NEW APPROACHES FOR THE GENERATION AND ANALYSIS OF MICROBIAL TYPING DATA

Edited by

Lenie Dijkshoorn
Department of Infectious Diseases C5-P
Leiden University Medical Center
P.O. Box 9600
2300 RC Leiden
The Netherlands

Kevin J. Towner
Public Health Laboratory
University Hospital
Queen's Medical Centre
Nottingham NG7 2UH
United Kingdom

Marc Struelens
Department of Microbiology
Hôpital Erasme
ULB
808, Route de Lennik
B-1070 Bruxelles
Belgium

ELSEVIER

Amsterdam - Boston - Heidelberg - London - New York - Oxford
Paris - San Diego - San Francisco - Singapore - Sydney - Tokyo

ELSEVIER B.V.
Sara Burgerhartstraat 25
P.O. Box 211, 1000 AE Amsterdam, The Netherlands

First edition: 2001
Second impression: 2003

Library of Congress Cataloging in Publication Data
A catalog record from the Library of Congress has been applied for.

ISBN: 0-444-50740-X

The paper used in this publication meets the requirements of ANSI/NISO Z39.48-1992 (Permanence of Paper).
Printed in the Netherlands.

Preface

Classification and identification methods for microorganisms have been based for many years on the phenotypic properties exhibited by individual isolates. However, increasing use of molecular identification and typing techniques has resulted in a re-evaluation of the whole process of classification and identification, with the recognition of new relationships and the revision of some previously accepted taxonomic schemes. Organisms can be classified at different levels, of which the most immediately important to microbiologists is the concept of the bacterial species. It is now accepted by microbial taxonomists that a complete genomic DNA sequence should form the reference standard for determining phylogeny, and hence taxonomy and species, but rRNA relatedness has also been used increasingly (and more controversially) to make assumptions about the phylogeny of microorganisms and to delineate new species. Once individual species have been defined, they can be arranged on a hierarchical system into genera and families. The different philosophical approaches underpinning microbial classification are outlined in Chapter 1 of this book, but whichever approach is used, it is important that molecular relationships should be depicted accurately in classification schemes.

Once accurate identification to the species level has been achieved, many branches of microbiology, particularly medical microbiology, require the investigation of diversity below the species level, a process referred to as *typing*. As with identification to the species level, many well-established typing methods, developed over many years, are based on the study of phenotypic properties. However, most of these methods are restricted to particular groups of organisms and are not capable of rapid development to deal with emerging problems. While a complete DNA sequence would also form the ultimate reference standard for defining subtypes within a species, it is highly unlikely that routine microbiology laboratories will ever have the resources or capability to routinely sequence all of their isolates requiring investigation at this level. However, for many purposes, it is unnecessary to define the precise 'type' of an isolate. It is often sufficient to provide answers about the relatedness (or otherwise) of small collections of isolates. To this aim, a wide range of readily applicable comparative molecular fingerprinting methods has been developed over the past two decades, several of which have the potential to be used for studying microbial diversity in any competent microbiology laboratory. Details of the most important of these methods can be found in the succeeding chapters of this book. Visual comparison of the different fingerprints is often sufficient for day-to-day analysis (the intuitive pattern recognition capability of the human brain is still superior to all available forms of artificial intelligence for this purpose), but a computer-assisted strategy is required for the analysis of large sets of more complex fingerprints obtained over extended periods of time from different geographical locations. Such a strategy also enables the possibility of constructing reference databases of fingerprint patterns that can be contributed to and accessed by laboratories situated in many different countries. The aim of this book is to describe the novel methods that are currently available for the generation, analysis and classification of microbial typing fingerprints, and to indicate how they can be used to provide timely and accurate information of use to microbiologists working in a range of different specialisations.

The idea for this book originated from the discussions held among scientists from diverse backgrounds during workshops organised by the European Study Group on Epidemiological Markers (ESGEM) of the European Society of Clinical Microbiology and Infectious Diseases (ESCMID). We hope that more such initiatives will follow as the state of the art rapidly evolves in this field. We are indebted to all the contributing authors for their collaboration, friendship and enthusiasm. We would also like to thank our families and colleagues for their encouragement, patience and understanding.

Lenie Dijkshoorn
Kevin Towner
Marc Struelens

List of contributors

Antoon D.L. Akkermans
Laboratory of Microbiology
Wageningen University
Hesselink van Suchtelenweg 4
6703 CT Wageningen
The Netherlands
Tel: + 31 317 483486
Fax: + 31 317 483829
E-mail: antoon.akkermans@algemeen.micr.wau.nl

Dominique A. Caugant
Department of Bacteriology
WHO Collaborating Centre for Reference and Research on Meningococci
National Institute of Public Health
PO Box 4404 Nydalen
N-0403 Oslo
Norway
Tel: + 47 220 42311
Fax: + 44 220 42518
E-mail: dominique.caugant@folkehelsa.no
and
Insitute of Oral Biology
University of Oslo
PO Box 1052 Blindern
N-0316 Oslo
Norway

Willem M. de Vos
Laboratory of Microbiology
Wageningen University
Hesselink van Suchtelenweg 4
6703 CT Wageningen
The Netherlands
Tel: + 31 317 483100
Fax: + 31 317 483829
E-mail: Willem.deVos@algemeen.micr.wau.nl

Ariane Deplano
Centre for Molecular Diagnostic Microbiology
Department of Microbiology
Hôpital Erasme
Free University of Brussels
808, Route de Lennik
B-1070 Brussels
Belgium
Tel: +32 2 555 45 18
Fax: +32 2 555 31 10
E-mail: adeplano@ulb.ac.be

Lenie Dijkshoorn
Department of Infectious Diseases C5-P
Leiden University Medical Center
P.O. Box 9600
2300 RC Leiden
The Netherlands
Tel: + 31 (0)71 5263582
Fax: + 31 (0) 71 5266758
E-mail: L.Dijkshoorn@lumc.nl

Francine Grimont
Unité des Entérobactéries
INSERM U389
Institut Pasteur
28 Rue du Docteur Roux
F-75724 Paris Cedex 15
France
Tel: + 33 145 68 83 44
Fax: + 33 1 45 688837
E-mail: Fgrimont@pasteur.fr

Patrick A.D. Grimont
Unité des Entérobactéries
INSERM U389
Institut Pasteur
28 Rue du Docteur Roux
F-75724 Paris Cedex 15
France
Tel: + 33 145 68 83 40
Fax: + 33 1 45 688837
E-mail: Pgrimont@pasteur.fr

Hajo Grundmann
Division of Microbiology
University of Nottingham
University Hospital
Queen's Medical Centre
Nottingham NG7 2UH
UK
Tel: + 44 115 9709163
Fax: + 44 115 9422190
E-mail: Hajo.Grundmann@nottingham.ac.uk

John H. Hauman
8 Oxford Street
St. Clair
Dunedin
New Zealand
Tel: +64 3 455 9205
Fax: +64 3 455 9205
E-mail: jhhauman@es.co.nz

Herre F. Heersma
Division of Public Health Research
Management Team Computerisation and Methodological Consultancy
National Institute of Public Health and the Environment
PO Box 1
3720 BA Bilthoven
The Netherlands
Tel: + 31 (0) 30 2742067
Fax: + 31 (0) 30 2744456
E-mail: herre.heersma@rivm.nl

Marc Heyndrickx
Department for Animal Product Quality
Center for Agricultural Research
Brusselse Steenweg 370
B-9090 Melle
Belgium
Tel: + 32 9 252 1861
Fax: + 32 9 252 5085
E-mail: M.Heyndrickx@clo.fgov.be

Paul J.D. Janssen
EMBL Outstation – Hinxton
European Bioinformatics Institute
Wellcome Trust Genome Campus
Hinxton
Cambridge CB10 1SD
UK
Tel: + 44 (0) 1223 494 418
Fax: + 44 (0) 1223 494 468
E-mail: paul.janssen@advalvas.be

Kristin Kremer
Diagnostic Laboratory for Infectious Diseases Perinatal Screening
National Institute of Public Health and the Environment
PO Box 1
3720 BA Bilthoven
The Netherlands
Tel: +31 (0) 30 2742282
Fax: +31 (0) 30 2744418
E-mail: kristin.kremer@rivm.nl

Arjen van Ooyen
Nederlands Institute for Brain Research
Meibergdreef 33
1105 AZ Amsterdam
The Netherlands
Tel: + 31 20 5665500
Fax: + 31 20 566 5483
E-mail: A.van.Ooyen@nih.knaw.nl

Raf De Ryck
Centre for Molecular Diagnostic Microbiology
Department of Microbiology
Hôpital Erasme
Free University of Brussels
808, Route de Lennik
B-1070 Brussels
Belgium
Tel: +32 2 555 45 17
Fax: +32 2 555 64 59
E-mail: Raf.De.Ryck@ulb.ac.be

Nicholas A. Saunders
Molecular Biology Unit
Hepatitis and Retrovirus Laboratory
Central Public Health Laboratory
61 Colindale Avenue
NW9 5HT London
UK
Tel: +44 20 8200 4400 Ext. 3072
Fax: +44 20 8200 1569
E-mail: nsaunders@phls.nhs.uk

Dick van Soolingen
Diagnostic Laboratory for Infectious Diseases and Perinatal Screening
National Institute of Public Health and the Environment
PO Box 1
3720 BA Bilthoven
The Netherlands
Tel: +31 (0) 30 2742363
Fax: +31 (0) 30 2744418
E-mail: D.van.soolingen@rivm.nl

Marc J. Struelens
Centre for Molecular Diagnostic Microbiology
Department of Microbiology
Hôpital Erasme
Free University of Brussels
808, Route de Lennik
B-1070 Brussels
Belgium
Tel: + 32 2 555 4519
Fax: + 32 2 5556459
E-mail: marc.struelens@ulb.ac.be

Kevin J. Towner
Public Health Laboratory
University Hospital
Queen's Medical Centre
Nottingham NG7 2UH
UK
Tel: + 44 115 9709163
Fax: + 44 115 9422190
E-mail: Kevin.Towner@nott.ac.uk

Mario Vaneechoutte
Department of Clinical Chemistry, Microbiology & Immunology
University Hospital Ghent
De Pintelaan 185
B-9000 Ghent
Belgium
Tel: + 32 9 2403692
Fax: + 32 9 2403659
E-mail: mario.vaneechoutte@rug.ac.be

Erwin G. Zoetendal
Laboratory of Microbiology
Wageningen University
Hesselink van Suchtelenweg 4
6703 CT Wageningen
The Netherlands
Tel: + 31 317 483486
Fax: + 31 317 483829
E-mail: Erwin.Zoetendal@algemeen.micr.wau.nl

Contents

1 An Introduction to the Generation and Analysis of Microbial Typing Data

Lenie Dijkshoorn[1] and Kevin Towner[2]
[1]*Department of Infectious Diseases, Leiden University Medical Center, PO Box 9600, 2300 RC Leiden, The Netherlands;* [2]*Public Health Laboratory, University Hospital, Queen's Medical Centre, Nottingham NG7 2UH, UK*

CONTENTS

1.1 TYPING, WHAT DO WE MEAN?

The term typing is used frequently, but its interpretation varies depending on the context in which it is used. For example, it can be used in daily life as a very general way to describe distinctive features of an object or organism. In microbiology it is used in a more strict sense to describe a microbial isolate in such a way that it can be distinguished from other isolates. Thus, typing can be an *identifying* method in which a given character, e.g., a serotype, is used to identify an isolate as one already included in an existing typing scheme. Typing can also be done *comparatively*, which implies that microbial isolates of a defined set are compared to each other for similarity, without any reference to other isolates not included in the set, and without reference to existing classification schemes. If isolates are compared on the basis of complex characters, such as the DNA fragments comprising an electrophoretic profile, the term *fingerprinting* can be used instead.

The taxonomic level at which typing is performed can give rise to confusion among microbiologists. Most commonly it is understood that typing – either comparative or identifying – deals with discrimination of isolates *below* the species level, and this concept is used in this book in most cases. However, the term typing is sometimes used as a synonym for species identification. To some extent this is understandable since the classification of organisms is a dynamic process, and the splitting up of species into novel species previously considered subspecies or variants is not uncommon. Nevertheless, it should be noted that 'identification' implies *a priori* identification of an object or organism to an existing classification, whatever the taxonomic level is. In this context, if a subdivision below species level exists, e.g., at strain level, the term identification can be used to allocate isolates to the units of such a classification. Thus, the term 'strain identification' may be appropriate.

The following sections of this chapter deal with general aspects of typing, including data management and analysis, and quality control.

1.2 THE IMPORTANCE OF MICROBIAL SPECIES IDENTIFICATION AND TYPING

A. Species identification in microbiology

Since the early days of microbiology, numerous organisms have been found to be the causative agent of a specific infectious disease. These organisms were described in detail on the basis of methods available at the time (Brock, 1988; Dubos, 1988). In the course of time, more organisms were isolated in pure culture and classified according to their characters. The groups delineated were given names, and characters were sought that allowed their rapid identification. As a result, their association with particular diseases could be clarified and further research performed on their epidemiology and pathogenicity. Hence, strategies could be developed to combat these organisms. Thus, for many infectious agents that were recognised in the first era of microbiology, their identification to a species was, and still is, the signal for the implementation of specific prevention or control measures. Consequently, the threat of severe diseases, including scarlet fever, tuberculosis and typhoid fever, has been reduced drastically in the Western world in the 20th century as compared to the 19th century.

Elucidation of the role of particular microbial species in infectious diseases and their impact on human health has long dominated microbiology. However, although at first sight perhaps less spectacular, the diversity and ecology of microorganisms in general has also been studied from the very beginning, as exemplified by the work of Beijerinck and followers of the Delft School of Microbiology (La Rivière, 1997). Thus, numerous microbial species have been delineated in the past century that have no apparent clinical significance, and many of these are still the subject of study in other fields of microbiology, including microbial genetics, physiology, environmental and food microbiology (Dubos, 1988). As a result, numerous species are exploited in different ways, e.g., *Streptomyces* spp. for their production of antibiotics (Korn-Wendisch & Kutzner, 1992), or thermophilic archaebacteria and eubacteria for their production of important enzymes (Bergquist *et al.*, 1989). Other organisms, e.g., *Lactobacillus* spp. (Vogel & Ehrmann, 1996), are indispensable in the food industry, or are important, e.g., *Sphingomonas* spp. (Ye *et al.*, 1996) for their capacity to degrade xenobiotic compounds. Exploration of the general microbial diversity and precise recognition of species and strains are important for the study and use of these organisms. Novel DNA-based techniques, a number of which are described in this book, are useful tools for exploration of the microbial world.

B. Discrimination below the species level

Exploration of the diversity and biology of microbes during the last few decades has shown that strains of the same species may vary in characters, including pathogenicity or epidemicity. Many infections and outbreaks of infections are caused by specific strains of species which are in general not particularly pathogenic. For example, bacteria of the species *Escherichia coli* are normal inhabitants of the human colon, but particular strains, e.g., serotype O157:H7, can give rise to severe infections and large-scale foodborne epidemics (Griffin *et al.,* 1995). In hospitals, the advancement of health care over the past decades has been coupled with an increase in the number of severely ill patients in hospitals, and also with the emergence of nosocomial pathogens. These nosocomial pathogens, including *Pseudomonas aeruginosa, Klebsiella pneumoniae, Serratia marcescens, Acinetobacter* spp. and *Staphylococcus epidermidis*, are relatively harmless to the healthy individual, but can give rise to infections in vulnerable hospitalised patients (Bergogne-Bérézin & Towner, 1996; Farmer, 1999; Archer, 2000; Pollack, 2000). Usually these organisms are resistant to multiple antibiotics, and there are indications that – at least for some species – particular strains are responsible for these problems (Pitt *et al*., 1989; Mulligan *et al*., 1993; Dijkshoorn *et al*., 1996).

The precise genetic factors that determine the virulence or epidemicity of particular strains within these species are usually unknown, but typing can be an aid in determining strain characters (e.g., a sero- or phage type, or a DNA fingerprint) that are indicative of these pathogenic and/or epidemic strains. Thus, typing can be a tool for tracing and controlling the spread of clinically important strains.

1.3 IDENTIFICATION AND TYPING IN TAXONOMIC PERSPECTIVE

A. The concepts of identification, typing and fingerprinting

The purpose and taxonomic levels of identification, typing and fingerprinting are summarised in Table 1.1. These concepts have been been discussed from a taxonomic point of view by Goodfellow & O'Donnell (1993a, b). *Identification* is both the act and the result of determining whether an unknown organism belongs to a previously defined group, and organisms are given names by recognising that they are members of previously described taxa. The term *typing* is usually used to denote the differentiation of strains at subspecies level or below, and this activity is often used to trace the spread of a single strain causing an outbreak of infection. *Fingerprinting* is the characterisation of organisms in such a way that multivariate data are obtained, including graphic curves resulting from mass or infrared spectrometry, or banding patterns resulting from electrophoretic separation of protein or nucleic acid fragments. These fingerprints can be used for calculating the simi-

Table 1.1. Purpose and taxonomic levels of identification, typing and fingerprinting

Activity	Purpose	Taxonomic level
Identification	Allocation of unknown strains to described taxa	Subspecies, species or higher
Typing	Allocation of unknown strains to described 'types'. Discrimination of strains (= comparison of strains on the basis of similarity in one or more characters; see fingerprinting)	Below species level
Fingerprinting	Comparison of organisms on the basis of multi-variate, variable, quantitative characterisation data, spectra or electrophoretic profiles	Species level and below

larity of strains, followed by their grouping on the basis of similarity.

B. *Definitions*

Organisms can be classified at different levels. The species level, and categories below species level, have practical value in applied microbiology. As an introduction to the chapters dealing with typing methods in daily practice, it is necessary to provide a brief summary of some terms used to denote the different categories.

(i) Species

The species concept and the criteria used to delineate bacterial species are a continuous source of research and discussion (Wayne *et al.*, 1987; Dykhuizen & Green, 1991; Pennington, 1994; Stackebrandt & Goebel, 1994; Dijkshoorn *et al.*, 2000). In *Bergey's Manual of Systematic Bacteriology,* the bacterial species has been described as a collection of strains that share many features in common and differ considerably from other strains (Staley & Krieg, 1984). An *ad hoc* committee of the International Committee of Systematic Bacteriology has recommended that the complete DNA sequence of the genome should be the reference standard for phylogeny, and that phylogeny should determine taxonomy (Wayne *et al.*, 1987). It was proposed that a species should be the only taxonomic unit to be defined in phylogenetic terms, and that it should include strains of *c.*70% or greater DNA-DNA relatedness and with a difference of 5ºC or less in thermal stability (ΔT_m). Furthermore, it was stressed that names should not be allocated to genomic groups that cannot be differentiated by phenotypic properties. The overall concern of the Committee was that phylogenetically-based taxonomic schemes must also show phenotypic consistency. However, some genomic groups that have been delineated by DNA-DNA hybridisation are phenotypically so similar that they cannot be differentiated phenotypically for the time being. In these situations it seems best to allow a *nomenspecies* to contain more than one genomic group, and to designate the genomic groups (*genomovars*) by numbers (Ursing *et al.*, 1995).

Recently, rRNA relatedness has increasingly been the basis for assumptions on the phylogeny of organisms (Woese *et al.*, 1987). In the course of the 1990s, 16S rRNA sequence similarity data, rather than DNA-DNA hybridisation data, have increasingly been used for the creation of novel species that contain only one or a few strains (see e.g., Ravot *et al.*, 1999). However, it has been pointed out by Palleroni (1993) that 16S rRNA may not reflect the evolution of host bacteria sufficiently. Thus, the variability of rRNA sequences can be limited in closely related organisms (Fox *et al.*, 1992), while 16S rRNA sequence variation within species has also been noted (Vaneechoutte *et al.*, 1995; Cilia *et al.*, 1996).

Recent studies have suggested that sequences of conserved proteins provide an alternative model for microbial evolution (Gupta, 2000). Data on the homogeneity or heterogeneity of conserved proteins, and of rDNA and other DNA sequences *within* species, are still limited, but it is expected that the rapidly expanding libraries of microbial DNA and protein sequences may provide information in the near future on the validity of particular DNA or protein sequences used for the inclusion of individual strains in a species. At present, the common practice is to delineate species on the basis of a variety of genotypic and phenotypic characters. Such a so-called *polyphasic approach* (Colwell, 1970; Vandamme *et al.*, 1996) can be considered as a consensus classification, and numerous examples of species thus recognised are described in the *International Journal of Systematic and Evolutionary Microbiology* (formerly the *International Journal of Systematic Bacteriology*). These descriptions usually also include a detailed presentation of the phenotypic characters that can be used for identification of the described taxa.

(ii) Subspecies

In *Bergey's Manual of Systematic Bacteriology* (Staley & Krieg, 1984), the subspecies is described as the lowest taxonomic rank with a nomenclatural standing, but clearcut criteria for this rank were not given by the authors, nor by the International Committee on Systematic Bacteriology (Wayne *et al.*, 1987). Intra-subspecies ranks (e.g., serovars or phagovars; see below) have been acknowledged to have great practical usefulness, but they have no official standing in nomenclature.

(iii) Var / type

It was recommended by *Bergey's Manual of Systematic Bacteriology* (Staley & Krieg, 1984), that the term 'type' should not be used to denote, for instance, serotypes or biotypes. Instead it was suggested that the term 'var' (derived from variety) should be used, since the term 'type' should only be used to denote an example of a species or genus. However, this recommendation has received little appreciation or attention, and in the different fields of applied microbiology it is common to describe entities below the species level as types. These so-called types can be distinguished within a set of isolates on the basis of one single typing method (e.g., serotyping or biotyping), or on the basis of a combination of typing methods (Sloos *et al.*, 1996; Bernards *et al.*, 1997; Van Pelt *et al.*, 1999).

(iv) Strain

In laboratory jargon, a strain is usually considered to be a culture of a specific microorganism which is on the bench of the laboratory worker or stored in the refrigerator. As such, this concept corresponds to the description of Staley & Krieg (1984), i.e., 'a strain is made up of descendants of a single isolation in pure culture and is usually made up of a succession of cultures ultimately derived from a single colony'. A strain in this sense can be stored over longer periods under a specific designation and has been denoted as *the strain in the taxonomic sense* (Dijkshoorn *et al.,* 2000). The strain in the taxonomic sense may change genetically over time, but will keep its designation. In addition, another type of strain was considered, i.e., a *strain in nature*, which might include (e.g.) different isolates from different body sites of the same patient which are assumed to have been derived from an initial single colony, but not under human control. The concept of the strain in nature also applies to isolates of the same species that have spread among patients by cross infection, and which are assumed to represent the same strain, based on the finding that they exhibit common phenotypic and genotypic traits that are distinctive from those of other isolates of the same species (Struelens *et al.*, 1996).

(v) Clone

In many biological text books, the term *clone* is used to indicate all cells that descended from a common ancestor. This concept is relatively straightforward when applied in hospital epidemiology to isolates in a direct chain of replication and transmission from host to host or from the environment to host (Struelens *et al.*, 1996). In this sense, the terms strain (in nature) and clone are synonyms, and their recognition can be derived from data obtained by one or several typing methods, combined with data on the epidemiological origin of the organisms (Dijkshoorn *et al.*, 2000).

The term *clone* is also used in a wider context 'to denote bacterial cultures isolated independently from different sources, in different locations, and perhaps at different times, but showing so many identical phenotypic and genetic traits that the most likely explanation for this identity is a common origin' (Ørskov & Ørskov, 1983). Numerous population studies have suggested that a variety of bacterial species contain clones with a wide geographic spread (e.g., Achtman *et al.*, 1983; Ochman & Selander, 1984; Musser & Selander, 1990), while other species are panmictic or superficially clonal (Maynard Smith *et al*, 1993; see also Chapter 12). During the numerous generations in a line of descent, the genetic diversity within a clone will increase, and there are no strict rules to decide on whether isolates belong to a clone or not. In practice, clones are delineated using methods developed by population biologists, including multilocus enzyme electrophoresis (MLEE) or multilocus sequence typing (MLST), often supplemented by genotypic and phenotypic methods currently used in applied microbiology. The current trend is that population biologists and applied microbiologists are combining their efforts to obtain a better understanding of the diversity or non-diversity within bacterial species and, in particular, on the emergence of specific virulent clones

(Selander & Musser, 1990).

(vi) Isolate

Any pure culture (or subculture) of a bacterial species on a solid or liquid culture medium can be denoted as an isolate.

C. Common strategies in taxonomy, population biology and typing

On first sight, the delineation of species is the field of the taxonomist, while the recognition of clones is the field of the population biologist, *cq* the applied microbiologist, including the clinical microbiologist or plant biologist. At a closer look, this distinction is arbitrary. Apart from DNA-DNA hybridisation, the strategies used in these fields are increasingly the same, since many studies use a polyphasic approach combining DNA sequence analysis of specific genes or genomic fingerprinting techniques, and one or several methods for phenotypic characterisation. Thus, the study of diversity below the species level does not basically differ from the approaches used in taxonomy, and might be considered as 'taxonomy below the species level', although there are no formal recommendations for this process and there is a great lack of uniformity in the use of methods and criteria for delineation of the groups.

1.4 CATEGORIES OF TYPING METHODS

Typing methods can be grouped in different categories, including *phenotypic* vs *genotypic* methods, and *comparative* vs *definitive* (*absolute* or *identifying*) methods. Recently another term, *library typing*, has been proposed.

A. Phenotypic vs genotypic methods

Most commonly used typing methods are denoted as phenotypic or genotypic, depending on the markers used. In the former case, the characters used for distinction of organisms are phenotypic, e.g., an epitope which can be detected with specific antibodies, or the susceptibility to a specific bacteriophage (see section 1.6). If DNA or RNA are the chemical structures to be used for discrimination, then the method is considered as 'genotypic'. The term *molecular typing,* if used as a synonym for genotyping, is somewhat misleading since it may refer not only to DNA- or RNA-based methods, but also to methods that investigate other chemical classes at the molecular level, including lipopolysaccharides, lipids or proteins.

Norris (1980) distinguished different levels of genetic information in living cells, and proposed a scheme for the study of information at these different levels. An updated scheme with currently applied methods is presented in Table 1.2. At the first level, the genetic information is represented by itself, and the methods that differentiate organisms at this level are genotypic methods. At the second level, the genetic information is expressed in the structure of proteins; at the third level, in

Table 1.2. Levels of expression of genetic information and typing/identification methods for each level

Level	Marker/character	Method
1	The genome	GC determination, DNA/DNA and DNA/RNA hybridisation, PFGE, PCR fingerprinting, DNA sequencing and methods derived from sequencing, RFLP analysis (including ribotyping), AFLP, ARDRA
2	Proteins	Gel electrophoresis including SDS-PAGE and MLEE, amino-acid sequencing, serology
3	Cell components	Determination of amino-acid pools and cell wall composition, lipid analysis, infra-red spectrometry, pyrolysis gas liquid chromatography and mass spectrometry, bacteriocin and phage-typing, serology
4	Morphology and behaviour	Microscopic structure, motility, enzyme tests, physiology, antibiotic susceptibility, nutritional requirements

Adapted from Norris (1980).

the structure of all other cell components and products; and at the fourth level, in the morphology and behaviour of cells. The methods that use markers at levels 2–4 are phenotypic methods. The complexity of interactions of gene products increases from level 1 to 4, and accordingly the genetic relatedness of microorganisms is more difficult to deduce.

B. Identifying vs comparative typing methods

Apart from subdivisions based on the level of expression of the genome, typing methods can be subdivided according to the category of result obtained. The two categories, i.e., identifying and comparative methods, have already been outlined in section 1.1. With identifying (determinative) typing, organisms can be allocated to an already described type in an existing classification (typing) scheme. This would be the case with (e.g.) serotyping or phage-typing. In comparative typing, a group of organisms are compared to each other and grouped according to their similarity. In this approach, a set of organisms is considered as a whole, and comparisons are within this set without reference to an existing classification scheme. This approach is followed when there is no existing classification scheme encompassing all known types. Most of the recently developed genotypic typing methods fall into this second category.

C. Library typing methods

A major advantage of some phenotypic typing methods is that the data can be compared between laboratories. Thus, provided that the tools are available, any microbiologist in the world can identify local isolates by reference to the existing classification scheme, and consequently the geographic spread of these types (vars) can be investigated. In contrast, most genotypic fingerprinting methods may allow local recognition of an epidemic strain, but because of the inherent inter-laboratory variation, conclusions as to the geographic distribution of these strains has generally not been possible, despite the vast increase in the use of these methods.

It is now realised that so-called library typing methods are essential for large-scale surveillance systems and the study of the prevalence of particular strains or clones in the population (Struelens *et al.*, 1998). These methods must be standardised between laboratories, must use a uniform nomenclature, and should have a high throughput. Apart from serotyping, some promising library methods described elsewhere in this book are ribotyping (Chapter 5), selective amplification of restriction fragments by AFLP (Chapter 8), PCR-RFLP including ARDRA (Chapter 9), and insertion sequence fingerprinting (Chapter 10). In addition, binary typing (Van Leeuwen *et al.*, 1999; Zadoks *et al.*, 2000), combined with DNA chip technology, is a promising method for the future. However, with most of these methods, the necessary high throughput remains to be achieved.

1.5 APPLICATIONS OF TYPING

Microbial typing is probably applied most frequently in clinical, veterinary and food microbiology, and its uses in these fields are diverse (e.g., Maslow *et al.*, 1993; Pitt, 1994; Farber, 1996; Maslow & Mulligan, 1996; Goering, 1998). There are also numerous applications in other fields of microbiology, and genomic fingerprinting methods are also used in eukaryotic genome research. Some of the most obvious applications in the different fields are listed below and many examples can be found in the following chapters.

A. *Applications in medical microbiology*

Some examples of typing applications in medical microbiology are:

- solving diagnostic problems, e.g., determining in a patient with suspected endocarditis whether serial blood isolates represent the same strain;
- determining whether multiple isolates from the same hospital department represent an epidemic or endemic strain;
- elucidating the ecology or geographical spread of particular strains;
- studying the population structure of certain bacterial species;
- tracing specific types with known pathogenic features, e.g., certain *Salmonella* or *E. coli* serovars;
- studying microbial communities.

B. *Applications in other microbiological fields*

Comparative analysis of the genotypic and phenotypic characters of microorganisms finds many applications in non-medical fields. Some examples of their use are:

- in veterinary microbiology to investigate infections and outbreaks among animals, much along the same lines as in medical microbiology;
- in plant biology to trace plant pathogens or organisms in the rhizosphere;
- in the food industry, where the tracing of specific pathogens such as *E. coli* O157 or certain *Listeria* strains in food products is not only relevant for the food industry and public health, but also has legal and financial implications;
- in environmental microbiology where, as is the case in the food industry, identification of microorganisms with pathogenic potential, e.g., *Legionella pneumophila*, in environmental sources is important for public health and also has legal implications; in addition, fingerprinting methods are increasingly used to assess the microbial diversity of communities in the environment and the changes exerted by (e.g.) pollution;
- in taxonomy, where numerous examples can be found in the *International Journal of Systematic and Evolutionary Microbiology*.

C. Applications in eukaryotic gene analysis

Numerous applications involving the generation and analysis of genomic fingerprints, can be found in studies of eukaryotic genomes. These include:
- mutation analysis in the study of genetic disorders;
- studies in evolutionary biology;
- comparative analysis of genomes in forensic research, human history or anthropology;
- quality control of plant seeds in agriculture.

1.6 AN INTRODUCTION TO COMMONLY USED TYPING METHODS

The previous sections have described how typing methods can be divided into a number of different categories, including the major sub-division into phenotypic as opposed to genotypic methods. A complete DNA sequence would form the ultimate reference standard for recognising sub-types within a species, but short of achieving this ideal, any typing technique relies on finding detectable differences between isolates. Many existing typing systems for bacteria, particularly in clinical microbiology, are based on the recognition of differences in specific phenotypic properties and have largely stood the test of time for their particular applications. Thus, the main phenotypic typing techniques of biotyping, phage-typing, serotyping and bacteriocin typing are well-established and have been applied to a wide range of microorganisms. In addition to these, some newer 'molecular' phenotypic methods, such as typing based on analysis of proteins or lipopolysaccharides (see Chapter 4), also lend themselves readily to analysis by the fingerprinting techniques that are the main subject of this book. While the newer genotypic methods may eventually supersede the existing phenotypic methods, this is unlikely to happen in the short-term for species that have well-established phenotypic typing systems. It is therefore important to briefly consider the main advantages and disadvantages of these established methods in comparison with the newer genotypic methods that may eventually replace them.

A. *Biotyping*

Initial differentiation within a newly-delineated species is often achieved by examining the cultural and biochemical characteristics of a large collection of individual strains belonging to the species. Such characteristics may include colonial morphology, growth requirements, fermentation ability, carbon source utilisation and antibiotic resistances. However, as will be seen below, a major disadvantage shared by many phenotypic typing methods is that such properties may be rather difficult to interpret or subjective in their determination.

Many laboratories use commercially available galleries of tests, such as the API 20E system (bioMérieux, Marcy l'Etoile, France) with 20 different tests, to provide a biochemical profile from which different biotypes can be identified. However, variations in the duration of incubation and the inoculum size may affect interpretation of the results, while strains that are freshly isolated may exhibit different reactions compared with strains that have been stored. Newer automated biochemical fingerprinting systems have been reported to give good reproducibility and discrimination for certain species, e.g., *Enterobacter cloacae* (Kühn *et al*., 1991), but in many cases there may only be a very limited number of biochemical types within a species, resulting partly from the fact that traditional taxonomic procedures, particularly in clinical microbiology, may have relied on minor biochemical differences to define the species in the first place. However, if a suitable level of discrimination can be achieved, results can be scored as positive or negative, and then assessed by computer software. In such cases, following the calculation of similarity coefficients between every possible pair of isolates, the software can use an unweighted pair group method (see Chapter 2) to generate a dendrogram from a matrix of similarity coefficient values as an illustration of the inter-relationships of a particular set of isolates (e.g., Webster *et al*., 1996).

B. *Antibiogram typing*

So far as routine clinical microbiology laboratories are concerned, the first suspicion that an outbreak of infection is occurring is often still based on the observation of an antibiotic susceptibility pattern that is shared by a number of isolates of the same species. Antibiotic susceptibility or resistance profiles are normally easy to determine, but may be associated with potentially unstable extrachromosomal R plasmids. The relatively small number of agents with the potential to give different results also means that discrimination can be poor. Thus, while it is useful to identify recurring resistance patterns in a local situation, such patterns sometimes have little value when considered in isolation in comparative studies involving several different centres.

Antibiograms of isolates are often expressed simply in terms of 'resistant' or 'sensitive' depending on the breakpoints used. Such an approach has the effect of 'smoothing-out' subtler differences that may exist between strains. An alternative approach is to use a standardised disk inhibition zone method (Horrevorts *et al.*, 1995) which makes full use of all the available information. This method can be used, at least in the short-term, to generate reproducible and epidemiologically useful results for some organisms (Horrevorts *et al.*, 1995; Webster *et al.*, 1996). In such cases, similarity coefficients can be calculated as described above from the diameters of the inhibition zones (Blanc *et al.*, 1994; 1996). However, in general, it can be concluded that the determination of biotypes or antibiotic resistance profiles within a collection of isolates, while possibly helpful in a limited short-term local investigation, will only be useful in association with other typing methods for long-term investigations covering different epidemiological outbreaks or ecological situations.

C. Serotyping

Serotyping is one of the oldest typing procedures and still represents an important tool for typing isolates belonging to many microbial species (Towner & Cockayne, 1993). The method has been developed in particular detail for members of the Enterobacteriaceae (Ewing, 1986). Some of the original serotyping methods, e.g., slide agglutination and immunofluorescence tests, are still used widely today. Such tests are technically simple, albeit time-consuming if large numbers of isolates are to be examined. Immunoblotting allows the generation of complex antigen-based fingerprints that can be analysed with computers in much the same way as results from the other molecular fingerprinting techniques described in this book. ELISA tests can be used in large-scale studies where polysaccharide or heat-labile protein antigens are important for the differentiation of isolates. All of these techniques can be used with either polyclonal or monoclonal antibodies.

Serotyping has the advantage that it can be applied to many different genera, although a given set of reagents can usually only be applied to a single species. Many polyclonal and monoclonal antibodies used for routine typing of clinically important organisms are available commercially or through reference laboratories, but the complexity of raising and cross-testing antibodies means that it may require a considerable time to develop a serotyping scheme for a novel application. In general, it seems that the serotype of a microbe is a relatively stable and reliable typing marker, although some possible changes in structural antigens, related particularly to lysogenic conversion, have been noted occasionally (Meitert & Meitert, 1978). The main disadvantages of serotyping seem to be associated with problems in antisera production, standardisation of methodology and, sometimes, subjective assessment of results. It should also be noted that, since the method depends on the production of a range of specific antisera, a task which is logistically and ethically difficult for a routine microbiology laboratory to perform, serotyping schemes for certain genera are normally only available at central reference laboratories.

D. Phage-typing

The basis of phage-typing is the variable sensitivity of isolates to defined collections of bacteriophages which have been selected to provide the maximum sensitivity for differentiating isolates within a particular species. Phage-typing schemes have been developed for numerous bacterial species (Towner & Cockayne, 1993), many of which had not previously been typed successfully by other methods. Phage-typing is the most widely recognised typing method for *Staphylococcus aureus* (Parker, 1972), and is also still used widely for sub-dividing serotypes of *P. aeruginosa* (Bergan, 1978) and *Salmonella* / *Shigella* spp. (Guinée & van Leeuwen, 1978; Bergan, 1979).

Phage-typing schemes are highly sensitive, but have a number of important limitations (Meitert & Meitert, 1978). First, phage-typing is a technically demanding procedure in which environmental conditions and other variables must be controlled carefully. Second, phage-types can change following lysogenic conversion, loss of prophages, or gain or loss of R plasmids, and this variability is coupled with the continuous need to maintain the typing set of bacteriophages in a viable state by regular serial passage. Third, the discrimination is somewhat variable in that some species may contain too few phage-types and other species may contain several phage-types within a single strain, while with certain species (e.g., *Acinetobacter* spp.; Bouvet, 1991), a phage-typing scheme may only be reliable with isolates from a particular geographical region. Finally, sets of phages for typing purposes have been developed and refined for particular species over several decades, and it is difficult to apply phage-typing to a novel application in response to a sudden emerging clinical problem.

Overall, it seems likely that large laboratories will still continue to perform phage-typing for certain bacterial species in the short to medium term, but smaller laboratories will usually find it necessary to send their isolates to a central reference facility. The phage-typing approach does not lend itself readily to computer-assisted analysis, and phage-typing methods are gradually being superseded by genotypic techniques.

E. Bacteriocin typing

Bactericidal substances, normally proteins, which are active against different strains of bacteria are termed bacteriocins. Typing based on bacteriocins is normally performed by testing, often by cross-streaking, the sensitivity of 'unknown' isolates to bacteriocins produced by a set of standard selected strains. In general, a strain that produces a particular bacteriocin is also resistant to its action. Although bacteriocin production can be encoded by transmissible R plasmids, production or sensitivity to bacteriocins seem to be relatively stable properties.

Bacteriocin typing has been applied to a range of different bacteria (Towner & Cockayne, 1993), particularly *P. aeruginosa* and members of the Enterobacteriaceae, but the method is relatively labour intensive and also requires considerable

Table 1.3. Development of genotypic typing methods

First generation	• Analysis of plasmid content • Plasmid DNA restriction digests
Second generation	• Total chromosomal restriction digests • Analysis of RFLPs by hybridisation with probes – Ribotyping
Third generation	• Pulsed-field gel electrophoresis (PFGE) • PCR-based amplification methods – RAPD – REP-PCR – AFLP – PCR-ribotyping
Fourth generation	• Multilocus sequence typing (MLST) • DNA sequencing

development work before it can be used for a novel application.

F. Overview of the development of genotypic typing methods

All phenotypic methods suffer from the disadvantage that the observable characteristics are only an expression of the underlying genotype, and may or may not reflect the actual state of genetic relatedness in a group of isolates. In recent years, the scientific community has been confronted by an avalanche of new genotypic fingerprinting techniques, often with confusing or overlapping names and terminologies (reviewed in detail by Vaneechoutte, 1996). Nevertheless, a retrospective examination allows the development of genotypic typing methods to be divided into a number of phases or 'generations' (Goering, 2000; Table 1.3). Thus, the 'first generation' approach evolved in the 1970s and was based on the analysis of bacterial cells for the carriage of plasmids, together with more detailed analysis of restriction fragment length polymorphisms (RFLPs) by restriction endonuclease digestion of plasmid DNA. However, the potentially transient nature of plasmids meant that this approach was largely superseded in the early 1980s by techniques that examined RFLPs in the whole bacterial chromosome. Initial efforts using restriction endonuclease digestion and/or DNA probing techniques can be viewed as the 'second generation' methods. The development of pulsed-field gel electrophoresis (PFGE) in the mid-1980s ushered in the 'third generation', and PFGE has proven itself to be an especially powerful epidemiological tool that allows global chromosomal comparisons. At the same time, the discovery of PCR led to the development of a plethora of amplification-based techniques. Most recently, 'fourth generation' approaches involving a direct comparison of chromosomal sequences have started to assess potential inter-relationships between individual isolates at the most fundamental level. Examples and more details of the most useful genotypic approaches to typing can be found in the subsequent chapters of this book.

Table 1.4. Desired properties for an ideal typing scheme

1.	Ability to type the vast majority of isolates of a particular species encountered (i.e., a very low proportion of untypable isolates)
2.	Good discrimination, with the ability to recognise a reasonable number of types
3.	Good reproducibility over a long period of time and in different centres
4.	Readily applicable to natural isolates in addition to laboratory collections
5.	Rapid
6.	Amenable to computerised analysis and comparison with electronic databases
7.	Not too complicated or expensive

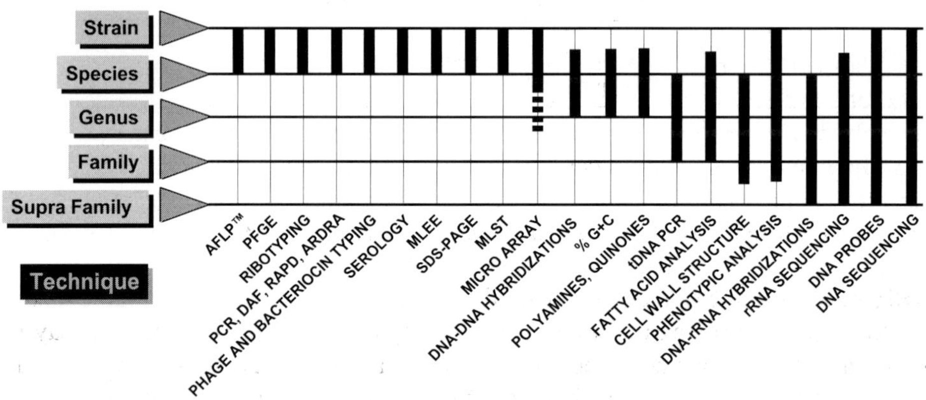

Fig. 1.1. Taxonomic resolution of current genotypic and phenotypic typing methods. (Courtesy of Dr Paul Janssen, Cambridge, UK.)

G. *Conclusions*

As will become apparent in the rest of this book, molecular fingerprinting methods are rapidly becoming the most commonly used approaches for assessing the relatedness of microorganisms in epidemiological studies. Such strategies are most often based on genotypic characteristics but, as described above, certain phenotypic properties can also be used. However, before using a particular strategy, it is important to ensure that the method distinguishes unrelated isolates, is capable of identifying the same strain in separate samples, and reflects genetic relatedness between epidemiologically linked isolates (see section 1.8). From a practical point of view, an ideal typing system should have a number of important characteristics (Table 1.4). Different methods are capable of varying levels of discrimination according to the precise organisms being studied, but an overall impression of the different levels of identification and typing achievable with different approaches is shown in Fig. 1.1.

From the above overview, it can be seen that none of the common phenotypic typing methods offers an ideal approach for the sub-dividing of microbial species. The importance and applicability of each of the methods may vary from one species to another, and also according to the precise geographical location in which they are being used. For some species, combined use of several different phenotypic methods may offer a reasonable approach, but such an approach may not be possible, or may take a long time to develop, for organisms which have not been well-studied previously. In contrast, genotypic methods have the potential to be used for studying diversity in any microbial species, with some genotypic methods also offering the possibility of providing a 'universal' approach to microbial typing in which the same basic methodology can be used to study isolates of any microorganism.

1.7 COMPUTER-ASSISTED DATA ANALYSIS

The human eye and brain are very efficient at recognising differences in electrophoretic patterns in two adjoining lanes of a gel and in correcting for distortions in gels due to technical problems. However, visual comparison and grouping of multiple profiles is extremely difficult, and the analysis of complex electrophoretic patterns is therefore dependent on the use of computer software. This is also the case for data analysis of the large number of phenotypic characters that can be generated by automated phenotypic identification systems such as the Biolog system (Biolog Inc., Hayward, CA, USA).

The most commonly used analytical method – cluster or pattern analysis – involves the pairwise comparison of isolates or objects, followed by their grouping on the basis of similarity, and their subsequent depiction in a tree-like structure termed a dendrogram. The idea of grouping organisms on the basis of their overall similarity in terms of large numbers of characters was originally suggested by the botanist Michel Adanson (1727–1806). However, it was not until the availability of high-speed computers for calculating these similarities that this concept could be applied in different disciplines. This approach, called *numerical taxonomy*, was developed for microbiology in the pioneering studies of Sneath (1957a, b), and was initially used for phenotypic characters that were scored as (+) or (–) (e.g., Baumann *et al.*, 1968). The underlying principles of numerical taxonomy are explained in detail elsewhere (Sneath, 1972; Sneath & Sokal, 1973; Bock, 1974; Aldenderfer & Blashfield, 1984; Everitt, 1998). A detailed description of how to score phenotypic characters, including morphological and physiological traits in numerical taxonomy, can be found in Lockhart & Liston (1970). An overview and practical details of applications in microbial fingerprinting are given in Chapters 2 and 3 of the present book.

The automated classification of microorganisms on the basis of protein electrophoretic fingerprints was first described by Kersters & De Ley (1975). A similar approach was also followed by Jackman (1985). Since the 1980s, there has been an explosive increase in the use of computer-assisted analysis of electrophoretic fingerprints in microbial taxonomy and epidemiology. The computer program originally developed by the group of Kersters and De Ley has evolved over the years into the highly advanced commercial software packages GelCompar and BioNumerics (Applied Maths, Kortrijk, Belgium). Other commercial software packages for the analysis of fingerprints include Taxotron (Institut Pasteur, Paris, France), Dendron (Solltech, Oakdale, IA, USA) and Bio Image Whole Band Analyser (Bio Image Inc., Ann Arbor, MI, USA).

As already discussed in previous sections, cluster analysis can be applied to both genotypic and phenotypic data, including DNA fragment electrophoretic patterns, profiles generated by mass spectrometry or gas chromatography, or biochemical profiles. A challenging development is the increase in availability of automated systems that generate complex phenotypic data sets. Examples are (i) fatty acid methyl ester analysis (FAME; MIDI Inc., Newark, DE, USA), (ii) 'fingerprint' analysis of the whole organism composition by mass spectrometry (Micromass, Manchester, UK), (iii) automated phenotypic profiling (Biolog Inc.), and (iv) automated antibiotic susceptibility testing based on diffusion zone image capturing by the use of (e.g.) the Sirscan system (i2a, Montpelier, France) or the Biomic system (Giles Scientific, Santa Barbara, CA, USA). These systems provide either spectra or complex data sets, and comparison of organisms on the basis of these character sets cannot be achieved without cluster or principal component analysis. The challenge is that these systems have a high throughput and that the data generated sometimes need to be analysed in combination with other data sets, e.g., genomic fingerprints or information on the epidemiological origin of the organisms obtained from a hospital's information system. The recently available BioNumerics software package allows combinations of all possible types of characters to be analysed. Such an integrated, polyphasic analysis of numerous genotypic and phenotypic characters, combined with automated data capturing, is a promising novel approach for research on microbial diversity at different taxonomic levels.

1.8 QUALITY ASPECTS OF MICROBIAL TYPING

A. *Overview*

When a typing system is being developed, a first critical question is whether the system provides distinctive fingerprints from different (i.e., genetically unrelated) isolates, and whether identical or highly similar fingerprints are generated from closely related organisms. If results are positive, further questions concerning the quality of the system become relevant, i.e., how reproducible is the method, how stable is the marker, and how large is the proportion of strains that can be typed? These quality criteria, together with more practical criteria that a typing system should meet, including cost, ease and speed, have been addressed in several previous reviews (e.g., Maslow & Mulligan, 1996; Struelens *et al.*, 1996; Tenover *et al.*, 1997). Some recommendations for typing microorganisms, most of which conform to the guidelines of the European Study Group on Epidemiological Markers (ESGEM) (Struelens *et al.*, 1996) are discussed in the following sections. However, these recommendations and quality criteria cannot be viewed apart from the basic question of *what is a type and how to delineate it*?

B. *Type definition and delineation*

With serotyping and phage-typing, a type is defined by its reactivity to an antiserum or phage. When fingerprinting methods are used, a limited number of fingerprints, e.g., PFGE profiles from a set of isolates in an outbreak, can be examined visually to determine whether or not they are identical. Patterns can be designated by letters or numbers for the purpose of such a study. Unambiguous discrimination of 'types' can still be difficult, since small differences in patterns between isolates may raise the question of whether or not the isolates are epidemically related. Suggestions on how to interpret visually the pattern variations generated by PFGE and other typing methods have been published (Tenover *et al.*, 1995; 1997). Nevertheless, a straightforward conclusion on the basis of just one or a few band differences is often difficult.

When large numbers of fingerprints are compared, this is usually done by computer-assisted pattern analysis. With this approach, organisms are grouped according to the similarity of their fingerprints. As is the case with visual analysis, it is often difficult to decide where to put the cutting level to delineate clusters that may represent groups of epidemiologically related isolates. If distinctive clusters contain isolates with a high similarity level, and different clusters are well separated, then these clusters may represent distinctive types. Knowledge on the natural discontinuities between strains within a species, as revealed by genotypic fingerprinting methods, is insufficient to define precise cutting levels for such an approach. Thus, the type delineation in such situations has to be based on as many features of the organisms being studied as possible, including the origin of the organisms in time and space, their ecology, and the correlation with other (genomic) typing methods. Furthermore, visual inspection of profiles of isolates in a cluster should intuitively corroborate the decision to use a particular cutting level.

C. *Discriminatory capacity*

Typing systems can differ in their discriminatory capacity depending on the marker used for discrimination. According to the ESGEM guidelines, a large collection (n = >100) of unrelated strains should be used to test the discriminatory capacity of a fingerprinting method and, ideally, each strain should give a distinctive fingerprint. Even larger collections should be tested for surveillance and population studies. The diversity resulting from such an exploration can be quantitated by the Simpson index of diversity (D) (Hunter, 1990). This index is a measure of the probability that a method will assign a different type to two unrelated strains sampled randomly and should generate a value of >0.95 (Blanc *et al.*, 1998). In theory, these guidelines are excellent but, in practice, smaller sets of strains are usually tested in epidemiological studies.

The strain collections used for assessing the discriminatory capacity should reflect the purpose of the typing system, but it is difficult to give detailed guidelines on the composition of the test collections while the diversity (within hospitals and in the community) and the evolutionary dynamics of the organisms being studied are unknown. With phage-typing it is usually recommended to use an international set of phages, supplemented with local phages, to cover the diversity of strains to be typed. Similarly, when using fingerprinting methods, it may be advisable to use strains from different countries, supplemented with local isolates, provided that the latter are not epidemiologically related.

D. Epidemiologic concordance

The capacity of a system to establish epidemiological relatedness between strains from the same outbreak has to be tested on sets from different outbreaks, each containing multiple (e.g., five) isolates per outbreak.

E. Typing system concordance

It is quite common to evaluate the performance of a typing system in relation to the performance of other typing systems (Dijkshoorn *et al.*, 1993; Van der Zee *et al.*, 1999; Vogel *et al.*, 2000). The underlying assumption for this approach is that there should be congruence between classification of organisms by different methods. With whole genomic fingerprinting methods this may be the case. However, markers that are subject to rapid evolutionary changes or are encoded by transmissable genetic elements e.g., serotypes or antibiotic susceptibility spectra, may show deviations from the pattern of grouping found by other methods. An example of such discordance is the variation in antibiotic susceptibility and biotype found in groups of *Acinetobacter baumannii* isolates that share the same ribotype and/or AFLP type (Dijkshoorn *et al.*, 1996; Nemec *et al.*, 1999). This lack of congruence does not necessarily imply that the methods showing rapid variation are useless. Rather, they can be used for subtyping isolates that otherwise show clonal relationship.

F. Reproducibility

Guidelines for evaluating the reproducibility of a method have been suggested by ESGEM (Struelens *et al.*, 1996). These comprise serial experiments that assess the influence of all possible steps in the process of type designation, and it is suggested that the reproducibility should ideally be >0.95. During use of a method, controls should always be included to monitor the procedure being used. For example, at least one strain with stable markers (see below) should always be processed in the same way as the test strains, including the initial steps of cultivation and sample preparation.

Ideally, a fingerprint generated by a particular typing method should be fully reproducible within and between laboratories. Unfortunately, many genotypic typing methods do not meet this requirement. In particular, early attempts at PCR fingerprinting were notorious for lack of reproducibility (Tyler *et al.*, 1997). The factors that influence the variability of the results of genotypic methods are numerous, including person-to-person variation, differences in sample preparation, in usage of reagents and equipment, in experimental conditions, or in data analysis. The impact of these sources of variation differs, depending on the setting in which the methods are being used. For example, if a small set of five isolates has to be typed during an outbreak, most factors can be kept constant and conclusions on whether the isolates are indistinguishable may be arrived at without any problems. However, even in the same locality, when many isolates have to be compared during endemic episodes, isolates have to be processed on different occasions and reproducibility is more difficult to control. It is even more difficult to control the many variables existing between laboratories, e.g., in multicentre studies on the geographic spread of specific pathogens.

Although reproducibility is a major problem in genotypic fingerprinting, recent developments have shown that these difficulties can be overcome if standardised reagents and/or uniform procedures are used. Thus, RFLP fingerprints of *Mycobacterium tuberculosis* have proven to be sufficiently reproducible to set up a large international database containing these profiles (see Chapter 3). In addition, the PCR fingerprints of a set of *Acinetobacter* strains generated in a multicentre study using standardised and quality controlled commercially available PCR reagents could be compared successfully between different laboratories (see Chapter 6; Grundmann *et al.*, 1997). Rigorous standardisation has also allowed PFGE patterns to be compared between laboratories through the National Molecular Subtyping Network (PulseNet; *http://www.cdc.gov/ncidod/dbmd/pulsenet/pulsenet.htm*). Perhaps the best example to date of a reproducible system is the fully automated Riboprinter (Qualicon Europe, Warwick, UK), which generates ribotypes for different microoorganisms according to a standardised protocol.

G. *The stability of strains*

A prerequisite of an epidemiological typing system is that the marker (e.g., the genomic fingerprint) used to type a strain is stable. However, the genetic material of microorganisms may undergo changes both *in vitro* (Arber *et al.*, 1994; Nakatsu *et al.*, 1998) and *in vivo*, and these changes may be reflected in the marker used. ESGEM has recommended that the in-vitro stability of at least 10 strains should be tested after 50 serial passages (Struelens *et al.*, 1996). Vogel *et al.* (1999) tested the in-vitro stability for three strains each of *P. aeruginosa*, *S. marcescens*, *K. pneumoniae*, and *K. oxytoca*. The strains were serially subcultured and stored at different temperatures, thereby simulating what happens to a strain in the laboratory when it is under investigation. No changes in fingerprints were observed during this test period.

In daily practice, the typing result obtained with a clinical isolate is frequently based on one subculture from a single colony, while the possible variability of the strain is unknown. Pitt (1994) suggested that such variability should be estimated when setting up a typing system by comparing various colonial variants of the same strains; isolates from different body sites; multiple colonies from primary platings of specimens; and antibiotic-sensitive and -resistant isolates from the same patient. The in-vivo changes that occur during the passage from one patient to another can be estimated by comparing multiple isolates from different patients during a clearcut outbreak (Struelens *et al.*, 1996).

H. Typability

Phenotypic typing methods such as serotyping or phage-typing do not always allow the characterisation of all isolates. In contrast, genotyping usually provides a characteristic fingerprint for most isolates. An exception is plasmid typing, since plasmids are accessory genetic elements that are not always present in all strains. Occasionally, other genomic fingerprinting methods, e.g., ribotyping or PFGE, do not produce a fingerprint with a specific isolate. A possible explanation is that such isolates are strong endonuclease producers, which results in breakdown of the DNA and prevents further processing of the specimens for typing. Typability can be calculated approximately when the discriminatory capacity of a typing system is tested on a large set of strains (see above). The actual typability will become apparent once the system is used in practice.

I. Standardisation of data analysis

The analysis of fingerprints, whether done visually or with computer assistance, can also contribute to bias in type delineation. For example, the visual type classification of PCR patterns in a multicentre study varied widely among participating centres (Van Belkum *et al.*, 1995; Deplano *et al.*, 2000). Computer-assisted data analysis may also result in different outcomes between scientists and laboratories. Sources of variables are: data capturing equipment, settings for data capture, software packages for analysis of fingerprints, and settings used with these packages. A severe source of bias can also be the visual designation of bands before comparisons are made (Burr & Pepper, 1997), which is a requirement of some software packages (e.g., Bio Image Whole Band Analyser). There is an urgent need to assess the impact of the many variables on computer-assisted data analysis. Once this is known, guidelines for standardising data capture and analysis can be agreed. As a first step, it would be helpful if it became a standard procedure for authors to give a detailed description in their publications of the precise analytical procedure used.

J. Conclusions

It is frequently stated that genomic fingerprinting methods are easy to perform, rapid, and can be applied to a wide range of organisms. However, setting up and validating these methods is a lengthy task and has to be performed for each species independently. Large sets of well-described strains are indispensable for this purpose. These requirements can only be met by specialised laboratories, or within frameworks of collaborating laboratories. Such activities have already resulted in the successful exploitation of MLST for *Streptococcus pneumoniae* (Enright & Spratt, 1998) and *Neisseria meningitidis* (Maiden *et al*., 1998), and the realisation of the *M. tuberculosis* database (see Chapter 3). Further to these collaborative studies, a number of other networks have been initiated for validating typing methods and investigating the diversity and spread of specific bacterial species. It is expected that these activities will have a great impact on the knowledge of the population structure and spread of many pathogens, and may provide an important tool to aid in their control (Stephenson, 1997).

1.9 STRAIN COLLECTION AND DATA MANAGEMENT

A. *Species identification*

It follows from the previous sections that specific strain collections are needed to test a typing method. First, the strains of these test sets must have been identified to a particular species according to the latest taxonomy. It should be noted that the taxonomy of many bacterial genera is revised from time to time due to new insights. Commercial identification systems – although easy to use – may not always have been validated with taxonomically well-described strains for all species, and may not use the most recent taxonomic scheme. In case of doubt, it is recommended that experts in the field should be consulted, i.e., specialists on the species being studied or curators of public culture collections.

B. *Culture collections*

It may be important to include reference strains in studies, perhaps to enable comparisons with the findings of other workers. Reference strains, e.g., strains from published outbreaks, can be requested directly from the authors or, if they have been deposited, from public culture collections. If specific clones are identified during large surveys, it is important that representatives of these clones are also deposited with culture collections under a specific designation, thus enabling direct comparisons to be made by the rest of the scientific community. As with type strains described in the *International Journal of Systematic and Evolutionary Microbiology*, the deposition and strain designation should be published in the same paper that describes the strain (clone).

C. *Storage of strains and data registration*

Strains should preferentially be stored at –145°C as there is no water movement at this temperature which can damage the cell structural components (Meryman, 1966), or otherwise at –70° or –80°C in glycerol broth. Other ways to store strains are by lyophilisation, or in 0.8% nutrient agar stab cultures (Kirsop & Doyle, 1991).

Table 1.5. Summary of data required for administration of epidemiological strain collections[a]

- Strain designation
- Eventual other designations
- Species name
- Received from
- Specimen (type)
- Patient code
- Date of isolation
- Hospital
- Department
- City
- Country
- Species identification (method)

[a] The list can be adapted to the purpose of the collection.

The records of the strains should contain relevant data on their origin, as summarised in Table 1.5. These are essential to decide whether strains are 'unrelated' or 'related'. It may be necessary for legal reasons to code the data so that the strains cannot be related to a specific hospital or patient. Since there is increasing interest in the wide geographical spread of particular epidemic clones, it is important that the strains in publications are retrievable by other researchers. Strains should therefore be designated by their collection number, and not just a serial number. Unfortunately, this important point has not yet been appreciated by the editors of most microbiological journals, with the significant exception of the *International Journal of Systematic and Evolutionary Microbiology.* There are several commercial software packages available for data administration, such as SPSS (SPSS Inc., Chicago, IL, USA) or Access (Microsoft, Redmond, WA, USA). These packages are also useful in sorting data, which can be useful for epidemiological purposes. Furthermore, the databases generated with these packages can be extended by the incorporation of additional data obtained over the course of time.

1.10 PROSPECTS FOR THE FUTURE

As mentioned earlier, a complete DNA sequence forms the ultimate reference standard for recognising sub-types within a species. It is clear that our increasing ability to rapidly sequence and compare the complete genomes of different isolates will bring many new scientific insights, and it is likely that epidemiology (and our insight into why certain strains have a propensity to become 'epidemic') is one of the areas that will benefit from this flood of sequence data. However, even with rapid automated sequencing techniques, it is highly unlikely that diagnostic microbiology laboratories will ever be in a position (or have the resources) to routinely sequence all their isolates of epidemiological interest.

One answer may be to concentrate on developing rapid sequencing methods for particular defined regions of the genome that can provide sufficient epidemiological and evolutionary information, and such techniques for sequencing 16S rDNA are already becoming available (see Chapter 11). An alternative advanced but simplified approach could eventually involve the use of DNA microarrays (sometimes called 'DNA chips'). Such microarrays normally consist of a very large number of evenly spaced spots of DNA fixed to a microscope slide. Each spot is a unique DNA fragment, typically a gene or part of a gene, transferred by a gridding robot from 96-well plates on to a slide. Standard microarrays may become commercially available and could then be hybridised in diagnostic laboratories with DNA extracted from 'unknown' isolates to generate different patterns of DNA hybridisation to the microarray. Each isolate would yield its own distinctive microarray pattern in a form that would be readily amenable to computerised analysis and comparison with electronic databases. However, for the time being, such technology remains in the future (but perhaps not too far away) and its cost effectiveness remains to be established.

In the meantime, progress in molecular biology has already resulted in the availability of molecular fingerprinting methods, several of which have the potential to be used in any competent microbiology laboratory for studying diversity in any microbial species. If only a few isolates are being compared, and the associated fingerprints are relatively simple, then visual comparison of pattern differences may be sufficient to assess the degree of relatedness. However, at present there is no general consensus as to the number of 'differences' (i.e., changes in fingerprint pattern) required for two isolates to be considered unrelated. For more complex fingerprints, and in cases where isolates from different geographical locations are being compared over significant time periods, a computer-assisted strategy is required that enables the formation of a database of fingerprint patterns. The challenge that is addressed in the rest of this book is to define how computer programs can be used to analyse molecular fingerprinting data and provide timely information of epidemiological importance or evolutionary significance.

REFERENCES

Achtman, M., Mercer, A., Kusecek, B., Pohl, A., Heuzenroeder, M., Aaronson, W., Sutton, A. & Silver, R.P. (1983). Six widespread bacterial clones among *Escherichia coli* K1 isolates. *Infection and Immunity* **39**, 315–335.

Aldenderfer, M.S. & Blashfield, R.K. (1984). *Cluster analysis (Sage University Press series on quantitative applications in the social sciences, No. 44).* Sage University Press, Beverly Hills.

Arber, W., Naas, T. & Blot, M. (1994). Generation of genetic diversity by DNA rearrangements in resting bacteria. *FEMS Microbiology Ecology* **15**, 5–14.

Archer, G.L. (2000). *Staphylococcus epidermidis* and other coagulase-negative staphylococci. In *Principles and practices of infectious diseases, 5th edn., vol. 2*, Mandell, G.L., Bennett, J.E. & Dolin, R., eds, pp. 2092–2100. Churchill Livingstone, Philadelphia.

Baumann, P., Doudoroff, M. & Stanier, R.Y. (1968). A study of the *Moraxella* group II. Oxidative-negative species (genus *Acinetobacter*). *Journal of Bacteriology* **95**, 1520–1541.

Bergan, T. (1978). Phage-typing of *Pseudomonas aeruginosa*. In *Methods in microbiology, vol. 10,*

Bergan, T. & Norris, J.R., eds, pp. 169–199. Academic Press, London.
Bergan, T. (1979). Bacteriophage typing of *Shigella*. In *Methods in microbiology, vol. 13*, Bergan, T. & Norris, J.R., eds, pp. 178–286. Academic Press, London.
Bergogne-Bérézin, E. & Towner, K.J. (1996). *Acinetobacter* spp. as nosocomial pathogens: microbiological, clinical, and epidemiological features. *Clinical Microbiology Reviews* **9**, 148–165.
Bergquist, P.L., Love, D.R., Croft, J.E., Streiff, M.B., Daniel, R.M. & Morgan, H.W. (1989). Genetics and potential biotechnological applications of thermophylic and extremely thermophylic archaebacteria and eubacteria. *Biotechnology & Genetic Engineering Reviews* **5**, 199–244.
Bernards, A.T., de Beaufort, A.J., Dijkshoorn, L. & van Boven, C.P.A. (1997). Outbreak of septicaemia in neonates caused by *Acinetobacter junii* investigated by amplified ribosomal DNA restriction analysis (ARDRA) and four typing methods. *Journal of Hospital Infection* **35**, 129–140.
Blanc, D.S, Lugeon, C., Wenger, A., Siegrist, H.H. & Francioli, P. (1994). Quantitative antibiogram typing using inhibition zone diameters compared with ribotyping for epidemiological typing of methicillin-resistant *Staphylococcus aureus*. *Journal of Clinical Microbiology* **32**, 2505–2509.
Blanc, D.S., Petignat, C., Moreillon, P., Wenger, A., Bille, J. & Francioli, P. (1996). Quantitative antibiogram as a typing method for the prospective epidemiological surveillance and control of MRSA: comparison with molecular typing. *Infection Control and Hospital Epidemiology* **17**, 654–659.
Blanc, D.S., Hauser, P.M., Francioli, P. & Bille, J. (1998). Molecular typing methods and their discriminatory power. *Clinical Microbiology and Infection* **4**, 61–63.
Bock, H.H. (1974). *Automatische klassifikation*. Vandenhoeck & Ruprecht, Göttingen.
Bouvet, P.J.M. (1991). Typing of *Acinetobacter*. In *The biology of Acinetobacter: taxonomy, clinical importance, molecular biology, physiology, industrial relevance*, Towner, K.J., Bergogne-Bérézin, E. & Fewson, C.A., eds, pp. 37–51. Plenum, New York.
Brock, T.D. (1988). Robert Koch, a life in medicine and bacteriology. In *Scientific revolutionaries: a bibliographic series*. Springer-Verlag, Berlin.
Burr, M.D. & Pepper, I.L. (1997). Variability in presence-absence scoring of AP PCR fingerprints affects computer matching of bacterial isolates. *Journal of Microbiological Methods* **29**, 63–68.
Cilia, V., Lafay, B. & Christen, R. (1996). Sequence heterogeneities among 16S ribosomal RNA sequences, and their effect on phylogenetic analyses at the species level. *Molecular Biology and Evolution* **13**, 451–461.
Colwell, R.R. (1970). Polyphasic taxonomy of bacteria. In *Culture collections of microorganisms. Proceedings of the international conference on culture collections, Tokyo, Oct. 7–11, 1968*, Iizuka, H. & Hasegawa, T., eds, pp. 421–436. University Park Press, Baltimore.
Deplano, A., Schuermans, A., Van Eldere, J., Witte, W., Meugnier, H., Etienne, J., Grundmann, H., Jonas, D., Noordhoek, G.T., Dijkstra, J, van Belkum, A., van Leeuwen, W., Tassios, P.T., Legakis, N.J., Van Der Zee, A., Bergmans, A., Blanc, D.S., Tenover, F.C., Cookson, B.C., O'Neil, G. & Struelens, M.J. (2000). Multicenter evaluation of epidemiological typing of methicillin-resistant S*taphylococcus aureus* strains by repetitive-element PCR analysis. *Journal of Clinical Microbiology* **38**, 3527–3533.
Dijkshoorn, L., Aucken, H.M., Gerner-Smidt, P., Kaufmann, M.E., Ursing, J. & Pitt, T.L. (1993). Correlation of typing methods for *Acinetobacter* isolates from hospital outbreaks. *Journal of Clinical Microbiology* **31**, 702–705.
Dijkshoorn, L., Aucken, H.M., Gerner-Smidt, P., Janssen, P., Kaufmann, M.E., Garaizar, J., Ursing, J. & Pitt, T.L. (1996). Comparison of outbreak and nonoutbreak *Acinetobacter baumannii* strains by genotypic and phenotypic methods. *Journal of Clinical Microbiology* **34**, 1519–1525.
Dijkshoorn, L., Ursing, B.M. & Ursing, J.B. (2000). Strain, clone and species: comments on three basic concepts of bacteriology. *Journal of Medical Microbiology* **49**, 397–401.
Dubos, R.J. (1988). Pasteur and modern science. In *Scientific revolutionaries: a bibliographic series*. Springer-Verlag, Berlin.
Dykhuizen, D.E. & Green, L. (1991). Recombination in *Escherichia coli* and the definition of biological species. *Journal of Bacteriology* **173**, 7257–7268.

Enright, M.C. & Spratt, B.G. (1998). A multilocus sequence typing scheme for *Streptococcus pneumoniae*: identification of clones associated with serious invasive disease. *Microbiology* **144**, 3049–3060.

Everitt, B. S. (1998). *Cluster analysis, 3rd edn*. Arnold, London.

Ewing, W.H. (1986). *Edwards and Ewing's identification of Enterobacteriaceae, 4th edn*. Elsevier Science, New York.

Farber, J.M. (1996). An introduction to the hows and whys of molecular typing. *Journal of Food Protection* **59**, 1091–1101.

Farmer, J.J. (1999). *Enterobacteriaceae*: introduction and identification. In *Manual of clinical microbiology, 7th edn*., Murray, P.R., Baron, E.J., Pfaller, M.A., Tenover, F.C. & Yolken R.H., eds. ASM Press, Washington D.C.

Fox, G.E., Wisotzkey, J.D. & Jurtshuk, P. (1992). How close is close: 16S rRNA sequence identity may not be sufficient to guarantee species identity. *International Journal of Systematic Bacteriology* **42**, 166–170.

Goering, R.V. (1998). The molecular epidemiology of nosocomial infection. An overview of principles, application and interpretation. In *Rapid detection of infectious agents*, Specter, S. & Friedman, H., eds, pp. 131–157. Plenum, New York.

Goering, R.V. (2000). The molecular epidemiology of nosocomial infection: past, present and future. *Reviews in Medical Microbiology* **11**, 145–152.

Goodfellow, M. & O'Donnell, A.G. (1993a). Roots of bacterial systematics. In *Handbook of new bacterial systematics*, Goodfellow, M. & O'Donnell, A.G., eds, pp. 3–54. Academic Press, London.

Goodfellow, M. & O'Donnell, A G., eds (1993b). Glossary of taxonomic terms. In *Handbook of new bacterial systematics*, pp. 525–549. Academic Press, London.

Griffin, P.M. (1995). *Escherichia coli* O157:H7 and other enterohemorrhagic *Escherichia coli*. In *Infections of the gastrointestinal tract*, Blaser, M.L., Ravdin, J.I., Greenberg, H.B. & Guerrant, R.L., eds, pp. 739–761. Raven Press, New York.

Grundmann, H.J., Towner, K.J., Dijkshoorn, L., Gerner-Smidt, P., Maher, M., Seifert, H. & Vaneechoutte, M. (1997). Multicenter study using standardized protocols and reagents for evaluation of PCR-fingerprinting reproducibility with *Acinetobacter* spp. *Journal of Clinical Microbiology* **35**, 3071–3077.

Guinée, P.A.M. & van Leeuwen, W.J. (1978). Phage-typing of Salmonella. In *Methods in microbiology, vol. 11*, Bergan, T. & Norris, J.R., eds, pp. 157–191. Academic Press, London.

Gupta, R.S. (2000). The phylogeny of proteobacteria: relationships to other eubacterial phyla and eukaryotes. *FEMS Microbiology Reviews* **24**, 367–402.

Horrevorts, A., Bergman, K., Kollée, L., Breuker, I., Tjernberg, I. & Dijkshoorn, L. (1995). Clinical and epidemiological investigations of *Acinetobacter* genomospecies 3 in a neonatal intensive care unit. *Journal of Clinical Microbiology* **33**, 1567–1572.

Hunter, P.R. (1990). Reproducibility and indices of discriminatory power of microbial typing methods. *Journal of Clinical Microbiology* **28**, 1903–1905.

Jackman, P.J.H. (1985). Bacterial taxonomy based on electrophoretic whole-cell protein patterns. In *Chemical methods in bacterial systematics*, Goodfellow, M. & Minnikin, D., eds, pp. 115–129. Academic Press, London.

Kersters, K. & De Ley, J. (1975). Identification and grouping of bacteria by numerical analysis of their protein patterns. *Journal of General Microbiology*, **87**, 333–342.

Kirsop, B.E. & Doyle, A., eds (1991). *Maintenance of microorganisms and cultured cells. A manual of laboratory methods, 2nd edn*. Academic Press, San Diego.

Korn-Wendisch, F. & Kutzner, H.J. (1992). The family Streptomycetaceae. In *The prokaryotes, 2nd edn, vol. 1*, Balows, A., Trüper, H.G., Dworkin, M., Harder, W. & Schleifer, K.-H., eds, pp. 921–995. Springer-Verlag, New York.

Kühn, I., Tullis, K. & Burman, L.G. (1991). The use of the PhP-KE biochemical fingerprinting system in epidemiological studies of faecal *Enterobacter cloacae* strains from infants in Swedish neonatal

wards. *Epidemiology and Infection* **107**, 311–319.
La Rivière, J.W.M. (1997). The Delft School of Microbiology in historical perspective. *Antonie van Leeuwenhoek* **71**, 3–13.
Lockhart, W.R. & Liston, J. (1970). *Methods for numerical taxonomy*. American Society for Microbiology, Bethesda, MD.
Maiden, M.C.J., Bygraves, J.A., Feil, E., Morelli, G., Russell, J.E., Urwin, R., Zhang, Q., Zhou, J., Zurth, K., Caugant, D.A., Feavers, I.M., Achtman, M. & Spratt, B.G. (1998). Multilocus sequence typing: a portable approach to the identification of clones within populations of pathogenic microorganisms. *Proceedings of the National Academy of Sciences of the United States of America* **95**, 3140–3145.
Maslow, J. & Mulligan, M.E. (1996). Epidemiologic typing systems. *Infection Control and Hospital Epidemiology* **17**, 595–604.
Maslow, J.N., Mulligan, M.E. & Arbeit, R.D. (1993). Molecular epidemiology: application of contemporary techniques to the typing of microorganisms. *Clinical Infectious Diseases* **17**, 153–164.
Maynard Smith, J., Smith, N.H., O'Rourke, M. & Spratt, B.G. (1993). How clonal are bacteria? *Proceedings of the National Academy of Sciences of the United States of America* **90**, 4384–4388.
Meitert, T. & Meitert, E. (1978). Usefulness, applications and limitations of epidemiological typing methods to elucidate nosocomial infections and the spread of communicable diseases. In *Methods in microbiology, vol. 10*, Bergan, T. & Norris, J.R., eds., pp. 1–37. Academic Press, London.
Meryman, H.T. (1966). Review of biological freezing. In *Cryobiology*, Meryman, H.T., ed., pp. 1–114. Academic Press, London.
Mulligan, M.E., Murray-Leisure, K.A., Ribner, B.S., Standiford, H.C., John, J.F., Korvick, J.A., Kauffman, C.A. & Yu, V.L. (1993). Methicillin-resistant *Staphylococcus aureus*: a consensus review of the microbiology, pathogenesis, and epidemiology with implications for prevention and management. *American Journal of Medicine* **94**, 313–328.
Musser, J.M. & Selander, R.K. (1990). Brazilian purpuric fever: evolutionary genetic relationships of the case clone of *Haemophilus influenzae* biogroup aegyptius to encapsulated strains of *Haemophilus influenzae*. *Journal of Infectious Diseases* **161**, 130–133.
Nakatsu C.H., Korona, R., Lenski, R.E., de Bruijn, F.J., Marsh, T.L. & Forney, L.J. (1998). Parallel and divergent genotypic evolution in experimental *Ralstonia* sp. *Journal of Bacteriology* **180**, 4325–4331.
Nemec, A., Janda, L, Melter, O. & Dijkshoorn, L. (1999). Genotypic and phenotypic similarity of multiresistant *Acinetobacter baumannii* isolates in the Czech Republic. *Journal of Medical Microbiology* **48**, 287–296.
Norris, J.R. (1980). Introduction. In *Microbial classification and identification*, Goodfellow, M. & Board, R.G., eds, pp. 1–10. Academic Press, London.
Ochman, H. & Selander, R.K. (1984). Evidence for clonal population structure in *Escherichia coli*. *Proceedings of the National Academy of Sciences of the United States of America* **81**, 198–201.
Ørskov, F. & Ørskov, I. (1983). Summary of a workshop on the clone concept in the epidemiology, taxonomy, and evolution of the Enterobacteriaceae and other bacteria. *Journal of Infectious Diseases* **148**, 346–357.
Palleroni, N.J. (1993). Structure of the bacterial genome. In *Handbook of new bacterial systematics*, Goodfellow, M. & O'Donnell, A.G., eds, pp. 57–113. Academic Press, London.
Parker, M.T. (1972). Phage typing of *Staphylococcus aureus*. In *Methods in microbiology, vol. 7B*, Norris, J.R. & Ribbons, D.W., eds, pp. 1–28. Academic Press, London.
Pennington, T.H. (1994). Molecular systematics and traditional medical microbiologists – problems and solutions. *Journal of Medical Microbiology* **41**, 371–3.
Pitt, T.L. (1994). Bacterial typing systems: the way ahead. *Journal of Medical Microbiology* **40**, 1–2.
Pitt, T.L., Livermore, D.M., Pitcher, D., Vatopoulos, A.C. & Legakis, N.J. (1989). Multiresistant serotype O12 *Pseudomonas aeruginosa*: evidence for a common strain in Europe. *Epidemiology and Infection* **103**, 565–576.

Pollack, M. (2000). *Pseudomonas aeruginosa*. In *Principles and practices of infectious diseases, 5th edn, vol. 2,* Mandell, G.L., Douglas, R.G. & Bennett J.E., eds, pp. 1673–1691. Churchill Livingstone, New York.

Ravot, G., Magot, M., Fardeau, M.L., Patel, B.K., Thomas, P., Garcia, J.L. & Ollivier, B. (1999). *Fusibacter paucivorans* gen. nov., sp. nov., an anaerobic, thiosulfate-reducing bacterium from an oil-producing well. *International Journal of Systematic Bacteriology* **49**, 1141–1147.

Selander, R.K. & Musser, J.M. (1990). Population genetics of bacterial pathogenesis. In *Molecular basis of bacterial pathogenesis (The bacteria, vol. II)*, Iglewski, B.H. & Clark, V.L., eds, pp. 11–36. Academic Press, San Diego.

Sloos, J.H., Dijkshoorn, L., Trienekens, T.A.M., Van Harsselaar, B., Van Dijk, Y. & van Boven, C.P.A. (1996). Multiresistant *Staphylococcus epidermidis* in a neonatal care unit. *Clinical Microbiology and Infection* **2**, 44–49.

Sneath, P.H.A. (1957a). Some thoughts on bacterial classification. *Journal of General Microbiology*, **17**, 184–200.

Sneath, P.H.A. (1957b). The application of computers to taxonomy. *Journal of General Microbiology*, **17**, 201–226.

Sneath, P.H.A. (1972). Computer taxonomy. In *Methods in microbiology, vol. 7A*, Norris, J.R. & Ribbons, D.W., eds, pp. 27–98. Academic Press, London.

Sneath, P.H.A. & Sokal, R.R. (1973). *Numerical taxonomy*. Freeman, San Francisco.

Stackebrandt, E. & Goebel, B.M. (1994). Taxonomic note: a place for DNA-DNA reassociation and 16S rRNA sequence analysis in the present species definition in bacteriology. *International Journal of Systematic Bacteriology* **44**, 846–849.

Staley, J.T. & Krieg, N.R. (1984). Classification of procaryotic organisms: an overview. In *Bergey's manual of systematic bacteriology, vol. 1*, Krieg, N.R., & Holt, J.G., eds, pp. 1–4. Williams & Wilkins, Baltimore.

Stephenson, J. (1997). New approaches for detecting and curtailing foodborne microbial infections. *Journal of the American Medical Association* **277**, 1337–1340.

Struelens, M.J. & the members of the European Study Group on Epidemiological Markers (ESGEM) of the European Society for Clinical Microbiology and Infectious Diseases (ESCMID) (1996). Consensus guidelines for appropriate use and evaluation of microbial epidemiologic typing systems. *Clinical Microbiology and Infection* **2**, 2–11.

Struelens, M.J., De Gheldre, Y. & Deplano, A. (1998). Comparative and library epidemiological typing systems: outbreak investigations versus surveillance systems. *Infection Control and Hospital Epidemiology* **19**, 565–569.

Tenover, F.C., Arbeit, R.D., Goering, R.V., Mickelsen, P.A., Murray, B.E., Persing, D.H. & Swaminathan, B. (1995). Interpreting chromosomal DNA restriction patterns produced by pulsed-field gel electrophoresis: criteria for bacterial strain typing. *Journal of Clinical Microbiology* **33**, 2233–2239.

Tenover, F.C., Arbeit, R.D., Goering, R.V. & the Molecular Typing Working Group of the Society for Healthcare Epidemiology of America (1997). How to select and interpret molecular strain typing methods for epidemiological studies of bacterial infections: a review for healthcare epidemiologists. *Infection Control and Hospital Epidemiology* **18**, 426–439.

Towner, K.J. & Cockayne, A. (1993). *Molecular methods for microbial identification and typing*. Chapman and Hall, London.

Tyler, K.D., Wang, G., Tyler, S.D. & Johnson, W.M. (1997). Factors affecting reliability and reproducibility of amplification-based DNA fingerprinting of representative bacterial pathogens. *Journal of Clinical Microbiology* **35**, 339–346.

Ursing, J.B., Rosselló-Mora, R.A., García-Valdés, E. & Lalucat, J. (1995). Taxonomic note: a pragmatic approach to the nomenclature of phenotypically similar genomic groups. *International Journal of Systematic Bacteriology* **45**, 406.

Van Belkum, A., Kluytmans, J., van Leeuwen, W., Bax, R., Quint, W., Peters, E., Fluit, A., Vandenbroucke-Grauls, C., van den Brule, A., Koeleman, H., Melchers, W., Meis, J., Elaichouni, A.,

$$D_{4,\{1,2\}} = \frac{D_{4,1} + D_{4,2}}{2}, \tag{3}$$

where $D_{4,\{1,2\}}$ is the similarity between object 4 and cluster {1,2}.

There are other ways to define the similarity between single objects and clusters of objects, and the method used to calculate the new similarity is what is called the clustering criterion or clustering method. In the new similarity matrix (Fig. 2.2d), the lowest value is again searched for, which is that between objects 3 and 4, and these objects are subsequently joined. Again a new similarity matrix is calculated, which now consists only of the similarity between cluster {1,2} and cluster {3,4} (Fig. 2.2e). Using the same clustering criterion as before, we obtain

$$D_{\{1,2\},\{3,4\}} = \frac{D_{\{1,2\},3} + D_{\{1,2\},4}}{2}. \tag{4}$$

The similarities $D_{\{1,2\},3}$ and $D_{\{1,2\},4}$ are given by equations (2) and (3), respectively (note that by definition $D_{\{1,2\},3} = D_{3,\{1,2\}}$ and $D_{\{1,2\},4} = D_{4,\{1,2\}}$).

The sequential union of points (groups) is now depicted in a dendrogram (Fig. 2.2f). First, objects 1 and 2 are joined. In the dendrogram, the level at which objects 1 and 2 are connected is the dissimilarity level in the similarity matrix that led to their union. Then, objects 3 and 4 are joined, and finally clusters {1,2} and {3,4}. In the dendrogram, the level at which the clusters are joined is the similarity value as calculated in equation (4); this is a measure for the similarity between *cluster* {1,2} and *cluster* {3,4}. Thus, the similarity between, for example, objects 2 and 4 is not shown in the dendrogram.

B. *General protocol for cluster analysis*

Keeping in mind the previous example, the general procedure for clustering is as follows (Fig. 2.3):

1. *Data set.* The starting point is a data set of objects that are described by the values they take on for a number of characters.
2. *Transformation.* Before calculating a similarity matrix, it may first be necessary to transform the data. This is necessary if the characters are qualitatively different or are expressed in different units. Transformation ensures that equal weight is given to all characters.
3. *Similarity matrix.* The next step is to choose a similarity measure and calculate the similarity between each pair of objects, yielding a triangular similarity matrix. Similarity measures are usually distance measures, but can also be derived from (e.g.) correlation coefficients. For electrophoresis data, the similarity between two objects can be expressed as the correlation between their banding patterns.
4. *Clustering.* Once the clustering method has been chosen – which is basically the formula that defines how to calculate the cluster-to-cluster similarities (and object-to-cluster similarities) from the basic object-to-object similarities – the

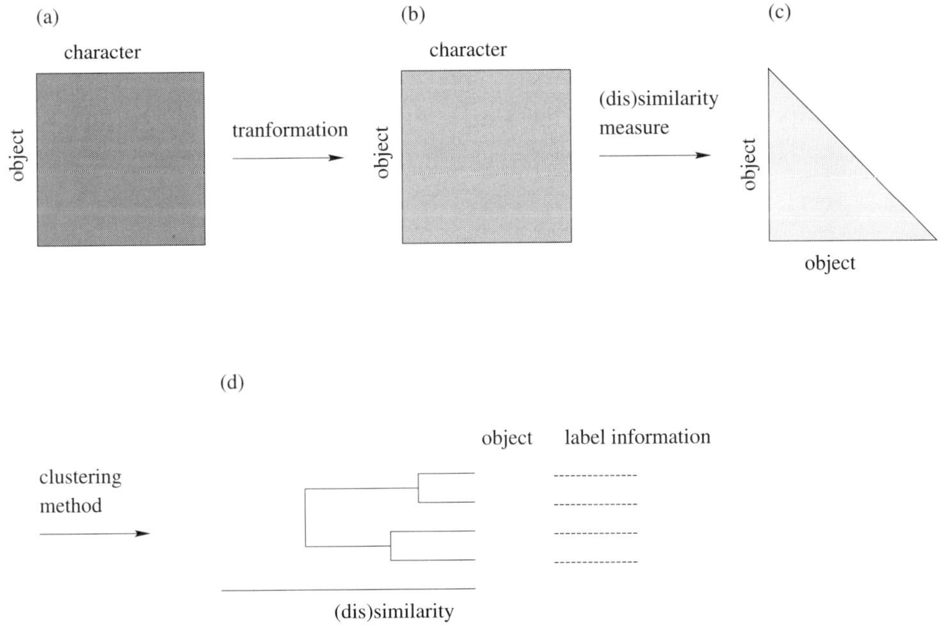

Fig. 2.3. The general protocol for cluster analysis (see text): (**a**) data set; (**b**) data set after transformation; (**c**) similarity matrix; (**d**) dendrogram.

similarity matrix can be used to form clusters.

5. *Dendrogram.* The result of this sequential joining of clusters is depicted in a dendrogram. In a dendrogram, the sequential union of objects and clusters is represented, together with the similarity value leading to this union. A dendrogram, therefore, does not define one partitioning, or grouping, of the set of objects, but contains many different partitionings of the set of objects. A particular partitioning can be obtained by "cutting" the dendrogram at some optimal value, defined relative to the dendrogram. For criteria to determine this cut-off value, see (e.g.) Blanc *et al.* (1994) and Hogeweg (1976b). In interpreting the groupings obtained, so-called label information can play an important role. Label information is basically all the information that is known about the objects which was not actually used in the clustering process itself (i.e., in determining the similarity between objects). Label information includes (e.g.) date of sampling, place of sampling, the date of analysis of the sampling, etc. It may be found – sometimes unexpectedly or unwanted – that the groupings obtained in the cluster analysis correlate with certain label information.

In the next sections, some of the most frequently used similarity measures and clustering methods will be briefly described.

C. *Similarity measures*

(i) *City-block distance*

The similarity measure used in the simple example, the city-block distance (or character difference), is given by

$$D_{i,j} = \sum_{k=1}^{N} \left| C_{k,i} - C_{k,j} \right| , \tag{5}$$

where $D_{i,j}$ is the dissimilarity between objects i and j, N is the total number of characters, and $C_{k,i}$ is the value that object i takes on for character k (index k runs from 1 to N). To calculate the mean city-block distance, the total number of characters is used as the denominator, i.e.,

$$D_{i,j} = \frac{1}{N} \sum_{k=1}^{N} \left| C_{k,i} - C_{k,j} \right| . \tag{6}$$

(ii) *Euclidean distance*

The distance between two objects in character space is used as a measure of their dissimilarity:

$$D_{i,j} = \sqrt{\sum_{k=1}^{N} (C_{k,i} - C_{k,j})^2} , \tag{7}$$

where $D_{i,j}$ is the distance between objects i and j, and $C_{k,i}$ is the value that object i takes on for character k (that $D_{i,j}$ represents distance can easily be seen for $N = 2$, using the Pythagorean theorem). To avoid the use of the square root, the value of the distance is often squared, and this expression is referred to as "squared Euclidean distance".

In comparing electrophoresis patterns, the matrix of similarities can be based either on the Pearson correlation coefficient or on one of the band-matching coefficients (Applied Maths, 1998).

(iii) *Pearson or product-moment correlation coefficient*

The similarity between two objects is calculated as the correlation between the two arrays of character values (typically densitometric values) taken on by the two objects:

$$S_{i,j} = \frac{\sum_{k=1}^{N} (C_{k,i} - \overline{C}_i)(C_{k,j} - \overline{C}_j)}{\sqrt{\sum_{k=1}^{N} (C_{k,i} - \overline{C}_i)^2 \sum_{k=1}^{N} (C_{k,j} - \overline{C}_j)^2}} , \tag{8}$$

where $S_{i,j}$ is the similarity (i.e., correlation coefficient) between objects i and j, $C_{k,i}$ is the value that object i takes on for character k, and $\overline{C}_i$ is the mean of all the character values of object i. The value of the correlation coefficient ranges from

+1 for perfect association to –1 for negative association; a value of 0 indicates that there is no association. That a correlation of 1 means perfect association can be seen by correlating object i to itself, i.e.,

$$S_{i,i} = \frac{\sum_{k=1}^{N}(C_{k,i}-\overline{C}_i)(C_{k,i}-\overline{C}_i)}{\sqrt{\sum_{k=1}^{N}(C_{k,i}-\overline{C}_i)^2\sum_{k=1}^{N}(C_{k,i}-\overline{C}_i)^2}} = \frac{\sum_{k=1}^{N}(C_{k,i}-\overline{C}_i)^2}{\sum_{k=1}^{N}(C_{k,i}-\overline{C}_i)^2} = 1. \qquad (9)$$

The correlation coefficient is a shape measure; i.e., it is sensitive to the pattern of dips and rises across the character values. Two profiles can have a correlation of +1 and yet not be truly identical (i.e., take on the same values). This occurs, for example, when the two profiles have the same pattern of dips and rises, but one profile is elevated compared to the other (see also Chapter 3).

(iv) Band-based similarity coefficients

(a) Coefficient of Jaccard. The similarity between two tracks of bands is the number of matching bands divided by the total number of bands in both tracks (i.e., the corresponding bands plus the track-specific bands):

$$S_{i,j} = \frac{n_{i,j}}{n_i + n_j - n_{i,j}}, \qquad (10)$$

where $S_{i,j}$ is the similarity between tracks i and j, $n_{i,j}$ is the number of corresponding bands for i and j, n_i is the total number of bands in i, and n_j is the total number of bands in j. So $n_i + n_j - n_{i,j}$ is the total number of bands in both tracks, not double counting the corresponding ones. If all bands in i match those in j, then $S_{i,j} = 1$.

(b) Area-sensitive coefficient. This is a more sophisticated similarity measure, which also takes into account the possible differences in areas of the matching bands:

$$S_{i,j} = \frac{A_{i,j}}{n_i + n_j - n_{i,j}}, \qquad (11)$$

where

$$A_{i,j} = \sum_{k=1}^{n_{i,j}} \frac{\alpha}{\alpha + \left|B_{i,k} - B_{j,k}\right|}, \qquad (12)$$

where α is a constant, and $\left|B_{i,k} - B_{j,k}\right|$ is the absolute difference between the areas of the k-th corresponding band in i and j, where k runs from 1 to $n_{i,j}$. Thus, differences in band areas of the corresponding bands are penalised. If the areas of all corresponding bands of both tracks are equal, this coefficient is reduced to the coefficient of Jaccard: if $B_{i,k} = B_{j,k}$ for all k, $A_{i,j} = \sum_{k=1}^{n_{i,j}} 1 = n_{i,j}$.

2. In statistics, attempts are made to make sample-independent statements. The data under consideration are assumed to be a random sample of the whole population, and the objective is to make statements about the whole population by looking at a representative sample of the population. Ideally, these statements should not change if a different random sample is taken from the population. In pattern detection, the data set under study is not considered a sample from a larger population but is considered all there is. A different structure may be found if new data is added (e.g., in taxonomy when new species are discovered).
3. In statistics, groups (and an underlying distribution) are pre-supposed and tests are made to determine whether these groups differ significantly from each other (i.e., more than can be expected on the basis of random fluctuations alone), while in pattern detection, groups are generated *per se*. In other words, concepts are *tested* in statistics (i.e., attempts are made to answer the question as to whether pre-supposed groups are different), while concepts (i.e., groupings) are *generated* in pattern detection. Descriptive statistics may be used in pattern detection for characterising the grouping obtained in cluster analysis.

Cluster analysis can best be seen as a heuristic, rather than a statistical, method for exploring the diversity in a data set by means of pattern generation. The result of a cluster analysis study can, and usually does, depend on the similarity measure used, the clustering method used, the set of objects in the study, the characters used to describe the objects, and the relative weight different characters are given in calculating the similarity between objects (see Hogeweg, 1976b; Van Ooyen & Hogeweg, 1990). Rather than trying to find the "right" pattern or classification, the differences in the patterns as revealed by the cluster analysis should be used to gain further understanding of the objects under study (see also Hogeweg, 1976a). Used in this heuristic way, cluster analysis is a powerful tool for data exploration in taxonomy and epidemiology, as well as in many other areas such as functional genomics.

REFERENCES

Aldenderfer, M.S. & Blashfield, R.K. (1984). *Cluster analysis.* Sage Publications, Newbury Park.

Applied Maths (1998). *GelCompar (comparative analysis of electrophoresis patterns) reference manual, version 4.1.* Applied Maths, Kortrijk.

Blanc, D.S., Lugeon, C., Wenger, A., Siegrist, H.H. & Francioli, P. (1994). Quantitative antibiogram typing using inhibition zone diameters compared with ribotyping for epidemiological typing of methicillin-resistant *Staphylococcus aureus*. *Journal of Clinical Microbiology* **32**, 2505–2509.

Blanc, D.S., Petignat, C., Moreillon, P., Wenger, A., Bille, J. & Francioli, P. (1996). Quantitative antibiogram as a typing method for the prospective epidemiological surveillance and control of MRSA: comparison with molecular typing. *Infection Control and Hospital Epidemiology* **17**, 654–659.

Bock, H.H. (1974). *Automatische klassifikation.* Vandenhoeck & Ruprecht, Gotingen.

Coenye, T., Falsen, E., Hoste, B., Ohlen, M., Goris, J., Govan, J.R.W., Gillis, M. & Vandamme, P. (2000). Description of *Pandoraea* gen. nov. with *Pandoraea apista* sp. nov., *Pandoraea pulmo-*

nicola sp. nov., *Pandoraea pnomenusa* sp. nov., *Pandoraea sputorum* sp. nov. and *Pandoraea norimbergensis* comb. nov. *International Journal of Systematic and Evolutionary Microbiology* **50**, 887–899.

Everitt, B. (1993). *Cluster analysis*, 3rd edn. Arnold, London.

Hogeweg, P. (1976a). *Topics in biological pattern analysis*. PhD Thesis, University of Utrecht.

Hogeweg, P. (1976b). Iterative character weighing in numerical taxonomy. *Computers in Biology and Medicine* **6**, 199–211.

Sloos, J.H., Horrevorts, A.M., Van Boven, C.P.A. & Dijkshoorn, L. (1998). Identification of multiresistant *Staphylococcus epidermidis* in neonates of a secondary care hospital using pulsed field gel electrophoresis and quantitative antibiogram typing. *Journal of Clinical Pathology* **51**, 62–67.

Sneath, P.H.A. &. Sokal, R.R (1973). *Numerical taxonomy*. Freeman, San Francisco.

Sokal, R.R. & Sneath, P.H.A. (1963). *Principles of numerical taxonomy*. Freeman, San Francisco.

Van Ooyen, A. & Hogeweg, P. (1990). Iterative character weighting based on mutation frequency: a new method for constructing phyletic trees. *Journal of Molecular Evolution* **31**, 330–342.

3 Setting-Up Intra- and Inter-Laboratory Databases of Electrophoretic Profiles

Herre F. Heersma[1], Kristin Kremer[2], Dick van Soolingen[2] and John Hauman[3]
[1]Management Team Computerisation and Methodological Consultancy, and [2]Laboratory for Infectious Diseases Surveillance, National Institute of Public Health and Environment, Bilthoven, The Netherlands; [3]Department of Microbiology, University of Otago, Dunedin, New Zealand

CONTENTS

3.1 INTRODUCTION

The microbial identification and typing techniques described in this book make use of a range of methodologies leading to the generation of a distinguishing fingerprint, normally combined with the use of electrophoresis as a separation strategy. During a long-term surveillance project at the Laboratory for Infectious Diseases at the National Institute of Public Health and Environment (RIVM), considerable experience has been acquired in constructing databases of typing results in the form of DNA fingerprints, in which laboratory, clinical and patient data are combined. This chapter describes the setting-up of databases of electrophoretic profiles and other laboratory data that can then be used for intra- and inter-laboratory studies, as well as national and international surveillance work investigating transmission routes, sources of infection, population based studies, etc. The work will be illustrated by reference to extensive studies that have been performed at RIVM with *Mycobacterium tuberculosis*.

The discovery of transposable elements and other repetitive DNA elements in strains belonging to the *M. tuberculosis* complex in the early 1990s led to various DNA fingerprinting methods being developed to differentiate between strains belonging to the complex. Some of these methods rely on PCR-based amplification of repetitive DNA sequences (Haas *et al.*, 1993; Plikaytis *et al.*, 1993; Friedman *et al.*, 1995; Kamerbeek *et al.*, 1997; Otal *et al.*, 1997). Other methods visualise restriction fragments containing particular repetitive DNA elements such as insertion sequences (IS) (Van Soolingen *et al.*, 1992; 1993; 1994a, b; Van Embden *et al.*, 1993; Wiid *et al.*, 1994). A comparison between PCR-based and restriction fragment length polymorphism (RFLP) typing methods revealed that IS*6110* RFLP typing is the most discriminative and reproducible typing method currently available (Kremer *et al.*, 1999).

This chapter will refer throughout to the results of international collaboration within the framework of the EU Concerted Action for Genetic Markers for the Epidemiology of Tuberculosis (CAonTB). Within this project, an inter-laboratory study has been carried out, with the goal of determining the limits of reproducibility of an international database of *M. tuberculosis* IS*6110* RFLP DNA patterns. This database was filled with results from over 30 laboratories worldwide. The laboratory work was already standardised with respect to restriction enzyme, probe and size markers. Based on this inter-laboratory study, standards and guidelines for computer-assisted analysis were established. This concerted effort to develop and standardise DNA fingerprinting and to develop computer-assisted methods has been very successful in establishing a robust infrastructure for studying the molecular epidemiology of TB in Europe. The experiences within this Concerted Action have led to exciting new possibilities for studying the transmission of tuberculosis. These experiences may also be applicable to other organisms. The analysis techniques described in this chapter are based on the application of GelCompar 2 and BioNumerics software (Applied Maths, Kortrijk, Belgium), although the techniques are also applicable for use with other software.

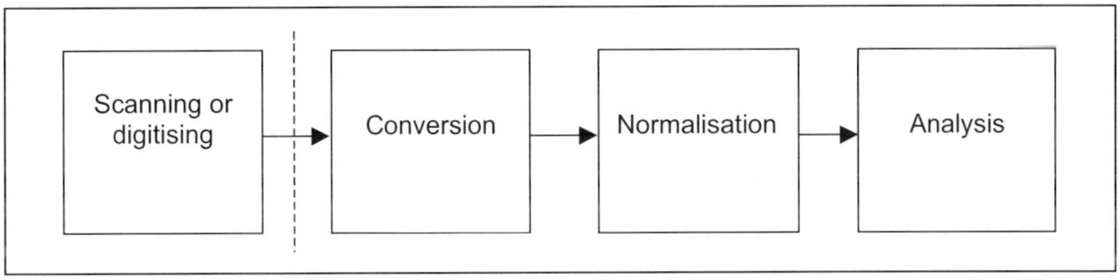

Fig. 3.1. The basic steps in computer-assisted analysis of electrophoretic profiles.

In brief, the application of electrophoresis for the computerised characterisation of an organism by its fingerprint involves two steps:

I. Electrophoresis of macromolecules (protein or DNA), resulting in a banding pattern on a gel, a film, an autoradiogram or a densitometric curve. Approaches leading to this first step (Van Soolingen *et al.*, 1999b), often referred to as 'the wet laboratory work', are discussed in the other chapters in this book.
II. Recording of the electrophoretic profiles by a camera, scanner or densitometer into a digitised form (bitmaps or densitometric curves), and subsequent analysis of the bitmap or densitometric curves with computer software.

The process of computer-assisted analysis of the electrophoretic profiles, based on one of a range of typing methods, basically consists of four distinct and sequential steps (Fig. 3.1):

1. creation of a digitised image that can be processed by a computer;
2. conversion of the digitised image into data that the computer analysis software can handle, and the recognition of the different lanes containing the images of the DNA fragments;
3. normalisation, involving interpolation of the different lanes to remove distortions;
4. analysis or comparison of the electrophoretic profiles processed in the previous phases; throughout the remainder of this chapter, electrophoretic profiles will be referred to as DNA fingerprints or DNA patterns.

If a DNA sequencer is being used to generate the data (e.g., see Chapter 6), steps I and II.1 are combined. The optical detector is located at a fixed position and the bands are recorded directly during electrophoresis while passing the detector.

The following sections discuss the basic steps involved and the factors that determine the comparability of DNA patterns and the reproducibility of databases of DNA patterns. The national database of *M. tuberculosis* IS*6110* RFLP patterns in The Netherlands and the international TBase database will be used as examples. Finally, results from an international inter-laboratory study based on the analysis of DNA patterns will be presented and some recommendations for setting-up intra- and inter-laboratory databases of DNA patterns are delineated.

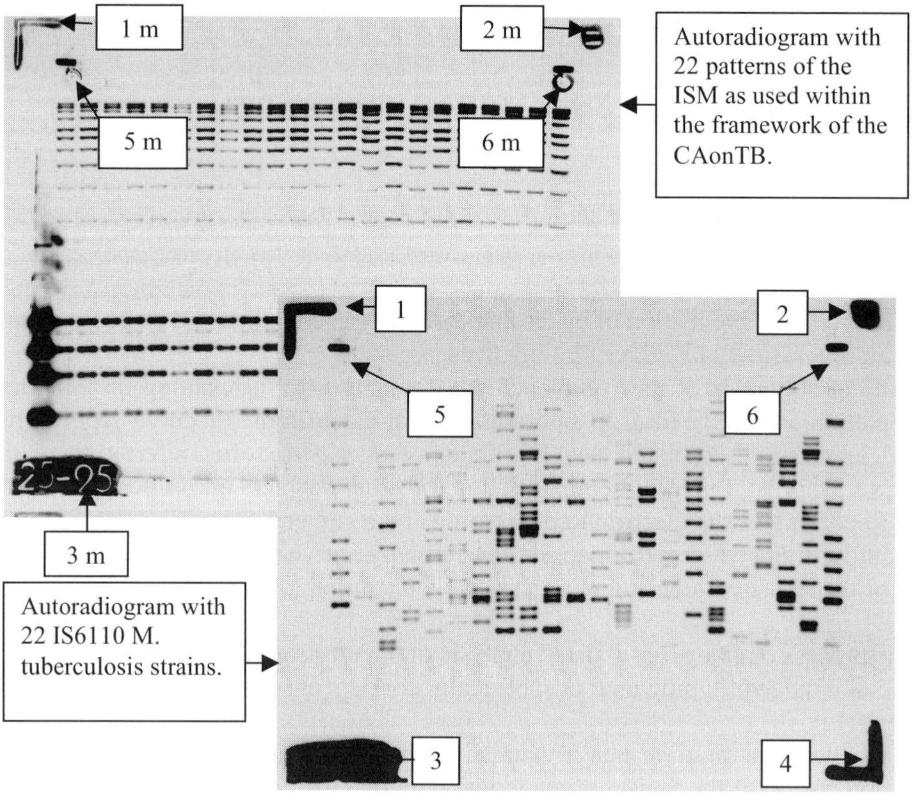

Fig. 3.2. Example of the analysis of two autoradiograms with ISMs. Marker points 1m, 2m, 3m and 4m (not visible here) on the ISM autoradiogram, and marker points 1, 2, 3 and 4 on the radiogram with the IS*6110* RFLP patterns are used for superimposition. To do so, both autoradiograms are put on a light box. After that step, small holes (5 and 6) are made with a sharp tool in both superimposed autoradiograms. These holes allow better alignment in the computer than the sometimes too dark markers.

3.2 PRINCIPLES OF COMPUTER-ASSISTED ANALYSIS OF DNA FINGERPRINTS

In this section, the recording of the patterns and computerised processing of the digitised information will be discussed. A more comprehensive description can be found in Heersma *et al.* (1998). The mobility of the DNA fragments in an agarose gel depends on numerous factors, including the DNA concentration, the type and concentration of the agarose, the voltage, temperature, buffer, electrophoresis time and temperature. Changes in any of these factors can lead to alterations in the profiles, which then influences the comparability of patterns, both within a single gel as well as between different gels.

Two basic techniques can be applied to improve the comparability of patterns (Van Embden *et al*, 1993; Friedman *et al*, 1995), namely the use of internal size markers (ISM) or external reference strains (ERS). The difference between these

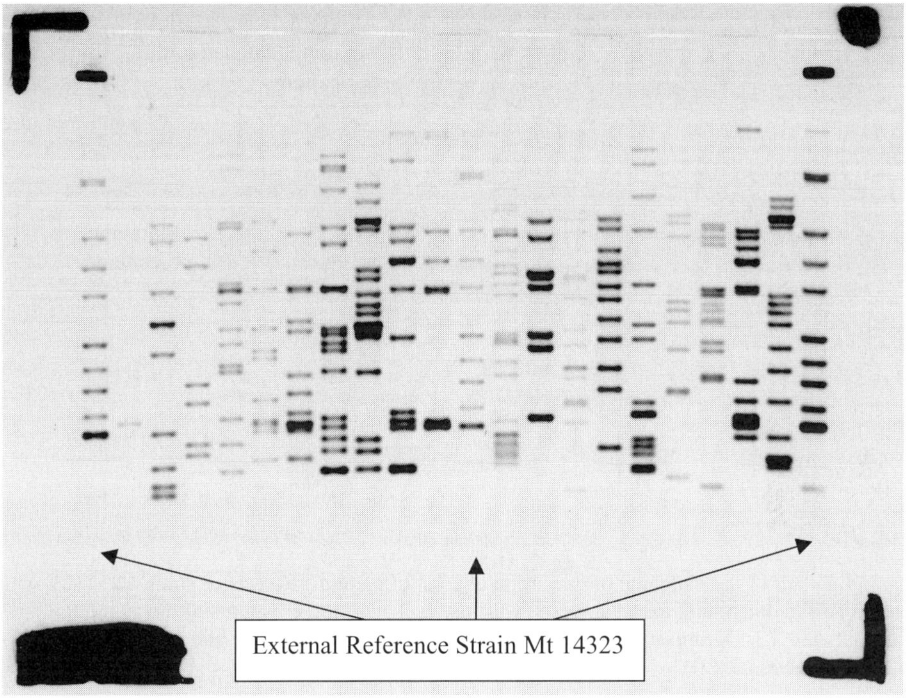

Fig. 3.3. Example of the use of ERS. In this autoradiogram, containing IS*6110* RFLP patterns from 22 *M. tuberculosis* strains, lanes 1, 12 and 22 contain the Mt14323 ERS. A disadvantage of this ERS is the sometimes very weak upper and lower bands. The marker points and the small holes (see Fig. 3.2) are not used in this case.

techniques can be seen in Figs. 3.2 and 3.3. When using ISM, the DNA patterns to be analysed and the reference patterns that enable compensation to be made for misalignments and errors are on different autoradiograms which have to be overlaid very accurately, whereas with ERS, the reference strains used to compensate for misalignments are on the same autoradiogram.

A. *Scanning the images*

Before further processing, electrophoretic patterns, whether on agarose gels or autoradiograms, have to be transformed into a digitised image. This can be done with a CCD camera or a scanner, usually a flatbed type. A CCD camera contains an objective lens and a light sensitive device, the charge coupled device (CCD), with a fixed resolution in pixels or dots per inch (dpi) in both directions.

A flatbed scanner usually consists of a glass cover, a light source and a computer-controlled array of light sensitive elements. In most cases, the resolution of a flatbed scanner is much higher than that of a CCD camera. With dedicated computer software, the image size and the resolution can be changed. Furthermore, properties of the digitised image, such as contrast and brilliance, can be adapted in

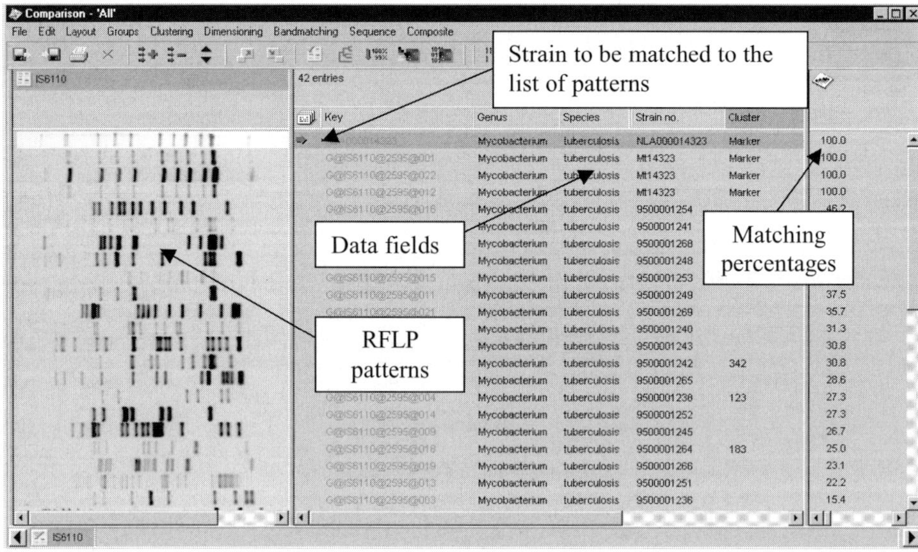

Fig. 3.4. Result of the matching of one strain to a list of patterns. Reference strain Mt14323, which was processed on another autoradiogram, was matched against the patterns of the autoradiogram in Figs. 3.2. and 3.3. Optimisation was enabled and both position tolerance and optimisation were set to 1% (see section 3.2.D).

various ways.

B. Conversion of images into the analysis software

In the next phase, the conversion or band pattern recognition phase, the digitised image is used as input, and this is converted to the BioNumerics or GelCompar 2 format. Pattern recognition software then gives the information necessary for the next steps, i.e., the overall borderlines of the desired parts of the image, the borderlines of the different 'raw' DNA fingerprints or 'Gelstrips' and, for the BioNumerics software, a densitometric curve for each DNA fingerprint. At the end of this step, each selected DNA pattern has its own individual entry in the computer database and is thereby separated from the rest of the gel or autoradiogram (Fig. 3.4).

In many cases, it can happen that the 256 grey levels are not fully used in the image file. This may result in a dark background and/or weak looking bands. The software offers the possibility of applying background subtraction and optimising use of the grey scale. This is based on the lightest and darkest optical density (OD) values, and the 'stretching' of these values to the full scale of 256 grey levels. In this way it is possible to remove the background 'noise' caused by the grey or blue photographic emulsion (Fig. 3.5).

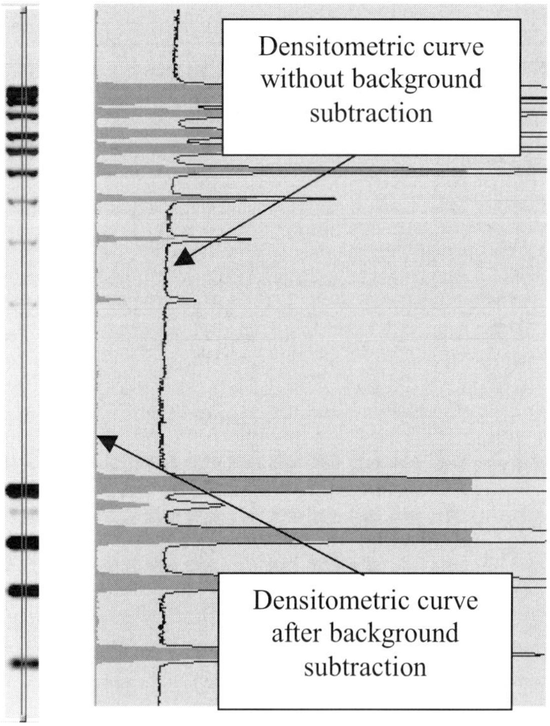

Fig. 3.5. Example of background subtraction to remove background noise caused by the photographic material.

C. Normalisation

The normalisation phase allows the alignment of patterns to be achieved by associating bands to predefined reference patterns. When using ERS, the alignment of reference patterns is done by aligning bands with corresponding predefined reference bands and by subsequent interpolation of the intermediate values. The other (non-reference) tracks are aligned gradually according to the closest neighbouring reference tracks on either side. If ISM are used, each track is interpolated separately and individually. It is possible to correct for distortions in the gel by combining any sample with any other, using one or more bands which are known to be the same in different patterns. In this way, it is possible to perform alignments in each track. This is the crucial difference between the two techniques: ERS can only be used to compensate for global distortions, while with ISM it is possible to minimise local distortions (Figs. 3.6 and 3.7).

D. The analysis phase

After the normalisation phase, the results can be analysed in two ways:

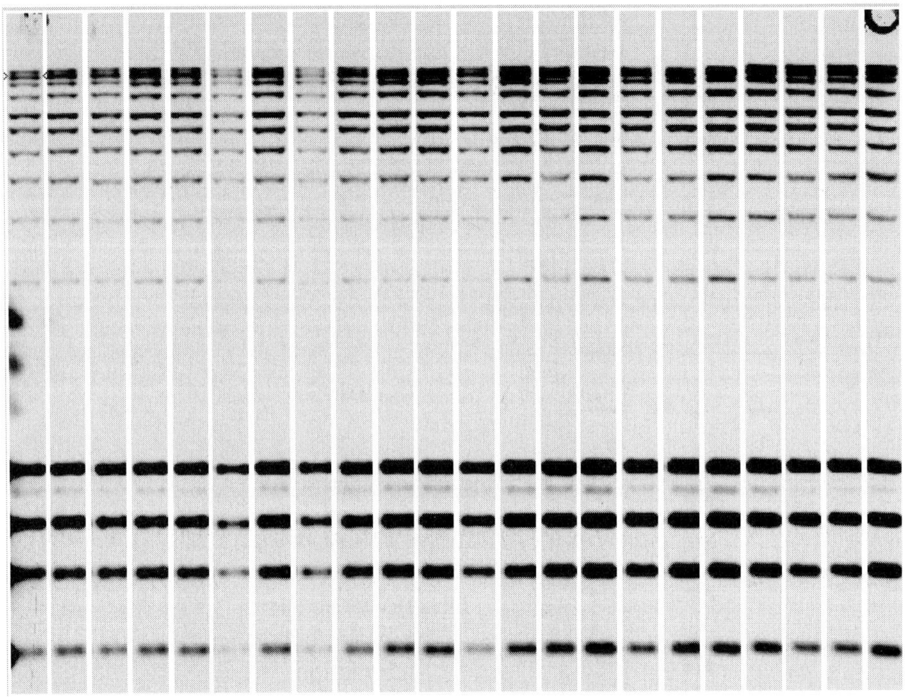

Fig. 3.6. The result of normalisation using ISMs. Each lane is used as a reference lane. Compare this to the result of the normalisation when using ERS in only three lanes (Fig. 3.7).

1. single patterns can be matched to a database or a list of patterns to reveal transmissions;
2. complete databases or lists of patterns can be compared and grouped to reveal the relatedness of patterns within a database.

This section discusses the different steps involved in the analysis phase. A comprehensive description of the theoretical background to pattern analysis can be found in Chapter 2.

(i) Matching single patterns against a list

When a single pattern is matched against a set (a database or a list) of *N* patterns, the similarity coefficient (SC) between each pattern in the set and the pattern to be matched is calculated. This SC is a number between 0 (no similarity at all) and 1 (100% identical patterns). The result is an array of *N* SCs. In the next step, the patterns are grouped with respect to their SC in decreasing order, and a list is depicted (Fig. 3.4).

(ii) Group and cluster analysis

Sets of DNA patterns, in databases or lists, can be compared mutually and grouped. This process is often called cluster analysis. A cluster is considered to be a group of

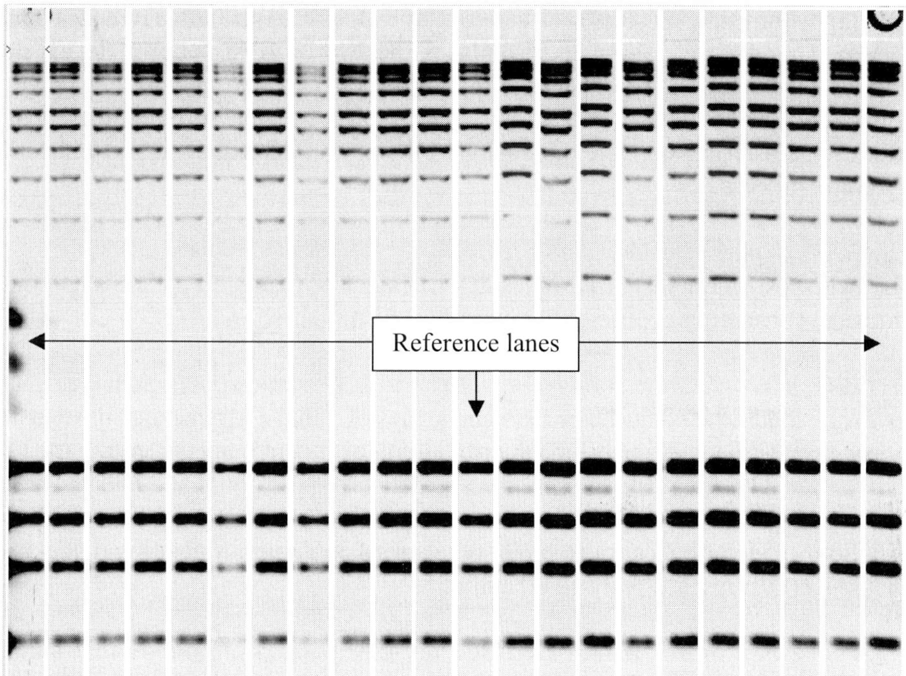

Fig. 3.7. The result of normalisation using ERS. In this case, lanes 1, 12 and 22 were reference lanes. It is clear that misalignments and errors, especially in lanes 15–21, were not well-compensated.

organisms with similar DNA patterns, and is delineated at a certain optimal ‘cutting’ level in a dendrogram (see Chapter 2).

As a first step, the SCs for all patterns (including a self-comparison) are calculated for each of the N patterns. This results in a symmetric $N \times N$ similarity matrix D (sometimes also called the dissimilarity or resemblance matrix); each element $D_{i,j}$ of the matrix represents the SC of patterns i and j. The elements $D_{i,i}$ on the main diagonal of the matrix are all 1 (= 100%), reflecting the SCs of each pattern when compared against itself. Again, for all other elements of the matrix, each SC value is a number between 0 (no similarity at all) and 1 (100% identical patterns). The ordering in this similarity matrix is decided by the way in which the patterns are grouped in the list or the database, i.e., randomly.

In the next step, the similarity matrix is transferred into a tree by using a clustering method. This is a series of steps that, theoretically, reduces the similarity matrix and, at the same time, develops a tree or dendrogram. In this example, a cluster is a set of DNA patterns that the computer considers to be identical according to the SCs. A cluster can contain as little as one pattern, indicating that the list contains no other pattern that matches at the defined cut-off level. The reorganised or optimised matrix is then kept in the computer.

The most frequently used algorithm for clustering DNA patterns is the unweighted pair-group method using arithmetic averages (UPGMA; see Chapter 2).

This algorithm first clusters two patterns by calculating the unweighted average pattern. This new average pattern replaces the two original patterns. In the next step, this average pattern clusters with another pattern, and again the average of both patterns is calculated, and a new pattern replaces the other two. Theoretically, after each clustering step, the dimension of the matrix decreases and the clustering is a finite process.

(iii) Use of similarity coefficients (SCs)
The two most frequently used measures to calculate SCs are Pearson's (product-moment) correlation coefficient or band-based comparisons.

(a) Pearson's correlation coefficient. Pearson's correlation coefficient (see Chapter 2) calculates the congruence between arrays of values, for instance the densitometric curves. Each element of the array contains a densitometric value, the OD, as interpreted by the scanner, optimised in the conversion phase and interpolated in the normalisation phase. This densitometric value can be anything between 0 and 255. So if the length of a track is 800 pixels (or BioNumerics positions), the array contains 800 greyscales between 0 and 255.

(b) Band-based SCs. Band-based SCs are calculated from arrays that contain only binary data, and reflect the fact that at a particular location on the autoradiogram, a DNA fragment is present (1) or not absent (0). Bands can be assigned manually or automatically. In this way it is possible to avoid or minimise the influence of all kinds of local errors, such as stain variations, non-linearities in the photographic material, differences in thickness of similar bands in different lanes, etc. Although most errors and non-linearities have been removed in the normalisation phase, differences in the morphology of bands can lead to differences in the interpretation of the positions of fragments on a gel. The band-based methods used most frequently to calculate SCs are the Dice and Jaccard methods. There is no difference between these measures with regard to the clustering, although Dice gives more weight to matching bands, whereas Jaccard gives more weight to differences (Figs. 3.8 and 3.9).

(c) Misalignments: tolerance settings and optimisation. Even in the best laboratory situation, the different steps will almost always lead to disturbances or misalignments, thereby influencing the reproducibility of the DNA patterns. When the Pearson correlation coefficient is used with GelCompar or BioNumerics software, clusters of 100% identical DNA patterns will be very rare because of differences in staining, background, or the morphology of bands, etc. Even for electrophoresis profiles of the same isolate on the same gel, small differences will always be present. The normalisation process also causes misalignments. *Optimisation* allows track-to-track corrections to compensate for the remaining misalignments by shifting one or both tracks with respect to the other until they reach their maximum correlation. The *Optimisation value* is the maximum shift between two pat-

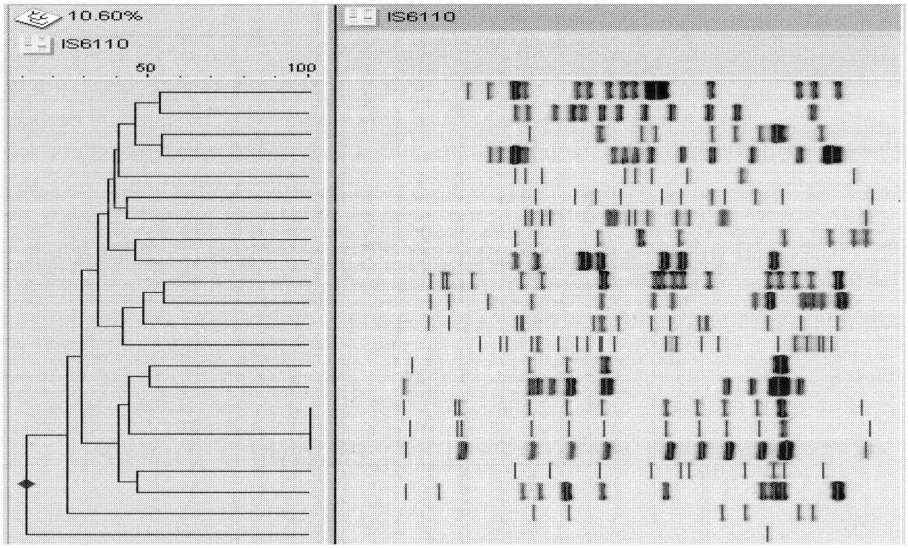

Fig. 3.8. Clustering of the *M. tuberculosis* IS*6110* RFLP patterns shown in Figs. 3.2 and 3.3. ISM were used with Jaccard and UPGMA, with tolerance and optimisation set at 1%. Compare this result with Fig. 3.9.

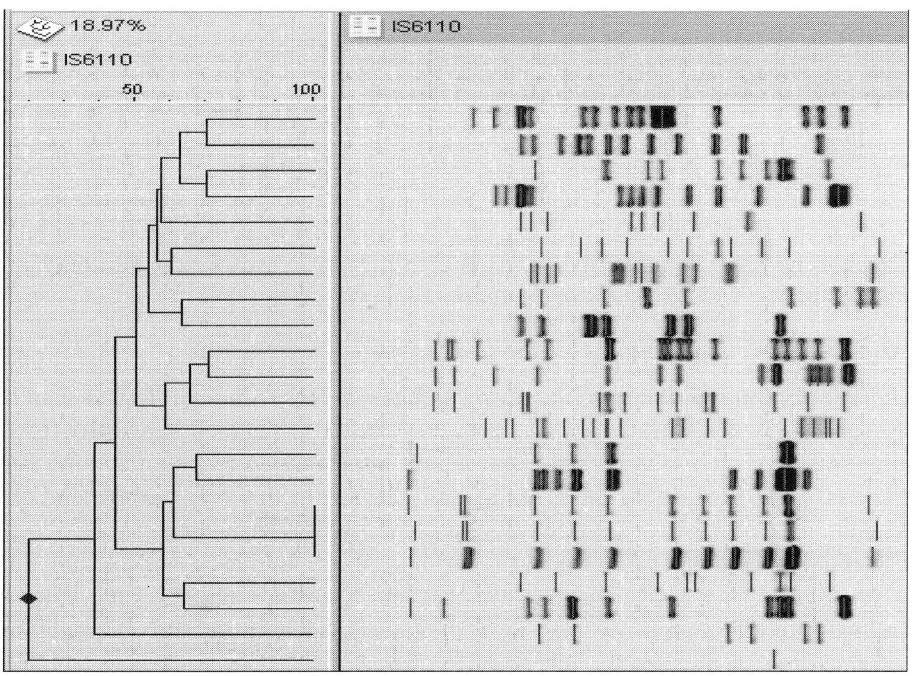

Fig. 3.9. Clustering of the *M. tuberculosis* IS*6110* RFLP patterns shown in Figs. 3.2 and 3.3 with the Dice coefficient. Other settings were identical to those used in Fig. 3.8.

terns, and can be adjusted by the user.

In cases where the physical and biological techniques lead in theory to binary results (i.e., there either is a fragment or there is not), such as with IS*6110* RFLP typing, it is recommended that band-based SCs are used. The lack of complete 100% reproducibility of the patterns again leads to small misalignments when assigning bands. *Position tolerance settings* enables the position tolerance to be specified, which is the distance allowed between the positions of two bands at which they will still be regarded as matching. This setting can be adjusted by the user. As when using the Pearson correlation coefficient, the *optimisation* feature then enables track-to-track corrections to be made to compensate for the remaining misalignments by shifting one or both tracks with respect to the other until they reach their maximum correlation; this value is the SC for these patterns.

3.3 FACTORS DETERMINING THE COMPARABILITY OF DNA PATTERNS

Each of the steps mentioned in the previous section has its own factors that determine the reproducibility of the final results. These factors are discussed in more detail below.

A. Resolution of the scanned images

A flatbed scanner can be used to scan polaroid photographs of a gel. When the patterns are reproduced on films or autoradiograms, the scanner should be extended with a transparency adapter in order to obtain the best results. Most modern flatbed scanners have a maximum resolution of at least 300 dpi and also at least 256 grey tones. This means that for a set of patterns with a total size of 10 × 15 cm (or about 4 × 6 inches), the computer image is about 1200 × 1800 pixels. This is an image size that can be processed by modern computers and gives very good results. It also leaves some room to adapt the scanned area. In most cases the resolution can be adapted to optimise file sizes in the computer.

Negatives and slides with a size of 24 × 36 mm can also be transferred into computer images, although an ordinary flatbed scanner with a hardware resolution of 300 or 600 dpi is inappropriate for scanning slides! Although flatbed scanners are supplied with software to increase the resolution of scanned images to 4800 or even 9600 dpi by applying optimisation and interpolation techniques, the results are very poor, at least for the goals of this chapter. In this case a dedicated slide scanner, with a higher resolution (at least 2000 dpi) has to be used.

A disadvantage of a CCD camera is the limited, and in most cases fixed, number of pixels, e.g., 768 × 480. Further, the absence of a zoom objective may limit the optimal use of the number of pixels, particularly if the relevant area is small compared to the area covered by the camera.

When DNA patterns from slides (24 × 36 mm), autoradiograms (e.g., with a format of 10 × 15 cm) and/or polaroids (e.g., with a format of 90 × 90 mm) have to be merged into one database, the resolutions have to be tuned to obtain an optimal result. In the example above, 300 dpi was used for the autoradiogram. Since a slide is about one-quarter the size of the autoradiogram used, a slide would have to be scanned in this case with a resolution of about 1200 dpi, whereas a polaroid would have to be scanned with a resolution of 10/9 × 200 = 220 dpi, assuming that the electrophoresis direction is parallel to the small edge of the autoradiogram.

Images can be saved in different file formats. The most commonly used formats are Bit Map (BMP), Tagged Image File Format (TIF or TIFF), and Graphical Image Format (GIF). Image formats like JPEG or JPG are based on compression techniques to decrease image file sizes drastically. These "lossy" formats are very useful for applications where the file sizes play an important role, e.g., Internet applications. However, the compression techniques result in a loss of information and the introduction of errors, and these formats should thus never be used for the goals described here. The GelCompar 2 and BioNumerics software programs only accept TIFF files.

In order to assign the positions of bands on computerised images of DNA patterns accurately, each band has to be covered by at least several pixels. At RIVM, the minimum band thickness of an IS*6110* RFLP DNA pattern on an autoradiogram with fingerprints of *M. tuberculosis* strains is about 1 mm. For an optimal result, initially at least 8 pixels have to cover the area of a band. Within the framework of the CAonTB project, this means that the optimal scanner setting is 200 dpi. If BioNumerics software is being used, the resolution is then transformed to about 160 dpi (700 pixels for the whole normalised image), which means that each band is covered by at least 6 pixels. By doing so, the maximum error, introduced by the number of pixels, is 1/700 or 0.14%. This error is relatively small and negligible, at least when it is compared to the other possible sources of error. Of course, values and numbers mentioned above are just examples, and may differ for each type of experiment.

B. Background subtraction and optimising the OD values

To optimise the quality of the DNA patterns, it is possible to apply background subtraction and optimise use of the OD scale in several of the various steps. The Pearson correlation coefficient is invariant to global shifts in the values of the arrays of elements and multiplication of the value with a factor. Pearson's correlation, *r*, is expressed by the following formula:

$$r = \frac{\sum(x-\bar{x})(y-\bar{y})}{\sqrt{\sum(x-\bar{x})^2(y-\bar{y})^2}}, \tag{1}$$

where $\bar{x}$ = average of n OD values x and $\bar{y}$ = average of n OD values y.

If a constant value A is then added to each array element of x (or y) (i.e., each element x is replaced with x + A), this shift does not influence the value of the correlation factor r (Fig. 3.10). In a similar way, it is easy to prove that if each

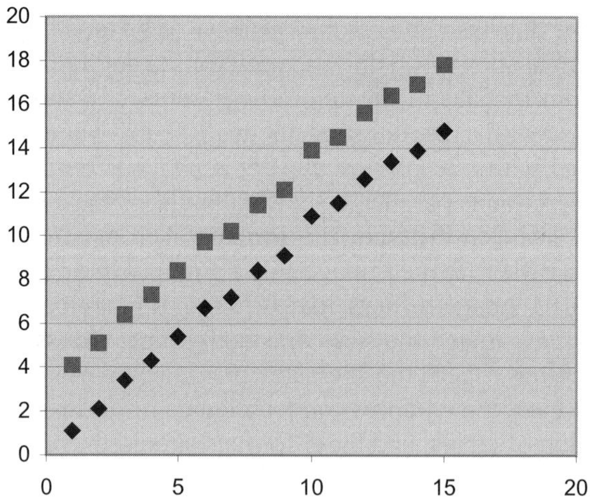

Fig. 3.10. Example of a series of 15 points x_i and a series of 15 points x_i +3. The points in both series have the same correlation (0.997). The correlation between both series is 1 !

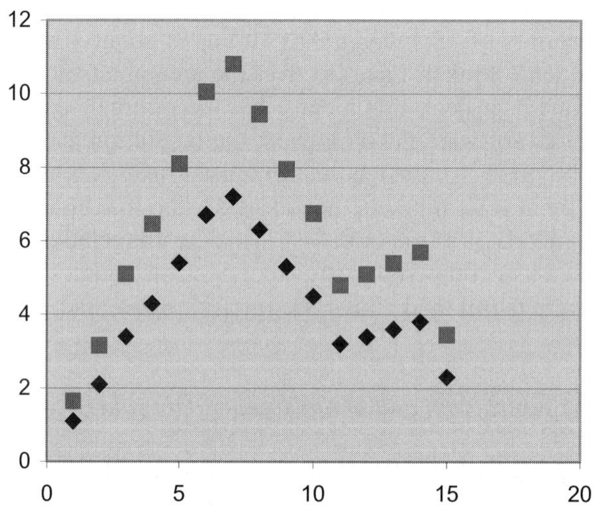

Fig. 3.11. Example of a series of 15 points x_i and a series of 15 points $1.5*x_i$. The points in both series have the same correlation (0.022). The correlation between both series of points is 1 !

OD value x (or y) is replaced with Ax (A is the constant factor, and in fact each array element is multiplied with A), that this factor can be erased from the formula and thereby does not influence the SC (Fig. 3.11). From this it can be concluded that both background subtraction and optimisation of the OD scale leave the SCs unaltered, and thus that they do not influence the final clustering. Theoretically, r can have each value between –1 and 1 (Everett, 1986; Romesburg, 1990; Armitage *et al.*, 1991; Norusis, 1993); however, due to the types of arrays (the OD values) for DNA patterns, r will never be 0 or less.

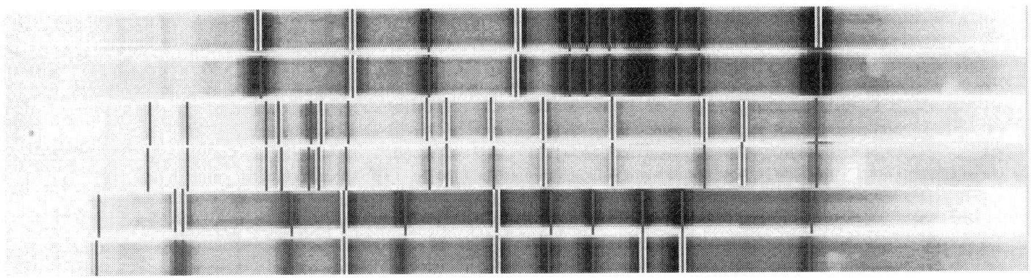

Fig. 3.12. Three ERS as used within the CAonTB. Each marker was processed twice on the same gel. Top, H37*Rv*; middle, TN2650; bottom, Mt14323. It is clear that H37*Rv* does not cover as much of the area as TN2650 and Mt14323.

C. Internal size markers (ISM) and external reference strains (ERS)

In order to be able to compare DNA patterns from different gels and from different laboratories in a computerised way, it is essential to use ISM and/or ERS. In this way it is possible to compensate for errors and misalignments. This is a necessary condition to be able to build databases with large numbers of DNA patterns for the purposes mentioned in the Introduction.

(i) ERS

By loading marker DNA in a few lanes of each gel, differences in migration between DNA fragments of a particular size on different gels and within gels can be corrected by interpolation techniques. Marker DNA should be run in at least three lanes when the gels used contain 22 lanes (i.e., lanes 1, 11 or 12 and 22), but preferably the ERS should be used after every four or five lanes (i.e., lanes 1, 6, 11, 16 and 22). The markers should cover at least the whole molecular size range of the fragments to be analysed. This means that, before selecting an ERS, a representative set of strains has to be fingerprinted. The markers can be a commercially available size ladder or a very specific strain that satisfies the conditions (Fig. 3.12). For similar reasons, this representative set of strains can be used to optimise the area to be scanned at the point when the standards are defined.

In cases where the ERS does not cover the whole "spectrum" of the DNA fragments, the results of the normalisation phase will be unpredictable for those bands that reflect fragments with a higher or a lower molecular size than the highest or lowest bands of the ERS. This is because the normalised band positions are calculated through extrapolation. As can be seen in Fig. 3.13, the ERS designated H37*Rv*, which was used for a long time as the ERS for analysis of *M. tuberculosis* strains, is not an optimal choice as about 15% of all bands belonging to IS*6110* RFLP DNA patterns of *M. tuberculosis* strains fall outside the area covered by the H37*Rv* bands. It should also be noted that if a pattern has only one neighbouring ERS, the alignments of the different patterns will also be the result of extrapolation. The normalisation process will then lead to unpredictable results.

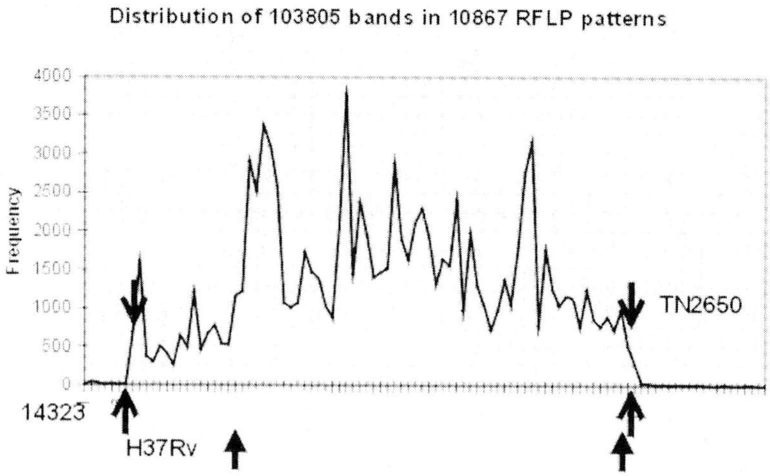

Fig. 3.13. The distribution of the bands of the three ERS in Fig. 3.12 plotted against the distribution of the bands of a large collection of *M. tuberculosis* strains. About 15% of the bands are not covered by H37*Rv*, and the results of the normalisation when using H37*Rv* for the bands representing higher molecular size fragments should be considered unreliable.

(ii) ISM

The use of ISM in each lane is a more accurate way to correct for inter- and intra-gel differences in DNA migration. The DNA of the strains to be analysed is mixed with a size marker DNA, which hybridises with a different probe, and two hybridisations are carried out: first to detect the DNA fragments to be analysed, and second to detect the marker DNA. Thus, two autoradiograms are obtained, one with the banding patterns of the strains to be analysed and one with ISM patterns. The basic idea behind this technique is that all disturbances during electrophoresis, leading to distortions in the patterns, affect the mobility of both the DNA fragments and the ISM fragments equally.

The most critical step when applying ISM is the superimposition of the two autoradiograms. The two autoradiograms are marked with marker points to enable the precise superimposition of the two images (this is the most critical step with regard to the introduction of errors). These marker points can be made by marking the membrane with a DNA mix, containing both strain and internal marker DNA. At the RIVM, a mix of commercially available DNA size markers is used as internal markers for the normalisation of the *M. tuberculosis* strains (super-coiled DNA-*Pvu*II/*Phi*X174-*Hae*III; Fig. 3.2). The application of ISM enables the elimination of most of the inaccuracies caused by heterogeneities in the gel and other influences for each individual pattern. Computer software allows both images to be superimposed by using the marker points (Heersma *et al.*, 1998; Van Soolingen *et al.*, 1999b).

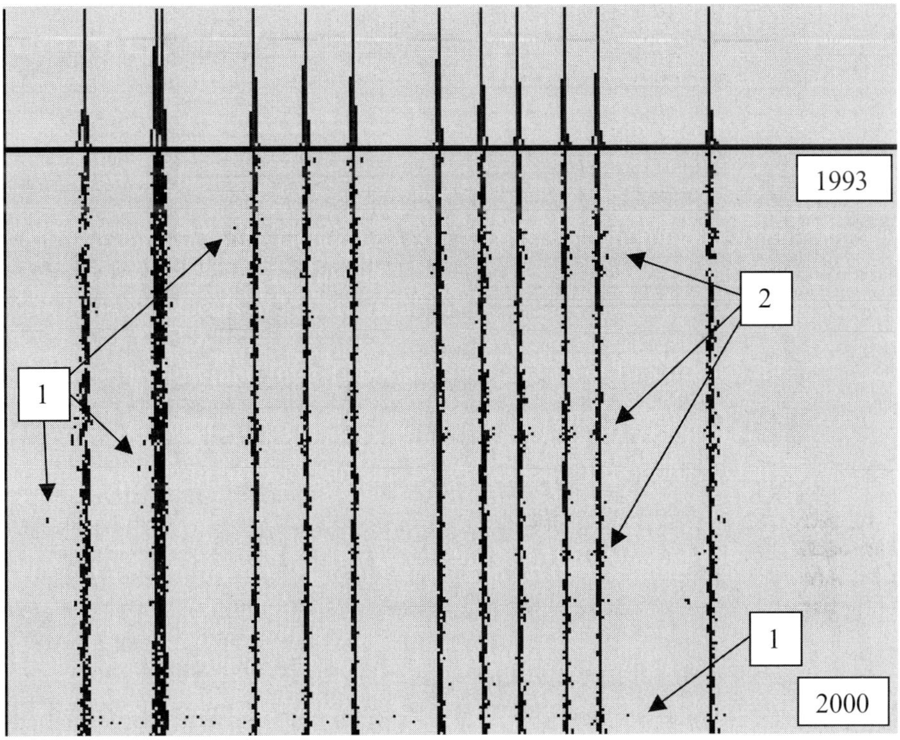

Fig. 3.14. The band positions of Mt14323 through the years, showing the ordering of 793 Mt14323 profiles as processed by the computer. Some of the results were rejected (1). Some misalignments were caused by tests with new marker techniques or new superimposition techniques (2) and these results were also rejected.

(iii) General recommendations for selection of ERS and ISM

Reference strains and markers suitable for setting up massive databases with many DNA patterns have to satisfy two important conditions:

- They must cover at least the whole spectrum of the DNA patterns from the visible fragments with the highest molecular sizes to the smallest visible fragments;
- The reference bands have to be distributed equally across the spectrum.

From Fig. 3.2, it can be seen that the marker used at RIVM (viz. super-coiled DNA-*Pvu*II/*Phi*X174-*Hae*III) does not fully satisfy the second condition. However, this has not influenced the reproducibility. This has been tested by comparing the matching of the patterns of the ERS Mt14323 through the years, because four of the 12 bands of Mt14323 are well-distributed in the area in the middle of this ISM (Fig. 3.14).

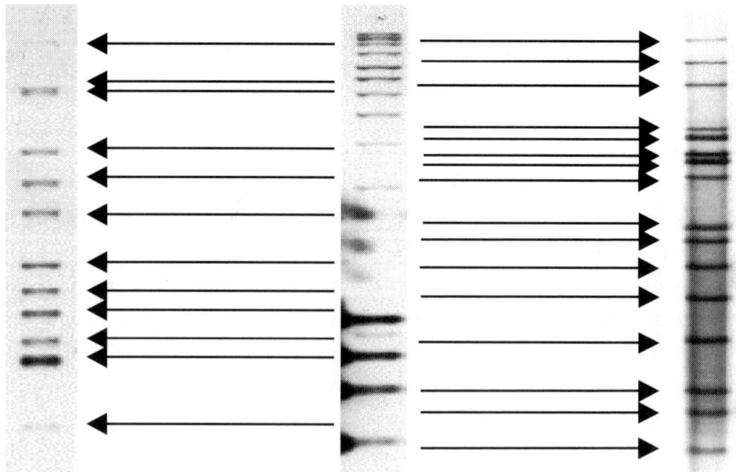

Fig. 3.15. Tuning ISM and ERS.

(iv) The use of different ISM and ERS within one project

Under well-defined circumstances, it is possible to use different ISM and ERS within one project. Within the framework of the CAonTB, all institutes use the same laboratory technique with regard to the choice of restriction enzyme, probe, etc. Although most of the participating institutes use one specific ISM, some use different ISM and/or ERS. At least three different ERS and two ISM were used within the ambit of the project.

One problem was that DNA patterns of identical strains, processed at different laboratories, did not cluster when using the standard tolerance settings. When comparing the fragment sizes of the ERS designated TN2650 (Fig. 3.15), which is considered as good an ERS as Mt14323, it appeared that there were small, but significant discrepancies between the molecular sizes of the ERS processed in different laboratories. These discrepancies influence the final analysis of DNA patterns dramatically. Originally, TN2650 was considered as a normal DNA pattern, and therefore the molecular sizes of TN2650 were based on a different ISM used within the CDC tuberculosis network in the USA, whereas the molecular sizes calculated in Bilthoven were based on the CAonTB "standard" ISM, super-coiled DNA-*Pvu*II/*Phi*X174-*Hae*III. Apparently, this reference strain and the size marker were not properly "tuned".

(iv) Tuning ISM and ERS

It was decided to use the ERS Mt14323, used at RIVM and in most of the laboratories collaborating in the CAonTB, as the 'gold standard' from which all other settings for different ERS and ISM could be derived. In a repeated experiment, all three ERS (Mt14323, H37*Rv* and TN2650) were processed on several autoradiograms with the same ISM, and then analysed by the computer. The result was a set of two derived standards (a set of reference band positions) based on Mt41323.

Next, the experiment was repeated for the other ISM (Marker X; Boehringer), again with the Mt14323 settings as the reference setting. The result was another derived standard, in this case an ISM. It appeared again that there was a significant discrepancy between the molecular sizes as supplied by the manufacturer and those calculated at RIVM. No explanation for this finding is readily available. From this and other experiments, it is clear that laboratory circumstances may differ, and that unexpected differences in databases may appear which influence the reproducibility.

Nevertheless, using this approach it was possible to ensure the interchangeability of databases based on different ISM and/or ERS, and to ensure the reproducibility of a central database. Each laboratory was supplied with a set of standard settings, based on the individual laboratory standards with respect to the ERS and/or ISM (Fig. 3.2).

D. Comparison of results when using ISM or ERS

There are pros and cons to the use of the ISM and ERS techniques. In most cases, applying ERS is cheaper for a number of reasons: e.g., no second film material is necessary, no extra marker DNA and extra hybridisation is required, less time consuming, etc. On the other hand, when applying ERS, at least two lanes, and preferably more, have to be filled with the ERS itself, which means a less economical use of the gels. A major drawback of applying ERS is the inability to remove all the distortions in profiles. Generally, the reproducibility of a profile is a reciprocal of the distance from the two neighbouring ERS lanes.

The most delicate part of the process when applying ISM is the superimposition of the autoradiogram containing the ISM and the second autoradiogram with the DNA in question. Simple techniques have been developed to ensure comparability (Heersma *et al.*, 1998). Each profile can then be aligned individually to remove distortions and misalignments (Fig. 3.2).

Within the framework of the CAonTB project, one of the challenges was to set up an international database of DNA fingerprints of *M. tuberculosis* strains, based on the analysis of RFLPs associated with IS*6110*. This database contains DNA patterns produced at about 30 different laboratories, spread throughout the world. The banding patterns usually consist of 1–24 bands of equal intensities, although some IS*6110*-containing *Pvu*II fragments migrate to similar positions on a gel, which may result in bands which appear broader and have a higher intensity. Each fragment is represented by a band position that is assigned either by the analysis software or manually. Broader and more intense bands are only considered as being two or more individual IS*6110*-containing fragments if the fragments can be distinguished by eye (Van Soolingen *et al.*, 1999b).

An inter-laboratory study was carried out to investigate the differences in the

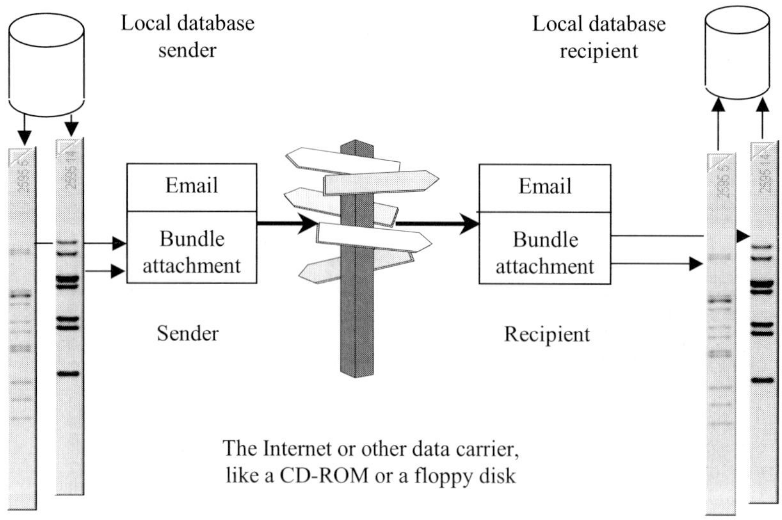

Fig. 3.19. The principles of a bundle. Typing results from different sources are grouped and can then be attached to an email or sent on a floppy disk or CD. The recipient laboratory can unpack or detach the bundle, add the results to the local database, and then apply a comparison or clustering.

population-based studies. In principle, if a fast typing technique is available, it is possible to set up "early warning" or "alert" systems, e.g., for surveillance of the transmission of specific phenotypes of an organism, such as multidrug-resistant (MDR) strains. The bundles concept, which is an open and well-documented standard developed by Applied Maths as part of the BioNumerics program, is a very flexible tool to support these goals. From a user-defined selection of information fields (often DNA patterns, but also other experimental types or data such as genus, species information, etc.) from a database, a bundle can be constructed. In the case of DNA patterns, the bundle contains information about the reference system and the molecular size settings. In this way it is very easy to remap or convert bundles to be compatible with databases based on different standards for the same organism. Bundles can be sent on different types of data carriers to other researchers, either as an attachment to an email or on a floppy disk or CD-ROM. The recipient can unpack the bundles in a very easy way (Applied Maths, 1999), add them to the relevant database, and then compare them with other DNA patterns (Fig. 3.19).

D. A=B and B=C does not mean A=C

In the world of mathematics, the transitive property says that if A=B and B=C, then A=C. However, this property does not always hold true for DNA fingerprints. As can be seen in Fig. 3.20 (1) and (2), the sets of strains A+B and B+C cluster at 100%. However, strain A does not cluster 100% with strain C, as shown in Fig. 3.20 (3). Moreover, if a dendrogram of the three patterns is created, one

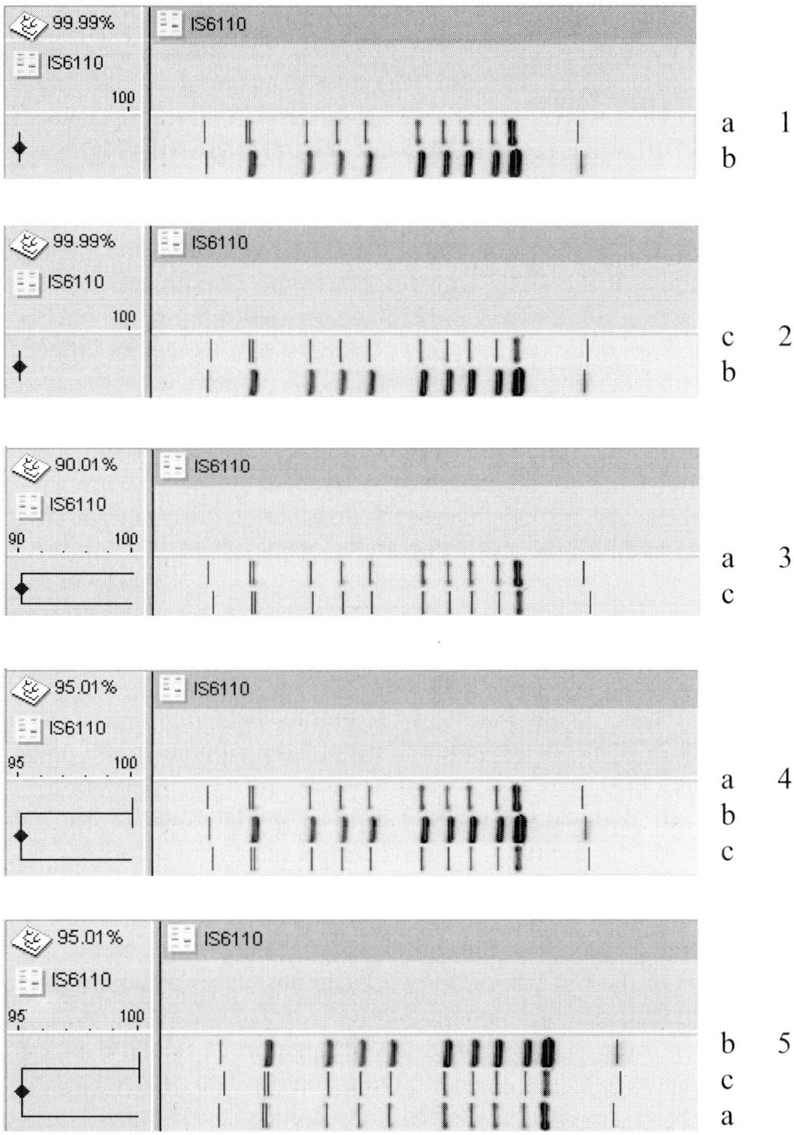

Fig. 3.20. The transitive property does not always apply to DNA fingerprints. In (1), strain *a* clusters 100% with strain *b*, and in (2), strain *b* clusters at 100% with strain *c*. However, in (3), strain *c* does not cluster with strain *a*. Further, if the ordering of the strains to be analysed is changed, the dendrogram may change, as shown in (4) and (5).

pattern does not cluster with the other two. The reason for this is the way that UPGMA clustering takes place. First, pattern one and pattern two are clustered, and the result is one average pattern, This new average pattern does not cluster 100% with the third pattern. If the order in which the clustering is calculated is changed, e.g., if patterns B and C are clustered first, pattern A will not cluster at

100% with the cluster of B and C, as shown in Fig. 3.20 (4) and (5). This is a disadvantage of the UPGMA method that has to be accepted.

3.5 SETTING UP INTRA- AND INTER-LABORATORY DATABASES OF ELECTROPHORETIC PROFILES: RECOMMENDATIONS

As a first stage, the laboratory work must be standardised with respect to the restriction enzyme, probe and sometimes also the size markers, and the fingerprinting technique must fit the goals with respect to the discriminative power. After this complicated process, a start can be made on analysing the laboratory results from different laboratories. This section considers various factors influencing the reproducibility of databases, as well as methods to compensate for these sources of error. It is very difficult to give a general "rule of thumb" for each type of experiment and organism. The following steps are suggested as preparations before setting up a live inter-laboratory database (it is assumed that one laboratory serves as the reference laboratory, responsible for the distribution of strains and fingerprints within the study, and also the collection and analysis of results from the different laboratories):

1. Inter-laboratory reproducibility study of both the laboratory and the computer analysis phases:
   - Collect a representative set of strains;
   - Select a reference strain that fits the requirements;
   - Select an ISM that fits the requirements (if the laboratory technique allows the use of an ISM);
   - Distribute an identical set of strains to different laboratories for fingerprinting;
   - Collect and analyse the results of the laboratory work;
   - Create an identical set of relevant fingerprints for each laboratory;
   - Each laboratory processes the strains and the sets of fingerprints;
   - The results of the wet laboratory work (in most cases an autoradiogram) are returned to the reference laboratory;
   - The results of the analysis of the sets of fingerprints can be returned as bundles, as an attachment to an email, or on another data carrier;
   - Both sets of results are analysed at the reference laboratory.

   In this way it is possible to get a good estimate of the errors introduced during the wet laboratory work and the computer analysis separately for each laboratory. At this point it is possible to tune the processes as accurately as possible. This is a necessary step to ensure the reproducibility of the database. Then:
2. If it is desired to use different ISM and/or ERS, one of the ISM or ERS should be selected as the "gold standard" for the project:
   - Create a database with the band reference positions of the different ISM and ERS;
   - Supply each laboratory with the settings.
3. In many cases, additional information on the isolates, such as epidemiological,

clinical, laboratory and patient data, is necessary. Before the start of the project, it is very important to have a commonly accepted description of the different database fields. Within the CAonTB, the conversion of these fields required a disproportionate amount of effort. The use of internationally accepted coding systems, such as ISO codes for countries, national codes for city of origin, ICD-10 codes for disease symptoms, codes for susceptibility and resistance levels, date fields (date of birth, date of isolation: yy-mm-dd or dd-mm-yy?), gender (male, female, unknown or M, F, ?), etc., are all basic things that can help to minimise the work.

4. Database management should be organised. A central approach, in which one laboratory is responsible for the final submission of DNA patterns and additional data to the general database, has proved to be a necessary condition. Apart from the DNA patterns, it is also necessary to exchange the ERS data for validation purposes.
5. If band-based comparison and clustering are used, the methods used to analyse and cluster DNA patterns should be decided, i.e., either Dice or Jaccard. Long-lasting discussions of the pros and cons should not be entered into as the final result will be the same with respect to clustering and branching of the dendrogram.

Sometimes compromises are necessary. Databases with DNA patterns based on applying different ISM and/or ERS and other deviant settings may already exist locally before setting up an inter-laboratory database of DNA patterns. Under certain circumstances it is possible to convert or remap the DNA patterns from one standard to another (Applied Maths, 1999).

3.6 FINAL REMARKS

The guidelines and standards presented in this chapter are based on experience gained during a project lasting from 1993 until the present. Many laboratories and institutes world-wide have collaborated in this project. A database with over 10,000 *M. tuberculosis* DNA patterns is available for use by scientists, both from within and outside the project. It is impressive, and often surprising, to see how DNA patterns from newly isolated strains cluster exactly to strains already in the database. In many cases, this indicates a cross-border infection with a particular *M. tuberculosis* strain. Many examples that underline the power of the approach have been published.

However, these guidelines and standards should not be interpreted as being universally applicable. Other organisms, different typing techniques, different laboratory circumstances, and many other factors, will determine the reproducibility of databases of DNA patterns. An inter-laboratory study as described in this chapter is therefore a necessary requirement. There is no general prescription on how to calculate or predict the different errors, misalignments, etc.

As an example of the possibilities of an international database of DNA patterns,

4 Fingerprinting of Microorganisms by Protein and Lipopolysaccharide SDS-PAGE

Lenie Dijkshoorn
Department of Infectious Diseases, Leiden University Medical Center, PO Box 9600, 2300 RC Leiden, The Netherlands

CONTENTS

4.1 INTRODUCTION

Characterisation of microorganisms on the basis of their protein electrophoretic profiles and, to a lesser extent, their lipopolysaccharide (LPS) electrophoretic profiles, has been performed in numerous studies since the 1970s. In the beginning, the primary aim of most studies was to investigate the chemical composition of cells or cell fragments, but it was soon realised that the profiles obtained could also be used to compare large numbers of strains for their relatedness. The two approaches used most commonly for comparative protein analysis of many strains are sodium dodecylsulphate-polyacrylamide gel electrophoresis (SDS-PAGE) and multilocus enzyme electrophoresis (MLEE). In particular, one-dimensional protein SDS-PAGE, once set-up and standardised, is a robust method that allows the construction of large databases of strains from numerous bacterial species. The same technology has also been found useful for typing microorganisms on the basis of their LPS composition.

Since the mid-1980s, nucleic acid-based technologies have advanced rapidly and a variety of DNA fingerprinting methods have now largely superseded protein profiling for typing bacterial strains. Even MLEE, which was for long the gold-standard method in population biology, is currently being replaced by a DNA sequence-based method, multilocus sequence typing (MLST; see Chapter 12). However, one-dimensional protein SDS-PAGE remains a powerful method for structural studies and for typing and classification of microorganisms. Similarly, LPS analysis is still an important method when screening organisms for their LPS phenotype, and is also still used to type organisms. The focus of the present chapter is on the comparative analysis of microorganisms on the basis of their protein and LPS profile as observed by SDS-PAGE. Further information on MLEE can be found in Chapter 12.

4.2 PROTEINS AND LIPOPOLYSACCHARIDES AS BACTERIAL CELL COMPONENTS

A. *The bacterial cell*

A general outline of bacterial cell composition can be found in major text books on microbiology (e.g., Brock *et al.*, 1994). A characteristic feature of the bacterial cell that distinguishes it from eukaryotic cells is the complex of layers surrounding the cytoplasm. This complex of layers, denoted the bacterial cell envelope, differs significantly between Gram-negative and Gram-positive bacteria. In Gram-negative bacteria, the basic structure comprises the cell (cytoplasmic) membrane, the periplasmic space which also includes the peptidoglycan layer, the outer membrane and, depending on the strain or species, capsule, slime and appendages such as flagellae or pili. The Gram-positive cell envelope contains a cell membrane surrounded by a complex matrix, including a dense three-dimensional peptidoglycan network interwoven with long, flexible chains of teichoic or teichuronic acids; there is no outer membrane, but capsule or slime may be present. The precise locations of many proteins in the Gram-positive cell wall are unknown; some proteins may be embedded in the matrix, whereas others, including protein A of *Staphylococcus aureus* are probably located at the periphery. An extensive overview of the cell surface of bacteria in general can be found in Hammond *et al.* (1984), while the architecture and function of the outer membrane of *Escherichia coli* and other Gram-negative bacteria has been reviewed by Lugtenberg & Van Alphen (1983).

B. *Proteins in the bacterial cell*

Proteins, being a major constituent of any living cell, occur both in the bacterial cytoplasm and in the cell envelope, and serve many functions, e.g., as catalysts, as structural components, as vehicles in transport, as organs of movement (flagellae), or as protective agents. Analytical studies using polyacrylamide gel electrophoresis (PAGE), with and without denaturing agents, have contributed considerably to the knowledge of certain proteins in the outer cell membrane of Gram-negative organisms and their biological function (see e.g., Lugtenberg, 1981; Lugtenberg & Van Alphen, 1983; Koebnik *et al.*, 2000). In comparative protein analysis for epidemiological or taxonomic purposes, microorganisms are grouped according to the protein profiles displayed by the use of SDS-PAGE, usually without any knowledge of the functions of the different proteins composing the profiles.

C. *Lipopolysaccharides in the Gram-negative cell envelope*

LPS is a major constituent of the external layer of the outer membrane in Gram-negative bacteria, with numerous biological functions (Rietschel & Brade, 1992;

E. Computer-assisted protein profile analysis

Protein profiles of microorganisms are relatively complex, with at least several tens of bands, which makes visual comparative analysis difficult. This was realised in the 1970s by Kersters & De Ley (1975), who consequently pioneered computer-assisted analysis of protein profiles. Their approach has been refined extensively over the course of time, and several different software packages (described elsewhere in this book) are now available for analysis of complex electrophoretic profiles, including DNA profiles. The basic principles and technical details of computer-assisted electrophoretic profile analysis are outlined in Chapters 2 and 3; some specific details for protein analysis are given in section 4.4.F.

4.4 PROTEIN SDS-PAGE IN PRACTICE

A. Fractions for investigation

The protein fractions used for comparative analysis of bacteria by SDS-PAGE belong to two major categories; i.e., whole-cell preparations, and cell envelopes or outer membrane preparations. Whole-cell preparations are the fractions used most commonly for protein profiling, both of Gram-positive and Gram-negative bacteria (Pot *et al.*, 1994). The preparations can be obtained easily by boiling cells in a reaction mix which causes lysis of cells, as well as denaturation and solubilisation of cell proteins; however, additional treatment, such as lysozyme or ultrasonic treatment, may sometimes be required. Whole-cell protein electrophoretic profiles are characterised by a large number of protein bands.

If Gram-negative bacteria are the subject of study, it can be advantageous to use cell envelopes or outer membrane preparations, since the profiles of these fractions comprise relatively few, densely stained bands and are more easy to interpret by eye as compared to whole-cell protein profiles. Methods for preparation of cell envelopes and outer membrane preparations have been reviewed by Lugtenberg & Van Alphen (1983). Cell envelopes can be obtained by ultrasonic disruption of cells combined with fractionated centrifugation, as described by Lugtenberg *et al.* (1975). Dijkshoorn *et al.* (1987a, b) used a slight modification of this procedure involving bench-top laboratory centrifugation rather than ultracentrifugation. Outer membrane fractions for comparative protein analysis are prepared most commonly by selective solubilisation of cell envelopes with detergents. However, the profiles of cell envelopes are, to a large extent, determined by the composition of the outer membrane proteins (Lugtenberg *et al.*, 1975); hence, in comparative analysis, it may suffice to use cell envelopes rather than outer membranes.

B. Sample preparation

Sample preparation is preceded by cultivation of the bacteria being studied, but culture conditions may differ depending on the precise organisms being investi-

Table 4.2. Sample buffer composition (Laemmli, 1970)

Tris-HCl	0.0625 M, pH 6.8
SDS	2% (w/v)
Glycerol	10% (w/v)
Bromophenol blue	0.001% (w/v)
β-Mercaptoethanol	5% (v/v)

gated. Once it has been established which medium and conditions will be used, care should be taken to adhere strictly to these conditions in order to obtain reproducible results.

(i) Preparation of whole-cell lysates

The modified method described here has been found useful for *Acinetobacter* (Dijkshoorn *et al.*, 1990) and *E. coli* (personal unpublished results). Bacterial stocks in glycerol broth at –80°C are subcultured on to blood agar and incubated overnight at the most suitable temperature for the organism being studied (e.g., 30°C for *Acinetobacter* or 37°C for *E. coli*). A subculture in 5 ml trypticase soya broth from four morphologically indistinguishable colonies is incubated overnight. From this culture, 20 drops are spread equally on to a trypticase soya agar plate which is incubated for 24 h. Next, cells are harvested by flooding the plate with several ml of phosphate buffered saline (PBS) and scraping off the cells. The cells are resuspended in PBS to a final volume of ≤30 ml, depending on the organism. The resulting bacterial suspension is centrifuged for 20 min at 10,000 *g*. The supernatant is discarded and the cells are first washed by resuspending them in ≤30 ml PBS, and then harvested by centrifuging for 10 min at 10,000 *g*. This washing and centrifugation step is repeated at least once. The supernatant is discarded after the final centrifugation step and quantitities of 100 mg wet weight of protein are dissolved in 2 ml of sample buffer (Table 4.2). The suspensions are then boiled for 10 min, followed by centrifugation for 10 min at 7000 *g* and 4ºC to sediment large cell fragments and cells that were not lysed. The supernates are then ready for use, or can be stored at –20°C for short periods or at –80°C for prolonged periods. If stored, they should be heated again for 5 min at 100°C before electrophoresis.

Some modifications of the procedure may be required for organisms that produce large amounts of polysaccharides or that are difficult to lyse, which may be the case for Gram-positive bacteria. Sedimentation of polysaccharide-producing organisms requires centrifugation for longer periods or at higher speed. The polysaccharides may interfere with the electrophoretic separation of proteins, and then have to be removed by repeated cycles of washing and centrifugation of cells (see above), sometimes preceded by mechanical disruption using rotating blades (Ultraturrax; Jahnke & Kunkel, Staufen, Germany). For Gram-positive bacteria, additional treatment before solubilisation in buffer may be required, including ultrasonic disruption or use of lysozyme to digest the peptidoglycan layer. Ultrasonic disruption of washed cells (see above) is done while cooling the cells on ice and fresh Coomassie Blue 0.15% (w/v) dissolved in 30:10 (v/v) methanol:acetic acid is added to the gel, which is stained for 1 h at room temperature with shaking. Excess stain is removed by replacing the staining solution with 25:10 (v/v) methanol:acetic acid and shaking. This step is repeated at least three times with fresh destaining solution.

Staining with Fast Green FCF has been used in studies with *E. coli* and *Acinetobacter* spp. (Lugtenberg *et al.*, 1976; Dijkshoorn *et al.*, 1990). The procedure involves fixation and staining simultaneously for 24 h at room temperature with shaking in 0.1% (w/v) Fast Green FCF dissolved in 50:10 (v/v) methanol:acetic acid, followed by destaining in 45:10 methanol:acetic acid. The final staining level of the gel can be standardised by choosing an acceptable level of stain (e.g., 0.0002% (w/v) Fast Green FCF in 45:10 (v/v) methanol:acetic acid), and equilibrating gels in this 'after-staining' solution before photographs or digital images are made. A sample of the 'after-staining' solution is kept as a control for a visual or photometric check during destaining. The dye in the 'after-staining' solution is not only important for standardisation of the staining level, but may also prevent certain bands losing dye during storage (Wilson, 1979).

E. Storage of gels

Gels can be stored for weeks or months in 'after-staining' solution. During such periods, the methanol may evaporate, which will cause swelling of the gels. This process is reversible, and the gels may regain their original size and aspect when they are re-equilibrated in 'after-staining' solution. For long-term preservation it is more practical to dry the gels. There are several systems available for drying gels, including conventional systems which dry gels with heating and/or vacuum. A more simple and convenient system is based on drying the gels in air with heating. For this purpose, gels are placed in a framework between sheets of cellophane and dried in a heated chamber. Once in the dry state, the gels can be stored easily, and subsequently reinspected or scanned at any time.

F. Computer-assisted profile analysis

(i) Sample and reference distribution

A prerequisite for successful pattern analysis is the preparation of high quality gels with a reference extract for normalisation in three or four lanes distributed equally over the gel. An example of a protein SDS-PAGE gel with 20 lanes is shown in Fig. 4.1. Proteins in the two outer lanes usually show a lower migration speed than proteins in other lanes, a phenomenon termed 'smiling'. These distortions may complicate pattern analysis, and therefore the outer lanes should not be used for test or reference extracts. The smiling effect can be reduced by running sample treatment buffer, of the same composition used for the other protein samples, in lanes 1 and 20 (see Fig. 4.1), or by loading a protein extract which is not part of the analysis. A reference extract for normalisation is included in lanes 3, 8, 13 and

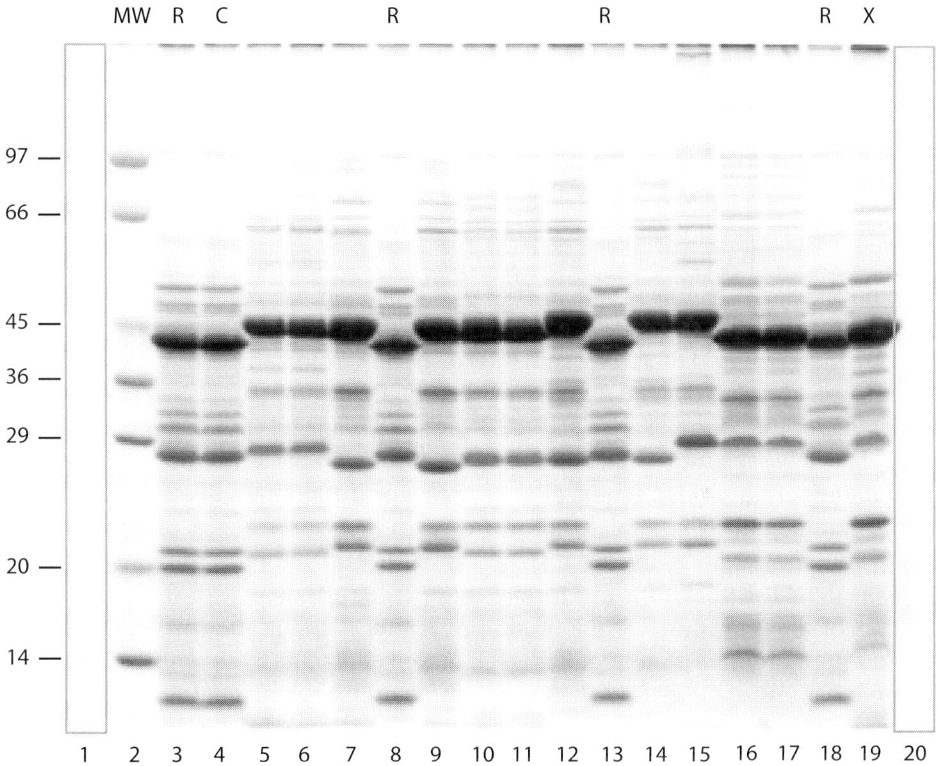

Fig. 4.1. Example of SDS-PAGE cell envelope profiles of *Acinetobacter* strains. Samples and reference extracts have been applied to the gel to conform with the requirements for computer-assisted gel analysis. Lanes 1 and 20, sample treatment buffer without protein (included on the gel in order to reduce smiling); R, reference extracts distributed equally for normalisation purposes; C, freshly prepared extract of the same strain used for normalisation (used as a control for the sample preparation procedure); X, sample extract with bands outside the reference area, and therefore not included in the analysis.

18 of the gel. This reference extract is selected on the basis that its profile contains multiple clear and equally distributed protein bands in the same size range as those of the unknown organisms being studied. A large stock of this reference extract in sample treatment buffer (Table 4.2) is prepared; this can be stored at –80°C for prolonged periods, or at –20°C for shorter periods. One lane (lane 4) contains another preparation from the reference strain, but this extract is freshly prepared at the same time as the test extracts, and serves as a control for the preparation method. In addition, one lane usually contains a calibration standard with proteins of known molecular sizes solubilised in the same buffer as the extracts being investigated. Thus, nine of the 20 lanes of the gel are used for quality and standardisation purposes, and 11 lanes remain for test extracts. The limited number of lanes available for test extracts is compensated for by the fact that extracts on high quality gels which include multiple reference lanes can be used to construct libraries of profiles and do not have to be run again.

(ii) Computer-assisted profile analysis

Images can be captured for computer-assisted analysis by flat-bed scanning of photos, by laser densitometry of dried gels, or by charge coupled device (CCD) video cameras. As discussed in Chapter 3, the resolution of most CCD cameras is lower than that of scanners, but CCD technology is rapidly improving, and resolutions of up to 1360 × 1024 pixels are now achievable. However, this section focuses on image capture by photography, followed by digitisation and analysis of protein gels with GelCompar or BioNumerics software (Applied Maths, Kortrijk, Belgium).

Moist or dried gels can be photographed with surface and transverse illumination by high quality laboratory cameras, or by Polaroid cameras. A red or yellow filter is used for gels stained with Fast Green FCF or Coomassie Blue, respectively. Black and white photographs (10 × 13 cm) are scanned with a flat-bed scanner (e.g., Hewlett Packard IICX), and images are stored in the TIFF format (see Chapter 3). The settings of the scanner should be standardised. Usually the file contains a maximum number of pixels to allow for saving on a 3.1" floppy disk.

Computer-assisted analysis with GelCompar or BioNumerics starts with conversion of the TIFF files to a file format that can be handled with these packages. Details of the procedure can be found in Chapter 3 and in the software manuals. For normalisation of protein patterns, an automated pattern recognition algorithm is used that aligns patterns by taking into account the whole contour of the curves rather than discrete peak positions (Vauterin *et al.*, 1993). Background subtraction is done by the rolling disk mechanism, with settings made according to the quality of the gel. For cluster analysis of protein profiles, the Pearson product moment correlation coefficient is preferred as a measure of dissimilarity (Kersters & De Ley, 1975). This coefficient is principally independent of the relative concentrations of patterns and is largely insensitive to differences in backgrounds (Vauterin & Vauterin, 1992). The Unweighted Pair Group Method using Arithmetic averages (UPGMA) is generally used to group protein profiles. Depending on the purpose of the study, proteins in specific size ranges can be used, while fragments in particular size ranges that are not considered relevant, or may even interfere, can be excluded from the analysis. It should be noted that computer-assisted analysis should always be accompanied by careful, visual inspection of profiles in order to check whether the grouping agrees with what can be seen intuitively by eye.

G. *Quality control and assessment*

It is sometimes stated that the computer-assisted normalisation process is a tool to compensate for variations between gels. However, the opposite is true since successful computer-assisted intra- and inter-gel analysis requires gels of a high quality that show minimal gel-to-gel variation. Thus, rigorous standardisation of the whole procedure is a prerequisite to computer-assisted data analysis. An overview of the variables that influence reproducibility is given in Table 4.3. Several aspects concerning the quality of protein electrophoresis, and its assessment, are discussed

Table 4.3. Variables that may influence the reproducibility of protein profiles and their analysis

Variable	Comments
Culture conditions	Test different media; use media that can be standardised as well as possible; determine growth conditions and growth phase that lead to reproducible results
Protein concentration of specimens	Standardise by determining the protein concentration or the wet weight of cells
Use of chemicals and solutions	Ammonium persulphate hydration may lead to too low a concentration and delayed polymerisation; limited shelf-life of acrylamide-Bis stocks because of acrylic acid formation; check pH and conductivity of buffers
Running conditions	Standardise current, cooling temperature and duration of run; check voltage, time and temperature of tank buffer
Staining, destaining and storage of gels	Some dyes do not dissolve well or do not form stable complexes with the proteins; see also Wilson (1979)
Photography, digitisation and computer analysis	Adhere to chosen settings

in the following sections.

(i) Determining the influence of culture conditions on the profiles of individual strains

Proteins are gene expression products, and their presence or absence in electrophoretic profiles may be influenced by environmental conditions (Lugtenberg *et al.*, 1976; Kawaji *et al.*, 1979). Therefore, when the epidemiological or taxonomic relatedness of organisms is being established on the basis of their protein profiles, it is important to first assess whether the protein profiles of these organisms are stable under different conditions. Introductory experiments on a few representative strains should be performed to test the influence of growth medium, growth temperature and growth phase. Hence, proteins that differ in expression in a strain can be recognised and eventually ignored during comparative analysis (although they may be interesting from the physiological point of view).

(ii) Testing the intra-strain stability of protein profiles

Once the growth conditions for the organisms being studied have been established, several representative strains should be investigated on different occasions for the stability of their profiles over time. In epidemiology, insight into the stability of a marker can also be obtained by investigating multiple colonies from the original culture, and by comparing multiple isolates from different body sites and different sampling times from the same patient. For well-defined single-strain outbreaks, a series of isolates from different patients can be investigated. If the profiles within and between patients are indistinguishable, then the profiles may be useful markers in epidemiology. If the patterns vary within and between patients, this may be of interest in the study of virulence factors.

(iii) Standardisation of protein electrophoresis and its analysis

Apart from the influence of culture conditions and growth phase, numerous experimental factors may influence the reproducibility of protein electrophoretic profiles, including the sample preparation procedure, the chemicals and solutions used during electrophoresis, the electrophoretic separation process, the staining and destaining procedure, and the process of digitisation and computer analysis.

With sample preparation, it is important that the protein concentration of specimens is standardised and that the samples applied to the gel do not contain too much protein. Large variations in the protein loading of samples in adjoining lanes lead to distortions in the migration patterns. Chemicals and the solutions used may also influence the results obtained, as was apparent from the finding of Lugtenberg *et al.* (1975) that use of SDS with contaminating C10 and C14 derivatives resulted in better separation of major outer membrane proteins than purified SDS. Some other reagents need specific attention, including ammonium persulphate, which should be kept dry to prevent hydration, and acrylamide, which has a limited shelf-life because of the formation of acrylic acid. The buffers of the electrophoresis system, particularly the separation gel buffer, have a great influence on the migration of proteins. Adjusting the pH of Tris solutions can be difficult, even with special 'Tris electrodes' (Costas *et al.*, 1990). This may result in the addition of too large an amount of HCl. High ionic strengths in buffers give slower rates of migration. To overcome this, it is recommended that the volume of HCl required should be determined precisely, and that both the conductivity and the pH of the solution should be checked. The process of staining is also subject to many variations, as shown in the detailed study of Wilson (1979) who compared Amido Black, Coomassie Blue and Fast Green dyes. Further trouble-shooting recommendations can be found in the reviews of Andrews (1986), Jackman (1987) and Costas *et al.* (1990), and also in the technical bulletins published by manufacturers (e.g., Bulletin 1156; Bio-Rad).

In conclusion, once a defined batch of chemicals shows acceptable results, it is recommended to adhere to their use, with solution preparation methods also being controlled as far as possible. Finally, digitisation and computer analysis are important sources of variation, but it appears that inter-laboratory standardisation of these processes can only be achieved in large-scale collaborative projects.

(iv) Control of the procedure

It should now be apparent that the whole procedure, from growth of organisms and sample preparation to final analysis, should be performed under carefully controlled conditions if it is intended to compare large numbers of strains and to generate databases for identification and large-scale surveys. Therefore, standard operating procedures (SOPs) should be used, and the preparation of solutions should be formally recorded so that possible irregularities in results can be linked to deviations in (e.g.) buffer preparation or new batches of chemicals.

To control the long-term reproducibility of the complete procedure, one representative strain should always be included as a *control strain* and processed in an identical way to the unknown samples being studied. The control strain should have a protein profile with a reasonable number of bands covering the molecular size range of fragments that can occur in the species being studied. Inclusion of this strain in each gel means that the reproducibility can be calculated following computer analysis by determining the clustering level of this control from different runs. If this strain is the same as that used to prepare reference extracts for normalisation of the protein profiles, it can be run on the gel in a lane next to the reference extract so that the pattern reproducibility can also be checked visually (see Fig. 4.1, lanes 3 and 4).

(v) Inter-laboratory variation

As with many fingerprinting methods, minor variations in performance between laboratories give rise to differences in patterns which hamper inter-laboratory comparisons. One inter-laboratory study compared the same set of *Campylobacter* strains by their whole-cell protein profiles (Costas *et al.*, 1990). The same pattern of grouping was obtained, although the profiles were clearly different due to the differences in procedures or reagents and, in particular, the different background subtraction methods of the computer analysis. Nevertheless, protein profiling is a robust method, and there is little doubt that protein profiles can be reproducible between laboratories if the procedures are standardised.

H. Applications of protein electrophoresis of microorganisms

Protein SDS-PAGE has been used with a great variety of microorganisms, including bacteria, viruses and fungi. Studies using the technology are often focused primarily on the proteins as cell constituents, with the aim of elucidating (e.g.) their chemical or antigenic character and their role in pathogenicity. For this purpose, a limited number of strains are usually investigated, and protein electrophoresis may be combined with various other techniques, including determination of enzymatic activities, immunoblotting to determine the presence of certain epitopes, or electron microscopy. In contrast, for epidemiology and taxonomy, many strains may be investigated in one run, and the protein profiles are used merely to determine relatedness between the strains. Hundreds of publications dealing with protein analysis of microorganisms are revealed by a simple search in the PubMed database on the Internet, and a comprehensive review is beyond the scope of this book. Table 4.4 lists some examples of studies in which protein SDS-PAGE has been applied and several of these are discussed in more detail below.

(i) Taxonomic applications

Grouping of organisms according to their similarity in protein profiles has been found to correlate well with grouping by DNA-DNA hybridisation (see e.g., Kersters & De Ley, 1975; Owen & Jackman, 1982). Consequently, protein SDS-PAGE has been found to be a powerful adjunct to DNA-DNA hybridisation and other methods, particularly when used in a polyphasic approach to resolve the taxonomy

Table 4.4. Examples of some organisms investigated by protein SDS-PAGE profiles

Organism	Reference
Aeromonas spp.	Kuijper *et al.* (1989)
Acinetobacter spp.	Dijkshoorn *et al.* (1987a, b; 1989; 1990)
Campylobacter spp.	Costas *et al.* (1990)
Coagulase-negative staphylococci	Dryden *et al.* (1992)
Clostridium difficile	Costas *et al.* (1994)
Enterobacter cloacae	Costas *et al.* (1989c)
Helicobacter spp.	Vandamme *et al.* (2000)
Lactobacillus acidophilus	Gancheva *et al.* (1999)
Legionella spp.	Verissimo *et al.* (1996)
MRSA	Costas *et al.* (1989a)
Providencia alcalifaciens	Holmes *et al.* (1988)
Providencia rettgeri	Costas *et al.* (1989b)
Viridans streptococci	Vandamme *et al.* (1998)

of bacterial genera at the species level. This usefulness is illustrated by Fig. 4.2, which shows the results of a cluster analysis of whole-cell protein profiles for strains belonging to species of the genus *Weissella* in which the different species are clearly separated. As is the case in this study, most taxonomic studies use whole-cell preparations rather than cell envelopes or outer membrane preparations. Computer analysis is an absolute requirement for grouping these profiles since they contain several tens of bands which makes grouping by eye impossible. One research group has set up a rigorously standardised system for protein SDS-PAGE combined with computer-assisted analysis, and uses the method as part of a polyphasic approach to investigate large collections of bacterial strains. As a result, important contributions have been made to understanding the taxonomy of a variety of bacterial genera (e.g., Vandamme *et al.*, 1998; Gancheva *et al.*, 1999; Heyndrickx *et al.*, 1999). By this standardised approach, libraries of protein profiles can be constructed with well-identified strains and then used to identify unknown strains and resolve misidentifications (Vandamme *et al.*, 2000).

(ii) Epidemiological typing

Another research group used highly standardised whole-cell protein analysis combined with numerical analysis during the 1980s and 1990s (Jackman, 1987; Costas *et al.*, 1990), and explored its usefulness for a variety of organisms, including *Clostridium difficile* (Costas et al., 1994), *Providencia alcalifaciens* (Holmes *et al.*, 1988), *Providencia rettgeri* (Costas *et al.*, 1989b) and methicillin-resistant *S. aureus* (Costas *et al.*, 1989a). It was recognised that predominant very dense protein bands could have a great impact on the outcome of the cluster analysis, particularly when Gram-negative organisms were being studied. This can be investigated by performing several different analyses, both of the whole profile and

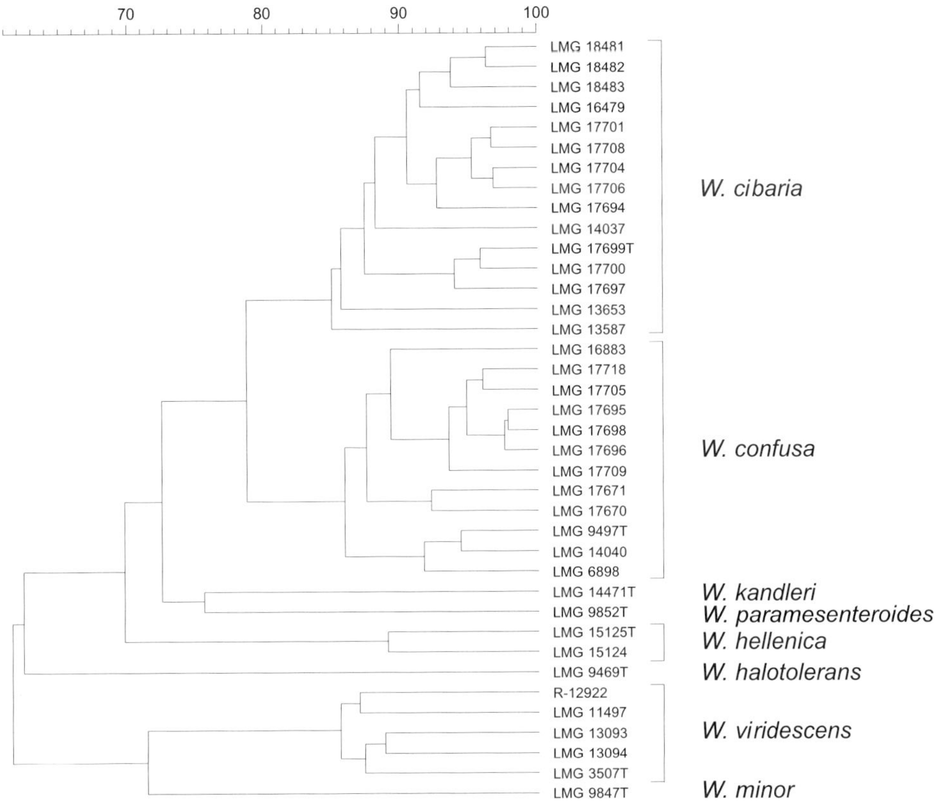

Fig. 4.2. Dendrogram showing cluster analysis of *Weissella* strains based on the unweighted pair-group average linkage of correlation coefficients between the protein profiles of the strains studied. The scale denotes the correlation coefficient as a percentage similarity for convenience (figure generously provided by Dr. Peter Vandamme).

of selected zones to distinguish between or within species. Many other studies on Gram-negative bacteria have used cell envelope or outer membrane profiles. If only a few strains are being compared, the profiles can be inspected easily by eye for relatedness since the number of bands is relatively low when compared to whole-cell profiles.

The usefulness of protein electrophoresis for both epidemiology and taxonomy is well-illustrated by studies with *Acinetobacter* spp. Initial studies (Dijkshoorn *et al.*, 1987a, b) used cell envelope protein SDS-PAGE profiles to type strains for epidemiological purposes. At the time, such profiles were analysed by eye. In later studies, the method was more standardised and numerical analysis of the profiles could be performed. In a comparative study of 120 strains belonging to different genomic species, a good correlation was found between the profiles of the 'minor' bands in the 55–97 kDa size range and the genomic species identification (Fig. 4.3). Thus, a small part of the overall profile could be used successfully for

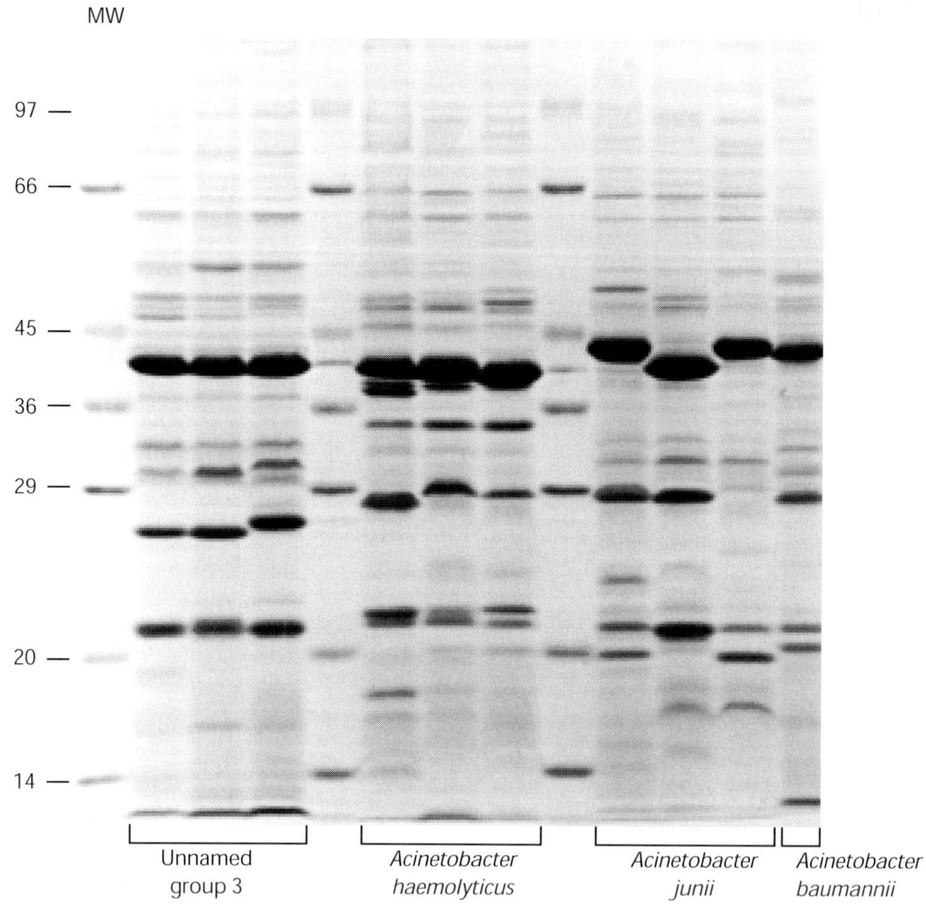

Fig. 4.3. Cell envelope protein profiles of *Acinetobacter* strains belonging to different genomic species.

(presumptive) species identification. More heavily stained bands in the 20–55 kDa size range showed considerable strain-to-strain variation and could be used for epidemiological typing (Figs. 4.1 and 4.3) (Dijkshoorn *et al.*, 1990). Whole-cell profiles appeared to be more difficult to interpret than cell envelope profiles (Fig. 4.4) and were less useful for typing purposes.

4.5 LIPOPOLYSACCHARIDE ANALYSIS

A. *Bacterial fractions for analysis*

LPS can be extracted from bacteria by the phenol-water procedure (Westphal & Jann, 1965), or by the phenol-chloroform, petroleum-ether extraction procedure (Galanos *et al.*, 1969) which is more commonly used for R-type LPS. Using SDS-PAGE, LPS can be separated into different moieties, and it has been shown with

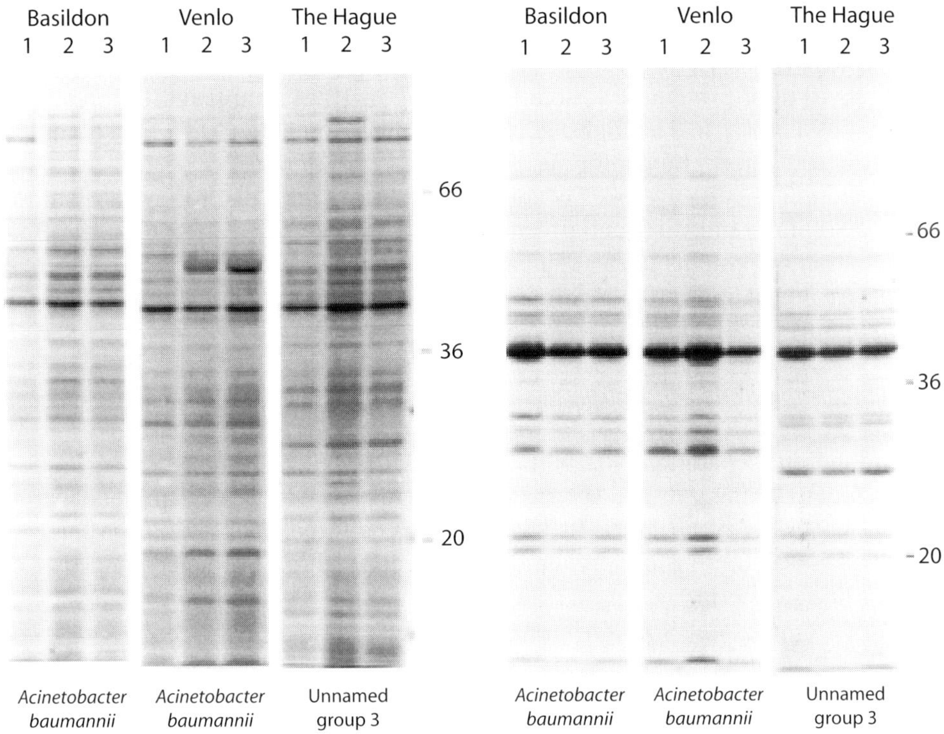

Fig. 4.4. Protein profiles of sets of *Acinetobacter* isolates from three outbreaks in hospitals. Left, whole-cell profiles; right, cell envelope profiles of the same isolates.

biochemically defined LPS fractions that differences in O-specific side chains, core oligosaccharides and lipid A are reflected by differences in the electrophoretic profiles obtained (Jann *et al.*, 1975). Tsai & Frasch (1982) developed a silver-staining method to visualise SDS-PAGE LPS profiles which was highly sensitive. In the comparative study of Hitchcock & Brown (1983), profiles of proteinase K-digested whole-cell lysates were found to be similar to profiles of purified LPS. Hence, it was concluded that proteinase K-treated whole-cell lysates could be used to provide preliminary data concerning the LPS phenotype. Once this was established, proteinase K-digested cell envelopes were also used for electrophoretic LPS analysis.

B. *Preparation of whole-cell lysates or cell envelopes followed by proteinase K digestion*

The modified method described here, as used by Vinogradov *et al.* (1997), involves first harvesting bacteria grown on solid medium (blood agar) with a sterile swab, resuspending the cells in 5 ml of 0.15 M NaCl, and then centrifuging at 7200 *g* for 10 min. The bacterial pellets are solubilised in 2 ml of 1 M Tris-HCl pH 6.8, 4% β-mercaptoethanol, 10% glycerol, and then stored at –20°C. For digestion with proteinase K, 15 μl of cell preparation are added to 15 μl sample buffer (Table 4.2) and heated to 100°C for 10 min. Following heating, 10 μl of sample extract is added to 10 μl of proteinase K solution (2.5 μg in 10 μl sample buffer) and incubated at 60°C for 1 h. The digested sample can be stored at –20°C, but should be heated at 100°C for 5 min before use.

Preparation of cell envelopes is done by harvesting the cells, followed by ultrasonic disruption and differential centrifugation as described in section 4.4.B.ii. Degradation of proteins in the cell envelopes is achieved by boiling for 5 min in sample treatment buffer (0.5 mg of protein/ml) and then incubating with proteinase K (50 μg/ml) for 1 h at 60°C.

C. Electrophoretic separation

Electrophoretic separation of LPS fragments by SDS-PAGE can be done using 5% acrylamide stacking gels and separation gels with 10 or 15% acrylamide, depending on whether S-type LPS or R-type LPS is to be visualised. Various buffer systems and concentrations of acrylamide have been used (e.g., Hitchcock & Brown, 1983; Maskell, 1991).

D. Silver-staining

The method outlined here corresponds to the method of Tsai & Frasch (1982), except that periodic acid is dissolved in water and ethanol is replaced as a fixative by isopropanol. The whole procedure is performed in a glass dish with gentle shaking, except for the overnight fixation step. All reagents are prepared in glass vessels and rinsed latex gloves are worn when handling the gel to prevent fingerprints. After use, the glass dish is cleaned with concentrated nitric acid, followed by careful rinsing with deionised or distilled water.

In the first step, following electrophoresis, the gel is fixed overnight in 250 ml 40% (v/v) isopropanol in 5% (v/v) acetic acid. Next, LPS is oxidised for 5 min in 200 ml 0.7% (w/v) periodic acid in water, and then washed three times with 500–1000 ml of water for 15 min, 30 min and 60 min, respectively. The water is drained and freshly prepared (5–10 min before use) silver-staining solution is added to the gel for 10 min. The staining solution contains sodium hydroxide, concentrated ammonium hydroxide and silver nitrate (Tsai & Frasch, 1982). It is important to maintain the 'high strength' of the ammonium hydroxide, and it is recommended that small aliquots from a freshly opened bottle should be stored in small vessels which can be tightly sealed (Hitchcock & Brown, 1983). After the staining step, the gel is washed three times as described above. Finally, the silver-stained profiles are visualised with 200 ml of a solution containing 10 mg citric acid and 100 μl 36% formaldehyde in water; however, 10–15 min may be required before an appropriate level of staining is obtained. It may be helpful to compare the results with a photograph showing the desired intensity of staining of a reference strain. Once this intensity is obtained, the gel is washed for 5 min with 500 ml of water and a photograph is taken immediately. Due to the washing procedure, the gel becomes fragile and care should be taken that it does not break into pieces.

E. Applications

LPS electrophoretic profiling has been used in numerous studies, frequently in combination with immunoblotting techniques to examine the antigenic characteristics of the separated LPS moieties. In many cases, these studies were performed to explore the correlation of LPS with pathogenicity or to study the interaction of LPS with the immune system. Exploitation of LPS diversity by SDS-PAGE, with or without immunoblotting, for the purpose of typing is relatively uncommon compared to the many pathogenicity studies, although many serotyping systems are based on the diversity of LPS in microorganisms. A drawback of LPS SDS-PAGE is that the method is rather laborious. However, it has undisputed potential as a typing method, as shown by Aucken & Pitt (1993) who applied the technique to 124 isolates belonging to 12 Gram-negative species and concluded that LPS profiling is useful for epidemiological investigations with small clusters of isolates in order to trace cross-infection between patients.

Another illustration of the usefulness of LPS profiling is shown by a series of recent studies on *Acinetobacter* LPS. Initial work with *Acinetobacter* strain NCTC 10305 had shown that its LPS resembled in many ways that of rough mutants belonging to the Enterobacteriaceae family (Brade & Galanos, 1982). However, ladder-type LPS (Fig. 4.5) can be observed in some strains and genomic species of *Acinetobacter* when proteinase K-treated cell envelopes of these organisms are investigated by SDS-PAGE combined with silver staining (personal unpublished results). This finding is suggestive of S-type LPS in the strains investigated. Occurrence of this type of LPS in *Acinetobacter* was also suggested by Smith & Alpar (1991) who used immunoblotting. The presence of S-type LPS in *Acinetobacter* has now been established by both chemical characterisation and immunoblotting of LPS separated by SDS-PAGE using whole-cell extracts (Vinogradov *et al.*, 1996; 1997; Haseley *et al.*, 1997). On the basis of these findings, monoclonal antibodies have been developed that are promising tools for rapid identification and typing of acinetobacters (Pantophlet *et al.,* 1999a, b).

4.6 CONCLUSIONS

The past three decades have shown that SDS-PAGE is a powerful and robust technique that can be used for analysis of proteins and LPS, either alone or in combination with other techniques. Thus, it is an important tool for taxonomy, epidemiology and more fundamental research. In general, the technique is considered to be rather laborious when compared to most DNA fingerprinting techniques. Nevertheless, once established in a standardised way, protein SDS-PAGE can be used to construct fingerprint libraries for the identification of species and strains of a variety of microorganisms. With the availability of commercially prepared reagents,

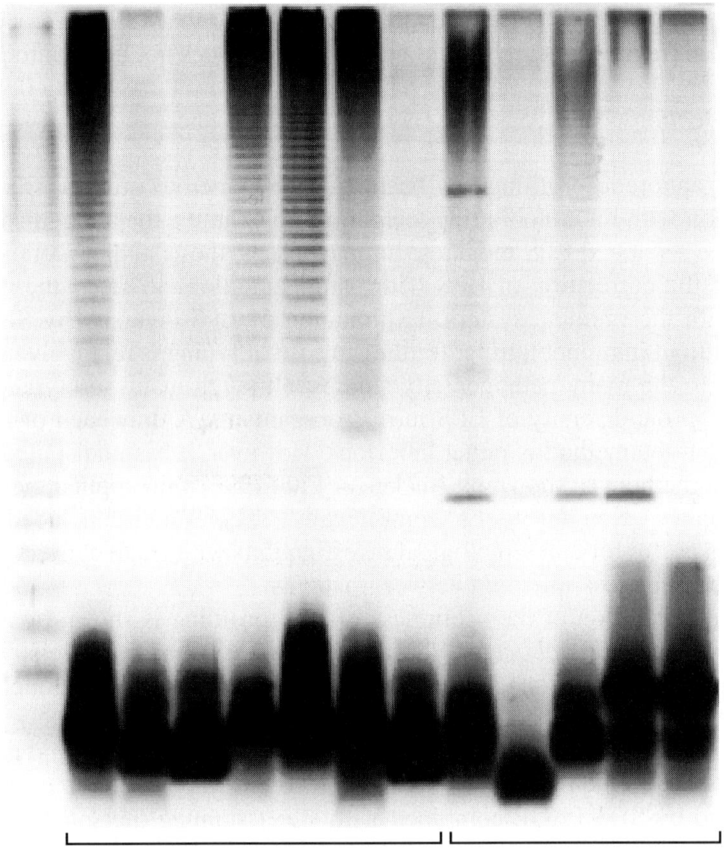

Fig. 4.5. Electrophoretic profiles of proteinase K-digested cell envelope fractions of *Acinetobacter* strains following visualisation by silver-staining.

the work load of the method is now comparable to that of other advanced genomic fingerprinting methods such as AFLP (see Chapter 8), and inter-laboratory standardisation has become feasible. Similarly, SDS-PAGE, with or without additional blotting procedures, can be useful for fingerprinting organisms on the basis of their LPS content, although such use will be confined largely to structural analysis rather than to typing. Overall, it can be expected that SDS-PAGE will retain its place for the immediate future on the list of major methods in biology and biochemistry that are suitable for fingerprinting microorganisms.

ACKNOWLEDGEMENTS

It is a pleasure to thank Dr. Peter Vandamme for providing Fig. 4.2, Mr. Jan Beentjes for processing the figures, and Dr. Ben de Jong, Mr. Dirk Dewettinck and Dr. Ralph Pantophlet for helpful advice and critical reading of the manuscript.

REFERENCES

Andrews, A.T. (1986). *Electrophoresis: theory, techniques, and biochemical and clinical applications*. Oxford University Press, New York.

Armstrong, S.K. & Parker, C.D. (1986). Heat-modifiable envelope proteins of *Bordetella pertussis*. *Infection and Immunity* **54**, 109–117.

Aucken, H.M. & Pitt, T.L. (1993). Lipopolysaccharide profile typing as a technique for comparative typing of gram-negative bacteria. *Journal of Clinical Microbiology* **31**, 1286–1289.

Brade, H. & Galanos, C. (1982). Isolation, purification, and chemical analysis of the lipopolysaccharide and lipid A of *Acinetobacter calcoaceticus* NCTC 10305. *European Journal of Biochemistry* **122**, 233–237.

Brock, T.D., Madigan, M.T., Martinko, J.M. & Parker, J. (1994). *Biology of microorganisms*, 7th edn. Prentice Hall, Englewood Cliffs, NJ.

Costas, M., Cookson, B.D., Talsania, H.G. & Owen, R.J. (1989a). Numerical analysis of electrophoretic protein patterns of methicillin-resistant strains of *Staphylococcus aureus*. *Journal of Clinical Microbiology* **27**, 2574–2581.

Costas, M., Holmes, B., Wood, A.C. & On, S.L. (1989b). Numerical analysis of electrophoretic protein patterns of *Providencia rettgeri* strains from human faeces, urine and other specimens. *Journal of Applied Bacteriology* **67**, 441–452.

Costas, M., Sloss, L.L., Owen, R.J. & Gaston, M.A. (1989c). Evaluation of numerical analysis of SDS-PAGE of protein patterns for typing *Enterobacter cloacae*. *Epidemiology and Infection* **103**, 265–274.

Costas, M., Pot, M., Vandamme, P., Owen, R.J. & Hill, L.R. (1990). Interlaboratory comparative study of the numerical analysis of one-dimensional sodium dodecyl sulphate-polyacrylamide gel electrophoretic protein patterns of *Campylobacter* strains. *Electrophoresis* **11**, 467–474.

Costas, M., Holmes, B., Ganner, M., On, S.L., Hoffman, P.N., Worsley, M.A. & Panigrahi, H. (1994). Identification of outbreak-associated and other strains of *Clostridium difficile* by numerical analysis of SDS-PAGE protein patterns. *Epidemiology and Infection* **113**, 1–12.

Dijkshoorn, L., Michel, M.F. & Degener, J.E. (1987a). Cell envelope protein profiles of *Acinetobacter calcoaceticus* strains isolated in hospitals. *Journal of Medical Microbiology* **23**, 313–319.

Dijkshoorn, L., van Vianen, W., Degener, J.E. & Michel, M.F. (1987b). Typing of *Acinetobacter calcoaceticus* strains isolated from hospital patients by cell envelope protein profiles. *Epidemiology and Infection* **99**, 659–667.

Dijkshoorn, L., Wubbels, J.L., Beunders, A.J., Degener, J.E., Boks, A.L. & Michel, M.F. (1989). Use of protein profiles to identify *Acinetobacter calcoaceticus* in a respiratory unit. *Journal of Clinical Pathology* **42**, 853–857.

Dijkshoorn, L., Tjernberg, I., Pot, B., Michel, M.F., Ursing, J. & Kersters, K. (1990). Numerical analysis of cell envelope protein profiles of *Acinetobacter* strains classified by DNA-DNA hybridization. *Systematic and Applied Microbiology* **13**, 338–344.

Dryden, M.S., Talsania, H.G., Martin, S., Cunningham, M., Richardson, J.F., Cookson, B., Marples, R.R. & Phillips, I. (1992). Evaluation of methods for typing coagulase-negative staphylococci. *Journal of Medical Microbiology* **37**, 109–117.

Filip, C., Fletcher, G., Wulff, J.L. & Earhart, C.F. (1973). Solubilization of the cytoplasmic membrane of *Escherichia coli* by the ionic detergent sodium-lauryl sarcosinate. *Journal of Bacteriology* **115**, 717–722.

Galanos, C., Lüderitz, O. & Westphal, O. (1969). A new method for the extraction of R lipopolysaccharides. *European Journal of Biochemistry* **9**, 245–249.

Gancheva, A., Pot, B., Vanhonacker, K., Hoste, B. & Kersters, K. (1999). A polyphasic approach towards the identification of strains belonging to *Lactobacillus acidophilus* and related species. *Systematic and Applied Microbiology* **22**, 573–585.

Gordon, A.H. (1975). Electrophoresis of proteins in polyacrylamide and starch gels. In *Laboratory techniques in biochemistry and molecular biology, vol. 1,* 2nd edn, Work, T.S. & Work, E., eds.

Elsevier, Amsterdam.

Hammond, S.M., Lambert, P.A. & Rycroft, A.N. (1984). *The bacterial cell surface*. Croom Helm, Beckenham.

Hancock, R.E.W. & Carey, A.M. (1979). Outer membrane of *Pseudomonas aeruginosa*: heat- and 2-mercaptoethanol-modifiable proteins. *Journal of Bacteriology* **140**, 902–910.

Haseley, S.R., Pantophlet, R., Brade, L., Holst, O. & Brade, H. (1997). Structural and serological characterisation of the O-antigenic polysaccharide of the lipopolysaccharide from *Acinetobacter junii* strain 65. *European Journal of Biochemistry* **245**, 477–481.

Heyndrickx, M., Lebbe, L., Kersters, K., Hoste, B., De Wachter, R., De Vos, P., Forsyth, G. & Logan, N.A. (1999). Proposal of *Virgibacillus proomii* sp. nov. and emended description of *Virgibacillus pantothenticus* (Proom and Knight 1950) Heyndrickx *et al.* 1998. *International Journal of Systematic Bacteriology* **49**, 1083–1090.

Hitchcock, P.J. & Brown, T.M. (1983). Morphological heterogeneity among *Salmonella* lipopolysaccharide chemotypes in silver-stained polyacrylamide gels. *Journal of Bacteriology* **154**, 269–277.

Holmes, B., Costas, M. & Sloss, L.L. (1988). Numerical analysis of electrophoretic protein patterns of *Providencia alcalifaciens* strains from human faeces and veterinary specimens. *Journal of Applied Bacteriology* **64**, 27–35.

Holst, O., Ulmer, A.J., Brade, H., Flad, H.-D. & Rietschel, E.T. (1996). Biochemistry and cell biology of bacterial endotoxins. *FEMS Immunology and Medical Microbiology* **16**, 83–104.

Jackman, P.J.H. (1987). Microbial systematics based on electrophoretic whole-cell proteins. In *Methods in microbiology, vol. 19*, Colwell, R.R. & Grigorova, R., eds. Academic Press, London.

Jacques, M. (1996). Role of lipo-oligosaccharides and lipopolysaccharides in bacterial adherence. *Trends in Microbiology* **4**, 408–410.

Jann, B., Reske, K. & Jann, K. (1975). Heterogeneity of lipopolysaccharides. Analysis of polysaccharide chain lengths by sodium dodecyl-sulphate polyacrylamide gel electrophoresis. *European Journal of Biochemistry* **60**, 239–246.

Kawaji, H., Mizuno, T. & Mizushima, S. (1979). Influence of molecular size and osmolarity of sugars and dextrans on the synthesis of outer membrane proteins O-8 and O-9 of *Escherichia coli* K12. *Journal of Bacteriology* **140**, 843–847.

Kersters, K. & De Ley, J. (1975). Identification and grouping of bacteria by numerical analysis of their electrophoretic protein patterns. *Journal of General Microbiology* **87**, 333–342.

Koebnik, R., Locher, K.P. & Van Gelder, P. (2000). Structure and function of bacterial outer membrane proteins: barrel in a nutshell. *Molecular Microbiology* **37**, 239–253.

Kuijper, E.J., van Alphen, L., Leenders, E. & Zanen, H.C. (1989). Typing of *Aeromonas* strains by DNA restriction endonuclease analysis and polyacrylamide gel electrophoresis of cell envelopes. *Journal of Clinical Microbiology* **27**, 1280–1285.

Laemmli, U.K. (1970). Cleavage of structural proteins during the assembly of the head of bacteriophage T4. *Nature* **227**, 680–685.

Lugtenberg, B. (1981). Composition and fuction of the outer membrane of *Escherichia coli*. *Trends in Biochemical Sciences* **2**, 262–265.

Lugtenberg, B. & Van Alphen, L. (1983). Molecular architecture and functioning of the outer membrane of *Escherichia coli* and other Gram-negative bacteria. *Biochimica et Biophysica Acta* **737**, 51–115.

Lugtenberg, B., Meijers, J., Peters, R., Van der Hoek, P. & Van Alphen, L. (1975). Electrophoretic resolution of the 'major outer membrane' protein' of *Escherichia coli* K12 into four bands. *FEBS Letters* **58**, 254–258.

Lugtenberg, B., Peters, R., Bernheimer, H. & Berendsen, W. (1976). Influence of cultural conditions and mutations on the composition of the outer membrane proteins of *Escherichia coli*. *Molecular and General Genetics* **147**, 251–262.

Maskell, J.P. (1991). The resolution of bacteroides lipopolysaccharides by polyacrylamide gel electrophoresis. *Journal of Medical Microbiology* **34**, 253–257.

Owen, R.J. & Jackman, P.J.H. (1982). The similarities between *Pseudomonas paucimobilis* and allied bacteria derived from analysis of deoxyribonucleic acids and electrophoretic protein patterns. *Journal of General Microbiology* **128**, 2945–2954.

Pantophlet, R., Brade, L. & Brade, H. (1999a). Use of murine O-antigen-specific monoclonal antibody to identify *Acinetobacter* strains of unnamed genomic species 13 *sensu* Tjernberg and Ursing. *Journal of Clinical Microbiology* **37**, 1693–1698.

Pantophlet, R., Brade, L. & Brade, H. (1999b). Identification of *Acinetobacter baumannii* strains with monoclonal antibodies against the O antigens of their lipopolysaccharides. *Clinical Diagnostics and Laboratory Immunology* **6**, 323–329.

Pot, B., Vandamme, P. & Kersters, K. (1994). Analysis of electrophoretic whole-organism protein fingerprints, p. 493–521. In *Modern microbial methods. Chemical methods in bacterial systematics.* Goodfellow, M. & O'Donnell, A.G., eds. J. Wiley and Sons, Chichester.

Preston, A., Mandrell, R.E., Gibson, B.W. & Apicella, M.A. (1996). The lipooligosaccharides of pathogenic Gram-negative bacteria. *Critical Reviews in Microbiology* **22**, 139–180.

Rietschel, E.T. & Brade, H. (1992). Bacterial endotoxins. *Scientific American* **267**, 26–33.

Smith, A.W. & Alpar, K.E. (1991). Immune response to *Acinetobacter calcoaceticus* infection in man. *Journal of Medical Microbiology* **34**, 83–88.

Tsai, C.-M. & Frasch, C.E. (1982). A sensitive silver stain for detecting lipopolysaccharides in polyacrylamide gels. *Analytical Biochemistry* **119**, 115–119.

Vandamme, P., Torck, U., Falsen, E., Pot, B., Goossens, H. & Kersters, K. (1998). Whole-cell protein electrophoretic analysis of viridans streptococci: evidence for heterogeneity among *Streptococcus mitis* biovars. *International Journal of Systematic Bacteriology* **48**, 117–125.

Vandamme, P., Harrington, C.S., Jalava, K. & On, S.L. (2000). Misidentifying helicobacters: the *Helicobacter cinaedi* example. *Journal of Clinical Microbiology* **38**, 2261–2266.

Vauterin, L. & Vauterin, P. (1992). Computer-aided objective comparison of electrophoresis patterns for grouping and identification of microorganisms. *European Microbiology* **2**, 37–41.

Vauterin, L., Swings, J. & Kersters, K. (1993). Protein electrophoresis and classification. In *Handbook of new bacterial systematics,* Goodfellow, M. & O'Donnell, A.G., eds, pp. 251–280. Academic Press, London.

Verissimo, A., Morais, P.V., Diogo, A., Gomes, C. & da Costa, M.S. (1996). Characterization of *Legionella* spp. by numerical analysis of whole-cell protein electrophoresis. *International Journal of Systematic Bacteriology* **46**, 41–49.

Vinogradov, E.V., Pantophlet, R., Dijkshoorn, L., Brade, L., Holst, O. & Brade, H. (1996). Structural and serological characterisation of two O-specific polysaccharides of *Acinetobacter. European Journal of Biochemistry* **239**, 602–610.

Vinogradov, E.V., Pantophlet, R., Haseley, S.R., Brade, L., Holst, O. & Brade, H. (1997). Structural and serological characterisation of the O-specific polysaccharide from lipopolysaccharide of *Acinetobacter calcoaceticus* strain 7 (DNA group 1). *European Journal of Biochemistry* **243**, 167–173.

Westphal, O. & Jann, K. (1965). Bacterial lipopolysaccharides. Extraction with phenol-water and further applications of the procedure. *Methods in Carbohydrate Chemistry* **5**, 83–91.

Williams, B.L. & Wilson, K. (1983). *A biologist's guide to principles and techniques of practical biochemistry*, 2nd edn., Willis, A.J. & Sleigh, M.A., eds, pp. 127–154. Arnold, London.

Wilson, C.M. (1979). Studies and critique of Amido Black 10B, Coomassie Blue R, and Fast Green FCF as stains for proteins after polyacrylamide gel electrophoresis. *Analytical Biochemistry* **96**, 263–278.

5 rRNA Gene Restriction Pattern Determination (Ribotyping) and Computer Interpretation

Patrick A.D. Grimont and Francine Grimont
Unité des Entérobactéries, INSERM U389, Institut Pasteur, 28 Rue du Dr Roux, 75724 Paris Cedex 15, France

CONTENTS

5.1 INTRODUCTION

Investigations of presumed outbreaks of bacterial infections require strain typing data to identify outbreak-related strains and to distinguish epidemic isolates from endemic isolates. It should be emphasised that precise identification at the species level is the primary epidemiological marker. As described in Chapter 1, traditional typing methods for bacteria were only devised after lengthy efforts and have been established for a limited set of bacterial species. Thus, serotyping schemes require a collection of rabbit sera, with each serum being collected after several immunisation steps and absorbed with cross-reacting bacterial suspensions. Such immunological reagents are normally applicable to only a single genomic species, but very powerful serotyping systems (such as the Kauffmann-White scheme for *Salmonella*) have been developed thanks to the efforts of several generations of workers.

Phage-typing has always been shown to have a better resolution than serotyping. However, a collection of well-chosen and pure phages is required and these phages generally have a limited host-range. A few months are necessary to devise a system for local strains, while several years of work are often needed to achieve an internationally useful system. For some genera (e.g., *Legionella)*, no phage have yet been isolated. Bacteriocin typing has the same limitations and requires the isolation of some bacteriocin-producing strains. Although a few classical typing systems have reached an impressive level of refinement (e.g., *Salmonella* serotyping), there is a growing need for universal typing methods (i.e., methods giving results with any bacterial species).

Early applications of a molecular approach to typing consisted of restriction analysis of plasmid DNA (Meyers *et al.*, 1976). However, this required the presence of a plasmid, its extraction and purification, and comparison of the same sort of plasmid in multiple plasmid-bearing strains. Comparison of restriction patterns of bacterial genomic DNA was subsequently proposed for Mollicutes (Bove & Saillard, 1979). However, for bacteria with larger genomes, restriction patterns were initially too complex to serve as fingerprints. The method proposed by Southern (1975) allowed visualisation of those restriction fragments in the overall complex pattern that hybridised with a probe. This method was soon used to analyse restriction fragment length polymorphisms (RFLPs) (Kan & Dozy, 1978).

The fact that genes coding for ribosomal ribonucleic acids (*rrs* genes coding for 16S rRNA and *rrl* genes coding for 23S rRNA) have large portions which have been conserved during evolution (Fox *et al.*, 1980) raised the possibility that these genes accounted for most of the DNA fragments shared amongst different bacterial species (Ostapchuk *et al.*, 1980). This idea was used to devise a general strategy for preparing species-specific probes (Grimont *et al.*, 1985). In the course of this work, it was observed that different species yielded different RFLP patterns when probed with labelled 16+23S rRNA. Other studies had previously reported the use of rRNA-RFLPs for different purposes (Ostapchuk *et al.*, 1980; Gottlieb & Rudner, 1985). However, the method called rRNA gene restriction pattern determination (Grimont & Grimont, 1986) was the first to propose a single probe mixture (16+23S rRNA from *Escherichia coli*) for specific and intra-specific identification of bacteria, irrespective of their phylogenetic position. This was the first attempt at universal molecular typing. The method was later named ribotyping (Stull *et al.*, 1988).

5.2 RIBOTYPING METHODOLOGY

A. *Manual method*

(i) DNA extraction and purification

In order to yield readable and reproducible rRNA gene restriction patterns after cleavage by restriction endonucleases, the extracted DNA must be of high molecular size and free from inhibitors which might interfere with endonuclease activity. Gram-negative bacteria, collected from a 10-ml stationary phase rich broth (e.g., trypto-casein soy broth) culture are easily lysed by detergents such as sodium dodecyl sulphate (SDS) in Tris-HCl, EDTA, NaCl buffer (Brenner *et al.*, 1982). Concomitant proteinase K treatment is likely to improve the yield. The precise protocol has been published previously (Grimont & Grimont, 1995).

Lysis of Gram-positive bacteria can be more tricky. It may be best to collect bacteria in the exponential phase of growth (40 ml) since lysis of some Gram-positive bacterial species in the stationary phase of growth can be very difficult. Different lysis protocols can be used, depending on the bacterial genus being studied (see references in section 5.5). For example, the protocol for coryneforms involves suspending bacterial cells in Tris-HCl, EDTA, NaCl, Triton X-100 buffer and treating first with lysozyme and mutanolysine, followed by proteinase K and SDS, while the protocol for staphylococci involves suspending bacterial cells in Tris-HCl, EDTA, NaCl, Triton X-100 buffer and treating first with lysostaphin, followed by SDS and proteinase K.

DNA purification involves shaking the bacterial lysate vigorously with an equal volume of 25:24:1 (v/v) phenol (saturated with Tris, EDTA, NaCl and protected against oxidation with 8-hydroxyquinoline):chloroform:isoamyl alcohol. After centrifugation, the upper (aqueous) phase containing the nucleic acids is collected and treated again with the phenol:chloroform mixture until there is no longer a white (proteinaceous) interface between phases. The nucleic acids are precipitated with two volumes of cold (–20°C) absolute ethanol and then dried in a vacuum dessicator. The dried precipitate is then dissolved in Tris-HCl, EDTA buffer.

Several kits are available for fast DNA extraction and purification. We have used the IsoQuick Nucleic Acid Extraction Kit (Orca Research, Bothell, WA, USA) and the QIAamp DNA Mini Kit (Qiagen, Valencia, CA, USA) successfully with Gram-negative bacteria, and the Wizard Genomic DNA purification kit (Promega, Madison, WI, USA) with lysed Gram-positive bacteria.

In all cases, it is useful to systematically submit a 100-µl portion of the DNA solution to microdialysis on Millipore type VS filters (diameter 2.5 cm, pore size

Meyers, J.A., Sanchez, D., Elwell, L.P. & Falkow, S. (1976). Simple agarose gel electrophoresis method for the identification and characterization of plasmid deoxyribonucleic acid. *Journal of Bacteriology* **127**, 1529–1537.

Moureau, P., Derclaye, I., Gregoire, D., Janssen, M. & Cornelis, G.R. (1989). *Campylobacter* species identification based on polymorphism of DNA encoding rRNA. *Journal of Clinical Microbiology* **27**, 1514–1517.

Oerter, K.E., Munson, P.J., McBride, W.O. & Rodbard, D. (1990). Computerized estimation of size of nucleic acid fragments using the four-parameter logistic model. *Analytical Biochemistry* **189**, 235–243.

Ostapchuk, P., Anilionis, A. & Riley, M. (1980). Conserved genes in enteric bacteria are not identical. *Molecular and General Genetics* **180**, 475–477.

Pelkonen, S., Romppanen, E.L., Siitonen, A. & Pelkonen, J. (1994). Differentiation of *Salmonella* serovar infantis isolates from human and animal sources by fingerprinting IS200 and 16S *rrn* loci. *Journal of Clinical Microbiology* **32**, 2128–2133.

Perolat, P., Grimont, F., Regnault, B., Grimont, P.A.D., Fourni, E., Thevenet, H. & Baranton, G. (1990). rRNA gene restriction patterns of *Leptospira*: a molecular typing system. *Research in Microbiology* **141**, 159–171.

Picard-Pasquier, N., Ouagued, M., Picard, B., Goullet, P. & Krishnamoorthy, R. (1989). A simple, sensitive method of analysing bacterial ribosomal DNA polymorphism. *Electrophoresis* **10**, 186–189.

Pignato, S., Giammanco, G.M., Grimont, F., Grimont, P.A.D. & Giammanco, G. (1999). Molecular characterization of genera *Proteus, Morganella* and *Providencia* by ribotyping. *Journal of Clinical Microbiology* **37**, 2840–2847.

Pitcher, D.G., Owen, R.J., Dyal, P. & Beck, A. (1987). Synthesis of a biotinylated DNA probe to detect ribosomal RNA cistrons in *Providencia stuartii. FEMS Microbiology Letters* **48**, 283–287.

Popovic, T., Bopp, C., Olsvik, O. & Wachsmuth, K. (1993). Epidemic application of a standardized ribotype scheme for *Vibrio cholerae* O1. *Journal of Clinical Microbiology* **31**, 2474–2482.

Popovic, T., Kombarova, S.Y., Reeves, M.W., Nakao, H., Mazurova, I.K., Wharton, M., Wachsmuth, I.K. & Wenger, J.D. (1996). Molecular epidemiology of diphtheria in Russia, 1985–1994. *Journal of Infectious Diseases* **174**, 1064–1072.

Popovic, T., Mazurova, I.K., Efstratiou, A., Vuopio-Varkila, J., Reeves, M.W., De Zoyza, A., Glushkevich, T. & Grimont, P. (2000). Molecular epidemiology of diphtheria. *Journal of Infectious Diseases* **181**, S168–S177.

Pourshafie, M.R., Grimont, F. & Grimont, P.A.D. (2000). Molecular epidemiological study of *Vibrio cholerae* isolates from infected patients in Teheran, Iran. *Journal of Medical Microbiology* **49**, 1085–1090.

Press, W.H., Flannery, B.P., Teukolsky, S.A. & Vetterling, W.T. (1986). *Numerical recipes*. Cambridge University Press, Cambridge.

Regnault, B., Grimont, F. & Grimont, P.A.D. (1997). Universal ribotyping method using a chemically-labelled oligonucleotide probe mixture. *Research in Microbiology* **148**, 649–659.

Romaniuk, P.J. & Trust, T.J. (1987). Identification of *Campylobacter* species by Southern hybridisation of genomic DNA using an oligonucleotide probe for 16S rRNA genes. *FEMS Microbiology Letters* **43**, 331–335.

Salloum, G., Meugnier, H., Reyrolle, M., Grimont, F., Grimont, P.A.D., Etienne, J. & Freney, J. (2001). Identification of *Legionella* species by ribotyping and other molecular methods. *Research in Microbiology,* in press.

Saunders, N., Harrisson, T.G., Kachwalla, N. & Taylor, A. (1988). Identification of species of the genus *Legionella* using a cloned rRNA gene from *Legionella pneumophila. Journal of General Microbiology* **134**, 2363–2374.

Schaffer, H.E. & Sederoff, R.R. (1981). Improved estimation of DNA fragment length from agarose gels. *Analytical Biochemistry* **115**, 113–122.

Schoonmaker, D., Heimberger, T. & Birkhead, G. (1992). Comparison of ribotyping and restriction enzyme analysis using pulsed-field gel electrophoresis of *Legionella pneumophila* isolates obtained during a nosocomial outbreak. *Journal of Clinical Microbiology* **30**, 1491–1498.

Sneath, P.H.A. & Sokal, R.R. (1973). *Numerical taxonomy*. Freeman, San Francisco.

Southern, E.M. (1975). Detection of specific sequences among DNA fragments separated by gel electrophoresis. *Journal of Molecular Biology* **98**, 503–517.

Southern, E.M. (1979). Measurement of DNA length by gel electrophoresis. *Analytical Biochemistry* **100**, 319–323.

Stanley, J., Baquar, N. & Threlfall, E.J. (1993). Genotypes and phylogenetic relationships of *Salmonella typhimurium* are defined by molecular fingerprinting of IS200 and16S rrn loci. *Journal of General Microbiology* **139**, 1133–1140.

Stull, T., LiPuma, J.J. & Edlind, T.D. (1988). A broad-spectrum probe for molecular epidemiology of bacteria: ribosomal RNA. *Journal of Infectious Diseases* **157**, 280–286.

Tamayo, M., Koblavi, S., Grimont, F., Castenada, E. & Grimont, P.A.D. (1997). Molecular epidemiology of *Vibrio cholerae* O1 isolates from Colombia. *Journal of Medical Microbiology* **46**, 611–616.

Tenover, F.C., Arbeit, R.D., Goering, R.V., Mickelsen, P.A., Murray, B.E., Persing, D.H. & Swaminathan, B. (1995). Interpreting chromosomal DNA restriction patterns produced by pulsed-field gel electrophoresis: criteria for bacterial strain typing. *Journal of Clinical Microbiology* **33**, 2233–2239.

Vandenesch, F., Eykyn, S.J., Bes, M., Meugnier, H., Fleurette, J. & Etienne, J. (1995). Identification and ribotypes of *Staphylococcus caprae* isolates isolated as human pathogens and from goat-milk. *Journal of Clinical Microbiology* **33**, 888–892.

Verger, J.-M., Grimont, F., Grimont, P.A.D. & Grayon, M. (1985). *Brucella*, a monospecific genus as shown by deoxyribonucleic acid hybridization. *International Journal of Systematic Bacteriology* **35**, 292–295.

Verger, J.-M., Grayon, M., Cloeckaert, A., Lefevre, M., Ageron, E. & Grimont, F. (2000). Classification of *Brucella* strains from marine mammals using DNA-DNA hybridization and ribotyping. *Research in Microbiology* **151**, 797–799.

Webster, J.A., Bannerman, T.L., Hubner, R.J., Ballard, D.N., Cole, E.M., Bruce, J.L., Fiedler, F., Schubert, K. & Kloos, W.E. (1994). Identification of the *Staphylococcus sciuri* species group with *Eco*RI fragments containing rRNA sequences and description of *Staphylococcus vitulus* sp. nov. *International Journal of Systematic Bacteriology* **44**, 454–460.

6 Generation and Analysis of RAPD Fingerprinting Profiles

Kevin Towner[1] and Hajo Grundmann[2]
[1]Public Health Laboratory and [2]Division of Microbiology, University Hospital, Queen's Medical Centre, Nottingham NG7 2UH, UK

CONTENTS

6.1 INTRODUCTION

Many of the bacterial typing methods described elsewhere in this book make use of the polymerase chain reaction (PCR) to amplify specific genes or spacer regions as a prelude to the generation of DNA fingerprints for typing purposes. However, although excellent and informative results can be produced, most of these techniques require several days to complete, usually involving several different molecular manipulations, and their very sophistication makes them less amenable to introduction in routine hospital microbiology laboratories. Accordingly, their use has tended to be confined to central research or reference laboratories. Unfortunately, central analysis of an ever-growing number of important microorganisms at a national or international level imposes an unmanageable workload on central reference facilities, with a concomitant delay in obtaining results, and involves the undesirable and increasingly expensive shipping of pathogens by air or surface carriers. Therefore, there is an increasingly urgent need for a simplified DNA fingerprinting method that can be used to generate meaningful and reproducible typing data rapidly, preferably within a single working day, at the local level.

In an attempt to address this problem, an increasing number of laboratories have investigated the use of two functionally interchangeable methods termed 'arbitrarily primed amplification' of chromosomal DNA (Welsh and McClelland, 1990; Williams *et al.*, 1990; Maslow *et al.*, 1993) and 'randomly amplified polymorphic DNA' (Caetano-Anollés *et al.*, 1991; Maslow *et al.*, 1993), otherwise known as AP-PCR and RAPD, respectively. Each of these techniques involves PCR amplification of 'random' fragments of genomic DNA with single primers, but the techniques are only actually 'random' in so far as a PCR primer(s) can be selected without regard to the sequence of the genome to be fingerprinted (Welsh & McClelland, 1990; Williams *et al.*, 1990). An alternative related approach, referred to as REP-PCR (Woods *et al.*, 1992), amplifies intervening sequences located between highly repetitive DNA motifs. These techniques (described in more detail in section 2A) have now been used successfully for localised epidemiological typing of an ever expanding range of bacteria. However, difficulties arise when attempting to compare the fingerprint banding patterns generated from a large number of isolates over an extended time period. Other important problems (see section 2B) include a reported lack of reproducibility and difficulties with the standardisation of equipment and reagents (MacPherson *et al.*, 1993; Meunier & Grimont, 1993; van Belkum *et al.*, 1995; Power, 1996; Tyler *et al.*, 1997).

The aim of this chapter is to review the current methods used for the generation and analysis of bacterial PCR fingerprints with 'random' primers and to consider whether appropriate experimental methods can be defined for particular groups of bacteria that will produce reliable, discriminatory and reproducible results at the practical level in different laboratories.

6.2 RAPD FINGERPRINTING APPROACHES

A. *Overview of the generation of RAPD fingerprints*

(i) *RAPD fingerprinting*

While conventional PCR assays can be used to amplify DNA sequences that are characteristic of a particular species or even strain, the method has the drawback that specific oligonucleotide primers are required, thereby making knowledge of the DNA sequence of the organism being studied an essential prerequisite. Analysis of RAPD fingerprints removes this requirement since a primer(s) can be

Table 6.1. Suggested classification of arbitrary primer fingerprinting methods according to primer length (Caetano-Anollés *et al.*, 1992a)

Method	Primer length
AP-PCR	*c.* 18–24 mer
RAPD	*c.* 10 mer
DAF[a]	*c.* 5–8 mer

[a] 'DNA amplification fingerprinting'.

chosen without regard to the sequence of the genome to be fingerprinted (Welsh & McClelland, 1990; Williams *et al.*, 1990). Thus, the method requires no previous knowledge of the molecular biology of the organisms to be investigated and can therefore, in principle, be applied to any species from which DNA can be prepared.

It is worth noting that the nomenclature used to describe this technique can be confusing (Vaneechoutte, 1996). Several different names have been used to describe methods involving the generation of RAPD fingerprints. An umbrella term of MAAP ('multiple arbitrary amplicon profiling') has been proposed (Caetano-Anollés *et al.*, 1992a), and it has also been suggested that individual methods could be classified according to primer length (Table 6.1). However, at the present time, none of this proposed nomenclature has received formal recognition, and most recent publications continue to refer to the general method as AP-PCR or RAPD typing regardless of the length of primer used. The technique will be referred to as RAPD fingerprinting for the remainder of this chapter.

The basis of the method is the original observation by Welsh & McClelland (1990) that a single arbitrarily-chosen primer, combined with two cycles of PCR at low stringency and many cycles at high stringency, generates a discrete and reproducible set of amplification products characteristic of particular genomes. The rationale for the phenomenon probably rests on the fact that, at a sufficiently low temperature, a primer can be expected to anneal to many sites on the chromosome, with a variety of mismatches, to initiate DNA polymerisation. Some of these sites will be within several hundred base pairs of each other and on opposite strands, such that the intervening sequence will be capable of amplification by PCR (Fig. 6.1). The molecular basis underlying the generation of RAPD fingerprints has been analysed in more detail by Venugopal *et al.* (1993), but arbitrary primers as short as 5 mer have been shown to generate complex genomic fingerprints following PCR amplification (Caetano-Anollés *et al.*, 1991). A key feature is that the initial amplification reaction in RAPD analysis is usually performed at low stringency, normally achieved by using a low annealing temperature during the PCR.

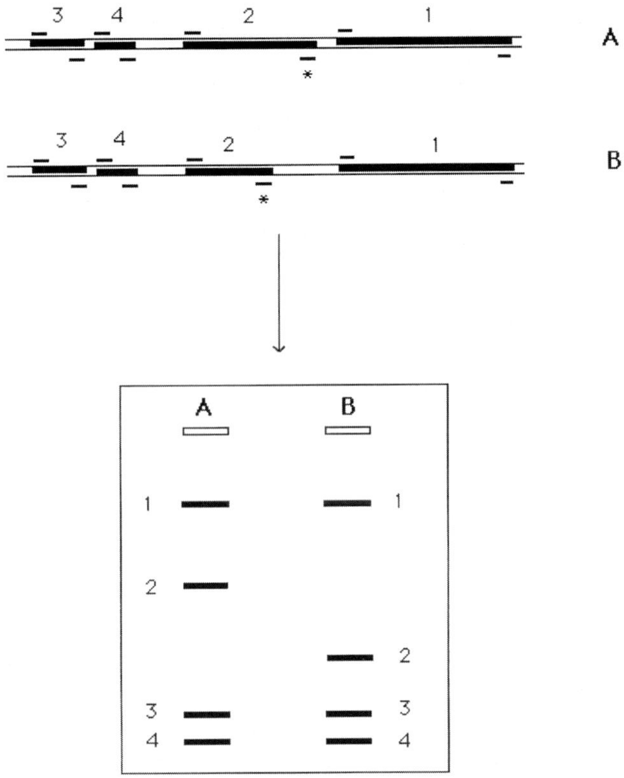

Fig. 6.1. Simplified representation of RAPD typing illustrating the binding of a single primer to both strands of two chromosomes (A and B). The resulting amplified DNA fragments are shown by the bold filled-in sections of the chromosomes. A single change in one annealing site (designated *) between the two chromosomes results in a significant difference in the corresponding RAPD fingerprint patterns as a result of the altered size of fragment 2.

(ii) REP-PCR

A related method, termed REP-PCR, is based on amplification from the sites of repetitive extragenic elements (REP) that are found in varying numbers and at different positions in the chromosomes of many different bacteria. As with RAPD, if two REP elements are located close enough to each other, they can act as priming sites to amplify the region of DNA between them. These intervening regions are termed inter-repeat fragments and, depending on the primer used, their number and sites (and consequently the number of fragments amplified) are variable from strain to strain.

Some individual techniques within this category have also been given specific names. Thus, a conceptually similar method known as ERIC-PCR is based on a family of conserved 'enterobacterial repetitive intergenic consensus' (ERIC) sequences found in the Enterobacteriaceae (Hulton *et al.*, 1991; Versalovic *et al.*, 1991). ERIC sequences resemble REP sequences, but their specific nucleotide sequences are quite different (Stern *et al.*, 1984; Hulton *et al.*, 1991). Similar techniques targeted against specific eukaryotic repeat sequences have also been described (Breukel *et al.*, 1990; Raich *et al.*, 1993).

The main characteristic distinguishing all these techniques from RAPD is that they produce distinct DNA fingerprints at relatively high stringency (high annealing temperatures similar to those employed in a standard PCR), although there is, of course, no reason why REP primers cannot be used in conjunction with lower annealing temperatures in classic RAPD reactions.

B. Factors affecting the generation of RAPD fingerprints

(i) Template:primer interactions

As described above, hybridisation of a single primer can occur to multiple sites on both DNA strands of a chromosome, particularly when the annealing step is carried out at a relatively low temperature (low stringency) to allow for slight mismatches. When two annealing sites are sufficiently close to each other on opposite strands, the intervening sequence is capable of amplification by PCR. Amplification products synthesised in the first round of PCR then become the preferred templates for subsequent amplification cycles (Caetano-Anollés *et al.*, 1992b) and result in the amplification of characteristic RAPD fingerprints of greater or lesser complexity, depending upon the precise primer used. However, it is worth noting that amplification products derived from a single primer will have palindromic ends that are capable of forming hairpin loop structures; these may, in turn, interfere with primer annealing in subsequent PCR rounds. As noted by Power (1996), the extent and stability of hairpin loop formation will vary between each individual amplification product, and therefore only some of the first round products will be amplified efficiently in subsequent cycles. This consideration is not a factor in amplification reactions (e.g., some REP-PCRs) that use two separate primers.

In subsequent rounds of PCR, decreasing efficiency of template denaturation tends to favour the formation of product:product duplexes rather than primer:template interactions, and the production of amplified product becomes linear rather than exponential. As different amplification products which act as templates are amplified individually, it might also be expected that templates present at low concentrations will be preferentially amplified during later PCR rounds. Therefore, on a theoretical basis, since the specificity of the amplification process depends on primer:template interactions, it might be expected that changes in the concentration of template DNA or primer will affect the precise nature of the PCR products amplified, and hence result in different RAPD fingerprints.

(ii) Reaction conditions

As described in section 6.2A, REP-PCR is distinguished from RAPD fingerprinting by the relatively high annealing temperatures used in the PCR, although REP primers can also be used with lower annealing temperatures in classical RAPD reactions. However, although production of adequate RAPD fingerprints generally requires an annealing temperature of 25–40°C, 10 to 15-mer primers have been shown to yield highly reproducible RAPD fingerprints at 45–50°C (Grundmann *et al.*, 1997). The use of a 'hot start' technique may be beneficial in terms of reproducibility at these higher annealing temperatures, but is not thought to influence RAPD fingerprint profiles at lower annealing temperatures (Power, 1996). Buffer components (e.g., Mg^{++} concentration) can be expected to influence RAPD fingerprints in much the same way as they influence standard diagnostic PCRs (Bassam *et al.*, 1992), but it seems that the precise number of cycles does not influence the qualitative nature of RAPD fingerprints provided that a minimum number of cycles (at least 30) is performed (Power, 1996).

(iii) Template DNA

This is an area in which it is the particular group of organisms being studied that seems to be critical. For some genera (e.g., *Acinetobacter* spp., *Klebsiella pneumonia, K. oxytoca, Pseudomonas aeruginosa*) it seems that reproducible RAPD fingerprints can be obtained simply by using heated cell suspensions (95°C for 15 min) as a DNA template (Grundmann *et al.*, 1997; Vogel *et al.*, 1999). This procedure has the major advantage of considerably simplifying the process and speeding-up the overall time required for RAPD analysis. However, for other genera (e.g., *Candida albicans*, *Serratia marcescens*) it seems that purified DNA is required for reproducible results (Power, 1996; Vogel *et al.*, 1999), perhaps because of the presence of heat-stable nucleases in the initial heated cell suspensions.

Observations regarding the influence of the template DNA concentration on RAPD fingerprints are rather divergent. A number of studies have concluded that the reproducibility of RAPD fingerprints is not strongly affected by the template DNA concentration (Welsh & McClelland, 1990; MacPherson *et al.*, 1993; Muralidharan & Wakeland, 1993; Van Belkum, 1994), with only variations in staining intensity being observed at moderately high DNA concentrations (>10 ng / 25-μL PCR). In contrast, other studies (e.g., Ellsworth *et al.*, 1993; Davin-Regli *et al.*, 1995) have reported that variations in the quantities of genomic template DNA alter the efficiency of amplification to such an extent that substantially different fingerprint patterns are generated. Variations seem to be greater at low DNA concentrations (<1 pg/μL) and could be explained by relative changes in the number of amplification events from perfect priming sites compared with rare sites, and the relative frequency of mismatch annealing events. Interestingly, excess template DNA can result in the production of a reduced number of amplification products (Power, 1996), perhaps because limiting concentrations of primer result in little or no amplification from sites at which annealing is less than optimum. In general, for most templates, 2.5 pmol of primer and 10 ng of template DNA in a 25-μL reaction mix seem to be optimal for generating sufficiently complex fingerprint patterns. However, providing a standardised DNA extraction method is used with cell suspensions of broadly equivalent initial cell densities, experience and published evidence suggests that reproducible fingerprints can be obtained with a range of bacterial genera without precise measurement of the template DNA concentration (Grundmann *et al.*, 1997; Vogel *et al.*, 1999).

An additional consideration is whether RAPD fingerprints should be performed with purified chromosomal DNA or total DNA (i.e., including any plasmid DNA present). Some authors have suggested that RAPD typing should be based strictly on purified chromosomal DNA alone (e.g., Power, 1996). However, total separation of chromosomal DNA from plasmid DNA can never be guaranteed, and the more lengthy extraction and purification procedures required have the effect of neutralising two of the major advantages of the RAPD approach, namely its simplicity and speed. A good compromise is to ensure that total DNA extracts are diluted as much as possible before amplification, thereby ensuring that the presence of plasmid DNA has a minimal effect.

(iv) Primer considerations

As described in section 6.2A, generation of RAPD fingerprints can be achieved with a primer chosen randomly without regard to the sequence of the genome being analysed. Nevertheless, it is clear that some arbitrary primers work better than others and may provide results that are more reproducible (Penner *et al.*, 1993). It also seems that the use of a combination of arbitrary primers in a single reaction can give more detailed and reproducible patterns (Tyler *et al.*, 1997). Consequently, primer selection is probably the single most influential parameter in the successful generation of RAPD fingerprints (van Belkum *et al.*, 1993), with primer sequence, primer length and primer concentration being the three most important considerations.

Primer sequence is generally not a major concern at the low annealing temperatures used in classical RAPD reactions, and many different arbitrary primer sequences have been used successfully. However, the precise number of fingerprint bands obtained will depend on the number and frequency of annealing sites on the template DNA being investigated. Since this is difficult to predict at low annealing temperatures, it is usually necessary to screen a number of different primers or primer combinations to obtain the optimum results for each individual species being investigated. Laboratories that do not wish to design, synthesise and test RAPD primers on an empirical basis may find that a useful starting point is a commercially available set of six 10-mer primers of arbitrary sequence (Amersham Pharmacia Biotech, Little Chalfont, UK) that have been specifically designed and tested for use in RAPD analysis with a variety of different species.

As with primer sequence, the effects of varying the primer length are also somewhat unpredictable. Clearly the optimum primer length should be related to the particular template DNA being investigated and, in theory, it might be expected that longer primers would provide a greater discriminatory ability than short primers (Power, 1996). Thus, MacPherson *et al.* (1993) found that 10-mer primers generated far fewer fragments than 20-mer primers in control experiments with *Toxoplasma gondii* DNA, while RAPD typing of methicillin-resistant *Staphylococcus aureus* (MRSA) and *Enterococcus faecium* was also more successful using prim-

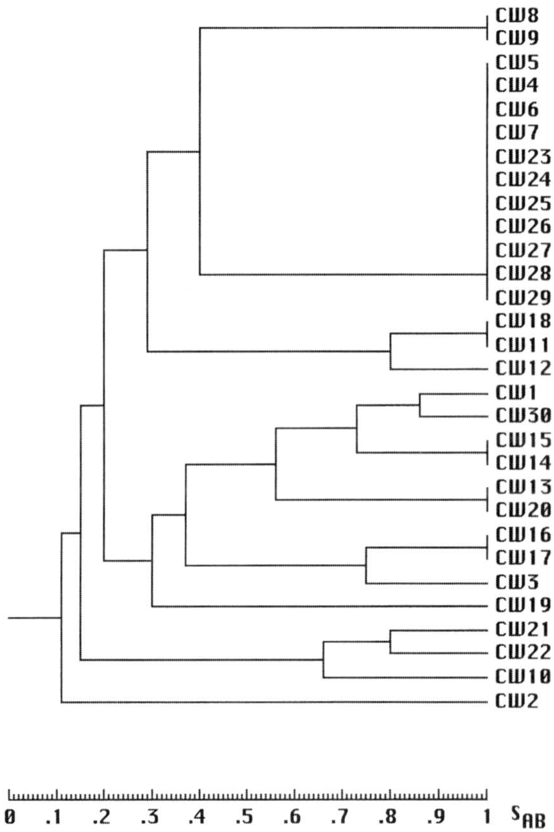

Fig. 6.4. Example of a dendrogram illustrating the relationships between 30 nosocomial isolates of *Acinetobacter baumannii*, based on analysis with the Dice coefficient of RAPD profiles generated with M13 core primer (Webster *et al.*, 1996). Within this particular system, a S_{AB} value of 0.7 (70%) has been shown to distinguish between genetically and epidemiologically unrelated isolates. Isolates with a S_{AB} value of >0.7 are considered to be closely related (Grundmann *et al.*, 1995, 1997; Webster *et al.*, 1996).

ing fluorescence density profiles generated by automated DNA sequencing equipment because of the PC-based computer platform that is generally used by such systems.

For PC users, the most widely used package for analysing RAPD fingerprints and digitised fluorescence density profiles is undoubtedly the BioNumerics program (Applied Maths, Kortrijk, Belgium) and its progenitor GelCompar. These programs can be used in the same way as DENDRON for examining scanned gel images and functions in a similar manner to DENDRON. When used for examining fluorescence density profiles, similarity analysis is usually carried out with Pearson's product moment correlation coefficient (as opposed to the Dice coefficient), followed by UPGMA cluster analysis (e.g., Grundmann *et al.*, 1995a, b; 1997; 1999). The Pearson coefficient calculates the correlation between arrays of values, in this case fluorescence density values for each point in the two profiles being compared. As the curves as a whole are compared, the Pearson correlation coefficient is less sensitive to differences in band intensities than the Dice coefficient and is ideally suited to PCR-based methods where intensity differences between bands matter more than with restriction fragment analyses. Crucially for the comparison of fluorescence density profiles, it is largely insensitive to relative concentrations, but it *is* sensitive to differences in background.

The question of whether to use the Dice or Pearson correlation coefficients to examine RAPD fingerprint profiles provides an interesting philosophical problem. As mentioned above, the Dice coefficient is based primarily on the presence or absence of bands in particular *positions*. Although the various programs available will give an automated interpretation of the positions of bands on a scanned gel image or fluorescence density profile, it is important that these positions are checked manually before final analysis, and this checking process is ultimately a question of subjective judgement. In contrast, the Pearson coefficient is more objective in that it depends on the variance between two fluorescence density values at each point in the curve pattern and consequently does not suffer from typical peak/shoulder mismatches. However, it relies totally on the computer's interpretation of the profile and accurate gel normalisation is therefore mandatory.

Seward *et al.* (1997) made a direct comparison of the overall robustness of the respective algorithms and the reproducibility of the cluster analysis results generated by the Dice and Pearson correlation coefficients by using both methods to examine the same set of RAPD fingerprints generated from isolates of known epidemiological relationships. Both methods were efficient at recognising groups of isolates with closely similar RAPD fingerprints (i.e., isolates which might be expected to have a close epidemiological or evolutionary relationship), but showed considerable variation in terms of the relationships suggested for less closely related isolates with comparatively low S_{AB} values. Overall it was concluded that both analysis methods were useful tools for examining genotypic relationships between individual isolates, but that assessment of the results and occasional intervention by a trained microbiologist is still essential to detect and correct artifacts introduced by the automated systems (Seward *et al.*, 1997).

Whatever the precise algorithm used to analyse the relationships between strains, the outcome is either expressed in terms of the percentage similarity between each individual pair of isolates (Dice coefficient) or as a correlation value (Pearson coefficient). There is a general consensus that several band differences (depending on the total number of bands) in the overall RAPD fingerprint profile are required before two isolates may be considered different (Struelens *et al.*, 1996). However, as outlined in section 6.3A, there are no hard or fast criteria that define the precise number of band differences that is required to separate two isolates, and it is clear that a 'cut-off' value should be defined empirically for each individual species:primer combination before epidemiological conclusions can be reached. To date, the only such studies with RAPD fingerprints have been concerned with strains of *Acinetobacter* spp., for which it seems that a cut-off similarity value of

70% is suitable for distinguishing unambiguously between genetically and epidemiologically unrelated isolates *with the particular sets of primers and amplification conditions used* (Grundmann *et al.*, 1995, 1997; Webster *et al.*, 1996). Such a cut-off value (which depends on the total number of bands in the profile) is quite helpful as it means that the minor variations in fingerprint profiles which seem to occur occasionally with most RAPD reactions have no influence on the final epidemiological conclusion. More studies need to be performed with RAPD fingerprints generated from epidemiologically defined collections of isolates belonging to other bacterial genera. It cannot be overemphasised that the absolute number of band differences is a measure that needs to be interpreted with extreme caution as its importance will be entirely proportional to the overall number of resolved RAPD fragments (Struelens *et al.*, 1996).

6.4 STANDARDISATION OF RAPD FINGERPRINTS

Section 6.2B described some of the factors that can affect the generation of PCR fingerprints. These have been discussed in more detail in previous reviews of this subject (Power, 1996; Tyler et al., 1997) and more details can be found in the references listed in Table 6.2. These factors can undoubtedly present major problems in terms of the reproducibility of the technique. Nevertheless, the RAPD technique offers some major advantages over traditional phenotypic typing methods (and the other molecular fingerprinting methods described in this book) by being rapid, relatively inexpensive, theoretically applicable to any organism, and technically feasible for most diagnostic microbiology laboratories. There is a general consensus amongst microbiologists that careful attention to detail and standardisation of techniques should be sufficient for an individual laboratory to use the technique *comparatively* for such purposes as recognising and tracing strains involved in an outbreak. The availability of specially designed and quality controlled RAPD analysis reagents has provided a considerable advance and, as with other PCR techniques, methods that remove or reduce non-specific annealing before amplification (e.g., the 'hot-start' technique) should help in the production of reproducible RAPD fingerprints.

The RAPD approach is being used increasingly in many microbiology laboratories for the epidemiological typing of an ever-expanding range of bacteria. However, variations between laboratories still present considerable difficulties and, to date, there has only been one study that has attempted to address this problem by combining the use of standardised methodology for the generation of RAPD fingerprints with automated laser fluorescence (ALF) analysis of the DNA fragments. In this study (Grundmann *et al.*, 1997), an exact measure of the inter-laboratory reproducibility of a given amplification profile was obtained by analysing the degree of similarity (pattern correlation) between RAPD fingerprints generated from DNA extracts prepared centrally or independently in seven European laboratories for the same isolates of *Acinetobacter* spp. The overall pattern correlation was calculated as the arithmetic mean of Pearson's product moment correlation

Table 6.2. Summary of important factors affecting the standardisation of bacterial RAPD fingerprints

	Parameter	Example references
Template DNA	Preparation	Elaichouni *et al.* (1994) McClelland & Welsh (1994) Micheli *et al.* (1994)
	Concentration	Bassam *et al.* (1992)
	Secondary structure	Caetano-Anollés (1993)
	RNA contamination	Ellsworth *et al.* (1993) Micheli *et al.* (1994)
Primer	Length	Caetano-Anollés *et al.* (1992b) Caetano-Anollés (1993)
	Concentration	Bassam *et al.* (1992) MacPherson *et al.* (1993)
	G+C content	Caetano-Anollés *et al.* (1992b) Caetano-Anollés (1993)
PCR reagents	Source / type	Bassam *et al.* (1992) Meunier & Grimont (1993) Schierwater & Ender (1993)
	Taq concentration	Bassam *et al.* (1992) Brikun *et al.* (1994)
	Magnesium concentration	Bassam *et al.* (1992) Ellsworth *et al.* (1993)
Equipment	Thermocycler model	MacPherson *et al.* (1993) Meunier & Grimont (1993) Penner *et al.* (1993)
Reaction and visualisation conditions	Annealing temperature	Caetano-Anollés *et al.* (1992b) Caetano-Anollés (1993) Ellsworth *et al.* (1993)
	Number of cycles	Bassam *et al.* (1992) Kangfu & Pauls (1992) MacPherson *et al.* (1993)
	Denaturing time	Kangfu & Pauls (1992)
	Extension time	Kangfu & Pauls (1992)
	Type of gel	Berg *et al.* (1994) McClelland & Welsh (1994)

coefficients of fingerprints from all identical isolates after parallel analysis of the RAPD fingerprints by ALF on sequencing gels. Table 6.3 shows the pattern correlation results obtained with each of the four individual primer sets investigated over the 120–800 bp size range. Overall, the pattern correlation was good (range: 83.3–86.6%), but was slightly better for the centrally prepared DNA extracts (87.1%) than the DNA extracts prepared in each individual laboratory (84.7%), although this difference was not significant. Since good discrimination between genetically and epidemiologically unrelated groups of *Acinetobacter* spp. could be

Table 6.3. RAPD pattern correlation results obtained following ALF analysis of data obtained in a multicentre trial in seven different European laboratories with isolates of *Acinetobacter* spp.

	Percentage pattern correlation (±SD)	
Primer(s)	Centrally prepared template DNA	Locally prepared template DNA
M13	88.6 (±7.8)	82.8 (±10.6)
DAF4	86.6 (±8.5)	87.9 (±8.5)
ERIC-2	86.3 (±12.5)	84.7 (±10.7)
REP1 + REP2	86.8 (±17.5)	83.5 (±8.6)
Mean	87.1	84.7

Data taken from Grundmann *et al.*, 1997.

achieved with the methods used at a similarity level of 70%, the study was successful in demonstrating that the minor variations in pattern correlations observed between the different participating laboratories would not be sufficient to alter the overall epidemiological conclusions. However, it was clear that such inter-laboratory comparisons are heavily dependent on rigorous normalisation of sample lanes on a gel to allow accurate intra- and inter-gel comparisons. This in turn necessitates the addition of internal molecular size markers to each and every lane, together with the inclusion of a set of molecular size markers spanning the entire range of DNA fragment sizes being analysed at regular intervals (at least every six lanes) on the gel.

6.5 CURRENT APPLICATIONS

A cursory examination of the scientific literature will reveal an ever-increasing number of articles describing the use of RAPD fingerprinting techniques for typing an ever-increasing range of bacterial species. However, rather fewer of these proposed applications have been rigorously assessed in terms of the performance criteria (typeability, reproducibility, stability, discriminatory power, epidemiological concordance and typing system concordance) recommended by the European Study Group on Epidemiological Markers (Struelens *et al.*, 1996). Nevertheless, it seems that the most obvious and valuable applications of RAPD fingerprinting lie in the relatively rapid and potentially discriminatory characterisation of isolates that is achievable on a day-to-day basis in individual diagnostic laboratories during outbreak investigations. Application of standardised RAPD fingerprinting techniques should provide rapid information at the local level and help with the initiation of effective infection control programmes.

Apart from outbreak investigations, it has also been suggested (Grundmann *et al.*, 1999; Webster & Towner, 2000) that RAPD fingerprinting techniques may have a role to play in the *routine* analysis of isolates in a diagnostic laboratory. Such routine typing of isolates would be a desirable component of infection control services in a setting such as an intensive care unit where high densities of cross-transmission might be expected as a result of the underlying illness of patients, the high frequency of invasive procedures, and the overall high levels of antibiotic selection pressure. At present, typing of bacteria is rarely performed in non-outbreak situations because most typing systems are time-consuming, not applicable to all bacteria, or not cost-effective. However, by using the relatively inexpensive and potentially universal RAPD fingerprinting technique in situations where bacteria of a single species are isolated from patients in conspicuous clusters, informative typing data can be made available to clinical staff in a timely fashion.

6.6 PROSPECTS FOR THE FUTURE

The spread of particular virulent, epidemic or antibiotic-resistant microorganisms poses a major and increasing threat to the health of populations throughout the world, with substantial additional costs associated with lengthened hospital stays, use of more expensive antibiotics, and increased morbidity and mortality. However, studies of the geographical epidemiology of particular microbial pathogens are often restricted mainly to individual research institutes because of the complicated or specialised nature of the typing systems that are currently being used. National Governments expect the scientific community to respond to the problem of cosmopolitan bacterial strains and their spread within the population, but tackling this problem requires more coordination of efforts. WHO has been taking the lead in developing networks for surveillance and identification of emerging and antibiotic-resistant pathogens in close collaboration with the EU and the Centers for Disease Control and Prevention in Atlanta, USA, but a mere report or database of accumulated susceptibility data and other associated information will not suffice for the timely recognition of particular virulent or epidemic organisms.

Identification and typing of bacteria is thus extremely important in efforts to monitor the geographical spread of pathogens. However, as outlined at the beginning of this chapter, central analysis of an ever-growing number of important microorganisms at a national or international level requires undesirable and potentially hazardous transport of pathogens between laboratories and imposes an unmanagable workload, with associated delays in obtaining results, on central reference facilities. Meaningful understanding and sensible intervention by local health authorities to prevent the national and international spread of disease can only be achieved by extensive and rapid communication of data between different local laboratories. Therefore, local isolate analysis combined with central data collection offers a more efficient and safer alternative. The design of new RAPD-based fingerprinting methods for analysis of microbial DNA means that such typing strategies can potentially be applied to many microorganisms of public health significance (Struelens *et al.*, 1996).

The RAPD fingerprinting technique, combined with the use of quality controlled reagents, standardised protocols and analysis on sequencing gels, has been shown in trial experiments to generate reproducible results of epidemiological significance with various groups of bacteria (Grundmann *et al.*, 1995a, b; 1997; Webster *et al.*, 1996; Vogel *et al.*, 1999), but the generation of numerous complex fingerprint profiles, particularly in cases where strains from different geographical locations are being compared over significant time periods, requires that a computer-assisted strategy is used to enable the formation of a database of fingerprint patterns. As described in section 6.3C, a number of software programs are available that successfully perform DNA fingerprint pattern analysis, clustering and comparison, but a consensus still needs to be reached for data generation, entry and retrieval protocols, and cut-off values for determining epidemiological relatedness.

Automated laser fluorescence-based analysis of RAPD fingerprints, combined with digital data communication via the Internet, may bring new opportunities for the comparison of typing data generated in different facilities. Such data could be made accessible to other laboratories via an Internet-based database. In places where automated sequencers are not available, DNA extracts could be easily prepared and shipped without accompanying risk to sentinel surveillance laboratories with appropriate equipment for immediate analysis. The increasing public health problem resulting from the national and international spread of pathogenic microorganisms means that it is now timely for further studies on establishing standardised methodologies and systems for use in conjunction with this technology.

REFERENCES

Bassam, B.J., Caetano-Annolés, G. & Gresshoff, P.M. (1992). DNA amplification fingerprinting of bacteria. *Applied Microbiology and Biotechnology* **38**, 70–76.

Berg, D.E., Akopyants, N.S. & Kersulyte, D. (1994). Fingerprinting microbial genomes using the RAPD or AP-PCR method. *Methods in Molecular and Cellular Biology* **5**, 13–24.

Blanc, D.S., Hauser, P.M., Francioli, P. & Bille, J. (1998). Molecular typing methods and their discriminatory power. *Clinical Microbiology and Infection* **4**, 61–63.

Breukel, C., Wijnen, J., Tops, C., van der Klift, H., Dauwerse, H. & Khan, P.M. (1990). Vector-Alu PCR: a rapid step in mapping cosmids and YACs. *Nucleic Acids Research* **18**, 3097.

Brikun, I., Suziedelis, K. & Berg, D.E. (1994). DNA sequence divergence among derivatives of *Escherichia coli* K-12 detected by arbitrary primer PCR (random amplified polymorphic DNA) fingerprinting. *Journal of Bacteriology* **176**, 1673–1682.

Caetano-Annolés, G. (1993). Amplifying DNA with arbitrary oligonucleotide primers. *PCR Methods and Applications* **3**, 85–94.

Caetano-Annolés, G., Bassam, B. J. & Gresshoff, P. M. (1991). DNA amplification fingerprinting using very short arbitrary oligonucleotide primers. *BioTechnology* **9**, 553–557.

Caetano-Annolés, G., Bassam, B. J. & Gresshoff, P. M. (1992a). DNA fingerprinting: MAAPing out a RAPD definition? *BioTechnology* **10**, 937.

Caetano-Annolés, G., Bassam, B. J. & Gresshoff, P. M. (1992b). Primer-template interactions during DNA amplification fingerprinting with single arbitrary oligonucleotides. *Molecular and General Genetics* **235**, 157–165.

Cotter, L., Daly, M., Greer, P., Cryan, B. & Fanning, S. (1998). Motif-dependent DNA analysis of a methicillin-resistant *Staphylococcus aureus* (MRSA) collection. *British Journal of Biomedical Science* **55**, 99–106.

Davin-Regli, A., Abed, Y., Charrel, R.N., Bollet, C. & de Micco, P. (1995). Variations in DNA concentrations significantly affect the reproducibility of RAPD fingerprint patterns. *Research in Microbiology* **146**, 561–568.

Elaichouni, A., Van Emmelo, J., Claeys, G., Verschraegen, G., Verelst, R. & Vaneechoutte, M. (1994). Study of the influence of plasmids on the arbitrary primer polymerase chain reaction fingerprint of *Escherichia coli* strains. *FEMS Microbiology Letters* **115**, 335–339.

Ellsworth, D.L., Rittenhouse, K.D. & Honeycutt, R.L. (1993). Artifactual variation in randomly amplified polymorphic DNA banding patterns. *BioTechniques* **14**, 214–217.

Grundmann, H.J., Schneider, C. & Daschner, F.D. (1995a). Fluorescence-based DNA fingerprinting elucidates nosocomial transmission of phenotypically variable *Pseudomonas aeruginosa* in intensive care units. *European Journal of Clinical Microbiology and Infectious Diseases* **14**, 1057–1062.

Grundmann, H., Schneider, C., Tichy, H.V., Simon, R., Klare, I., Hartung, D. & Daschner, F.D. (1995b). Automated laser fluorescence analysis of randomly amplified polymorphic DNA: a rapid method for investigating nosocomial transmission of *Acinetobacter baumannii*. *Journal of Medical Microbiology* **43**, 446–451.

Grundmann, H.J., Towner, K.J., Dijkshoorn, L., Gerner-Smidt, P., Mahar, M., Seifert, H. & Vaneechoutte, M. (1997). Multicenter study using standardized protocols and reagents for evaluation of reproducibility of PCR-based fingerprinting of *Acinetobacter* spp. *Journal of Clinical Microbiology* **35**, 3071–3077.

Grundmann, H.J., Hahn, A., Ehrenstein, B., Geiger, K., Just, H. & Daschner, F.D. (1999). Detection of cross-transmission of multiresistant Gram-negative bacilli and *Staphylococcus aureus* in adult intensive care units by routine typing of clinical isolates. *Clinical Microbiology and Infection* **5**, 355–363.

Hadrys, H., Balick, M. & Schierwater, B. (1992). Application of random amplified polymorphic DNA (RAPD) in molecular ecology. *Molecular Ecology* **1**, 55–63.

He, Q., Viljanen, M.K. & Mertsola, J. (1994). Effects of thermocyclers and primers on the reproducibility of banding patterns in randomly amplified polymorphic DNA analysis. *Molecular and Cellular Probes* **8**, 155–159.

Hulton, C. S., Higgins, C. F. & Sharp, P. M. (1991). ERIC sequences: a novel family of repetitive elements in the genomes of *Escherichia coli*, *Salmonella typhimurium* and other enterobacteria. *Molecular Microbiology* **5**, 825–834.

Kangfu, Y. & Pauls, K.P. (1992). Optimization of the PCR program for RAPD analysis. *Nucleic Acids Research* **20**, 2606.

Linz, U. (1990). Thermocycler temperature variation invalidates PCR results. *BioTechniques* **9**, 286–294.

MacPherson, J. M., Eckstein, P. E., Scoles, G. J. & Gajadhar, A. A. (1993). Variability of the random amplified polymorphic DNA assay among thermal cyclers, and effects of primer and DNA concentration. *Molecular and Cellular Probes* **7**, 293–299.

Maslow, J. N., Mulligan, M. E. & Arbeit, R. D. (1993). Molecular epidemiology: application of contemporary techniques to the typing of microorganisms. *Clinical Infectious Diseases* **17**, 153–164.

McClelland, M. & Welsh, J. (1994). DNA fingerprinting by arbitrarily primed PCR. *PCR Methods and Applications* **4**, S59-S65.

Meunier, J. R. & Grimont, P. A. D. (1993). Factors affecting reproducibility of random amplified polymorphic DNA fingerprinting. *Research in Microbiology* **144**, 373–379.

Micheli, M.R., Bova, R., Pascale, E. & D'Ambrosio, E. (1994). Reproducible DNA fingerprinting with the random amplified polymorphic DNA (RAPD) method. *Nucleic Acids Research* **22**, 1921–1922.

Munthali, M., Ford-Lloyd, B.V. & Newbury, H.J. (1992). The random amplification of polymorphic DNA for fingerprinting plants. *PCR Methods and Applications* **1**, 274–276.

Muralidharan, K. & Wakeland, E.K. (1993). Concentration of primer and template qualitatively affects products in random-amplified polymorphic DNA PCR. *BioTechniques* **14**, 362–364.

Penner, G.A., Bush, A., Wise, R., Kim, W., Domier, L., Kasha, K., Laroche, A., Scoles, G., Molnar, S.J. & Fedak, G. (1993). Reproducibility of random amplified polymorphic DNA (RAPD) analysis among laboratories. *PCR Methods and Applications* **2**, 341–345.

Power, E. G. M. (1996). RAPD typing in microbiology – a technical review. *Journal of Hospital Infection* **34**, 247–265.

Raich, T.J., Archer, J.L., Robertson, M.A., Tabachnik, W.J. & Beaty, B.J. (1993). Polymerase chain reaction approaches to *Culicoides* (Diptera: Ceratopogonidae) identification. *Journal of Medical Entomology* **30**, 228–232.

Sambrook, J., Fritsch, E.F. & Maniatis, T. (1989). Agarose gel electrophoresis. In *Molecular cloning, a laboratory manual*, 2nd edn., pp.6.3–6.19. Cold Spring Harbor Laboratory Press, New York.

Schierwater, B. & Ender, A. (1993). Different thermostable DNA polymerases may amplify different RAPD products. *Nucleic Acids Research* **21**, 4647–4648.

Seward, R.J., Ehrenstein, B., Grundmann, H.J. & Towner, K.J. (1997). Direct comparison of two commercially available computer programs for analysing DNA fingerprinting gels. *Journal of Medical Microbiology* **46**, 314–320.

Stegemann, J., Schwager, C., Erfle, H., Hewitt, N., Voss, H., Zimmermann, J. & Ansorge, W. (1991). High-speed online DNA sequencing on ultrathin slab gels. *Nucleic Acids Research* **19**, 675–676.

Stern, M. J., Ames, G. F., Smith, N. H., Robinson, E. C. & Higgins, C. F. (1984). Repetitive extragenic palindromic sequences: a major component of the bacterial genome. *Cell* **37**, 1015–1026.

Struelens, M.J., and the members of the European Study Group on Epidemiological Markers (ESGEM) of the European Society for Clinical Microbiology and Infectious Diseases (ESCMID) (1996). Consensus guidelines for appropriate use and evaluation of microbial epidemiologic typing systems. *Clinical Microbiology and Infection* **2**, 1–11.

Tenover, F.C., Arbeit, R.D., Goering, R.V., Mickelson, P.A., Murray, B.E., Persing, D.H. & Swaminathan, B. (1996). Interpreting chromosomal DNA restriction patterns produced by pulsed-field gel electrophoresis: criteria for bacterial strain typing. *Journal of Clinical Microbiology* **33**, 2233–2239.

Tyler, K. D., Wang, G., Tyler, S. D. & Johnson, W. M. (1997). Factors affecting reliability and reproducibility of amplification-based DNA fingerprinting of representative bacterial pathogens. *Journal of Clinical Microbiology* **35**, 339–346.

van Belkum, A. (1994). DNA fingerprinting of medically important microorganisms by use of PCR. *Clinical Microbiology Reviews* **7**, 174–184.

van Belkum, A., Bax, R., Peerbooms, P., Goessens, W., van Leeuwen, N. & Quint, W.G.V. (1993). Comparison of phage typing and DNA fingerprinting by PCR for discrimination of methicillin-resistant *Staphylococcus aureus* strains. *Journal of Clinical Microbiology* **31**, 798–803.

van Belkum, A., Kluytmans, J., van Leeuwen, W., Bax, R., Quint, W., Peters, E., Fluit, C., Vandenbroucke-Grauls, C., van den Brule, A., Koeleman, H., Melchers, W., Meis, J., Elaichouni, A., Vaneechoutte, M., Moonens, F., Maes, N., Struelens, M., Tenover, F. & Verbrugh, H. (1995). Multicenter evaluation of arbitrarily primed PCR for typing of *Staphylococcus aureus* strains. *Journal of Clinical Microbiology* **33**, 1537–1547.

Vaneechoutte, M. (1996). DNA fingerprinting techniques for microorganisms. A proposal for classification and nomenclature. *Molecular Biotechnology* **6**, 115–142.

Venugopal, G., Mohapatra, S., Salo, D. & Mohapatra, S. (1993). Multiple mismatch annealing: basis for random amplified polymorphic DNA fingerprinting. *Biochemical and Biophysical Research Communications* **197**, 1382–1387.

Versalovic, J., Thearith, K. & Lupski, J. R. (1991). Distribution of repetitive DNA sequences in eubacteria and applications to fingerprinting of bacteria genomes. *Nucleic Acids Research* **19**, 6823–6831.

Vogel, L., Jones, G., Triep, S., Koek, A. & Dijkshoorn, L. (1999). RAPD typing of *Klebsiella pneumonia, Klebsiella oxytoca, Serratia marcescens* and *Pseudomonas aeruginosa* isolates using standardized reagents. *Clinical Microbiology and Infection* **5**, 270–276.

Webster, C.A. & Towner, K.J. (2000). Use of RAPD-ALF analysis for investigating the frequency of bacterial cross-transmission in an adult intensive care unit. *Journal of Hospital Infection* **44**, 254–260.

Webster, C.A., Towner, K.J., Humphreys, H., Ehrenstein, B., Hartung, D. & Grundmann, H. (1996). Comparison of rapid automated laser fluorescence analysis of DNA fingerprints with four other computer-assisted approaches for studying relationships between *Acinetobacter baumannii* isolates. *Journal of Medical Microbiology* **44**, 185–194.

Welsh, J. & McClelland, M. (1990). Fingerprinting genomes using PCR with arbitrary primers. *Nucleic Acids Research* **18**, 7213–7218.

Williams, J. G. K., Kubelik, A. R., Livak, K. J., Rafalski, J. A. & Tingey, S. V. (1990). DNA polymorphisms amplified by arbitrary primers are useful as genetic markers. *Nucleic Acids Research* **18**, 6531–6535.

Woods, C. R., Versalovich, J., Koeuth, T. & Lupski, J. R. (1992). Analysis of relationships among isolates of *Citrobacter diversus* by using DNA fingerprints generated by repetitive sequence-based primers in the polymerase chain reaction. *Journal of Clinical Microbiology* **30**, 2921–2929.

7 Analysis of Microbial Genomic Macrorestriction Patterns by Pulsed-Field Gel Electrophoresis (PFGE) Typing

Marc J. Struelens[1,2], Raf De Ryck[1] and Ariane Deplano[1]
[1]*Centre for Molecular Diagnostic Microbiology, Department of Microbiology, Hôpital Erasme and* [2]*Infectious Diseases Epidemiology Unit, School of Public Health, Université Libre de Bruxelles, 808 route de Lennik, 1070-Bruxelles, Belgium*

CONTENTS

7.1 INTRODUCTION

Macrorestriction analysis of microbial genomes resolved by pulsed-field gel electrophoresis (PFGE) has emerged during the 1990s as the method of choice for studying the molecular epidemiology of bacterial pathogens, at least at the level of outbreak studies and hospital epidemiology (Tenover *et al.*, 1997; Struelens, 1998). Macrorestriction fingerprinting is a high resolution typing technique derived during the late 1980s (Grothues *et al.*, 1988) from the combination of two tech-

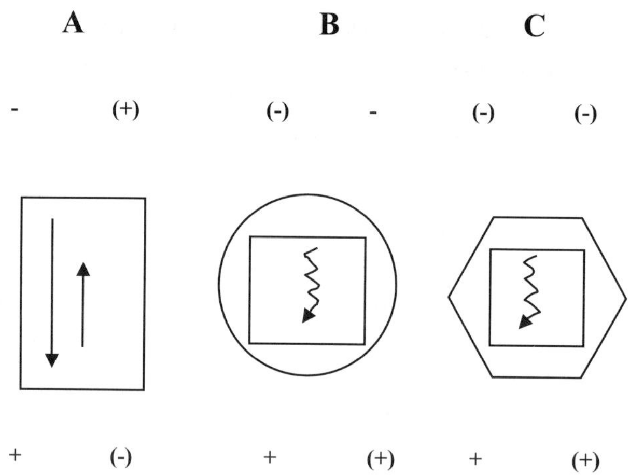

Fig. 7.1. Geometry of electrodes in commonly used PFGE equipment: **A**, FIGE; **B**, RFE; **C**, CHEF.

niques: chromosomal restriction fragment pattern analysis using low-frequency cutting enzymes (typically with less than 30 cleavage sites per genome) and PFGE, a modification of agarose gel electrophoresis initially introduced in 1984 for providing electrophoretic karyotypes of yeast (Schwartz & Cantor, 1984) and other eukaryotic microorganisms. PFGE is based on gel electrophoresis in which the electric field periodically changes direction and/or intensity. In contrast to unidirectional electrophoresis, which cannot separate DNA molecules larger than about 50 kb in size due to molecular trapping in the gel, PFGE can separate DNA molecules as large as 12 Mb.

7.2 PRINCIPLES OF PFGE

There are a number of physical models that have been proposed to predict the molecular separation of DNA in agarose gels by PFGE (Birren *et al.*, 1988; Kozulik, 1995). In response to changes in the orientation of the electric field, large DNA molecules migrate through the agarose matrix in a zig-zagging motion which has been visualised experimentally using fluorescent microscopy and labelled phage DNA (Gurrieri *et al.*, 1990). DNA molecules stretch out linearly in the direction of the electric field. When the field changes direction, the DNA molecules initially adopt a partially relaxed conformation, and then form multiple kinks in the direction of the new field until the predominant kink unfolds and the DNA molecule elongates again to migrate in the new direction. The larger the molecule, the longer the time required for relaxation and reorientation of its leading end. Thus, the time interval between changes of field direction, also called the *pulse time*, will primarily determine the size window for separation of DNA molecules by PFGE (Birren *et al.*, 1988).

During the 1980s, different instruments were developed for PFGE by using a diversity of geometry for placing the electrodes around the agarose gel. Fig. 7.1

schematically illustrates examples of systems commonly used for bacterial typing. The simplest and earliest instruments are based on field inversion gel electrophoresis (FIGE), in which a conventional gel chamber is exposed to periodic alternations in polarity of the electrodes. Net forward migration is obtained by increasing the ratio of forward to backward pulse times to 3:1. Zero integrated field electrophoresis is a modification of FIGE in which the electric field intensity is slightly higher forward than backward, thereby producing higher resolution, but at the cost of longer running-times. More recently, FIGE has been adapted to capillary electrophoresis systems (PFCE or pulsed-field capillary electrophoresis; Kim & Morris, 1995). Small amounts of DNA samples are injected into a capillary tubing containing a sieving matrix composed of long polymer chains. Under exposure to high frequency modulation of direct current and alternating voltage current, DNA molecules of up to 1.6 Mb in size can be separated within minutes, in contrast to the several hours or days required for PFGE (Kim & Morris, 1995). Another rapid method for separation of macrorestriction fragments has been described recently, based on ultra-sensitive flow-cytometry (Kim *et al.*, 1999).

Other PFGE instruments that allow co-linear separation of large DNA molecules include systems in which the electrode sets rotate around the gel chamber (RFE or rotating field electrophoresis; Fig. 7.1). Conversely, other types of apparatus rotate the gel periodically in a stationary field (RGE or rotating gel electrophoresis). However, due to mechanical constraints, these systems are fragile and do not allow very short pulse times. The most popular PFGE system used for bacterial typing is the contour-clamped homogeneous electric field (CHEF) apparatus, in which an hexagonal array of electrodes periodically alternate uniform fields with an angle of reorientation of 120°. CHEF instruments are available commercially from Bio-Rad (CHEF DR-II and III), Pharmacia (Pulsaphor) and BRL (Hexafield). A more sophisticated commercial model of the CHEF electrophoresis system with programmable electrodes (CHEF Mapper; Bio-Rad) allows versatile variation of electric field angle and shape that enables optimisation of resolution by molecular size range.

Several inter-dependent physical and chemical factors affect the separation of DNA molecules in PFGE. The most important parameter is the pulse time. CHEF electrophoresis performed using a single pulse time will produce a non-linear relationship between fragment size and migration distance. The longer the pulse time, the larger the size of the molecules that can be separated, but at the cost of lower resolution, i.e., separation of DNA molecules or fragments of similar size. Maximal resolution is seen just below the maximum size of non-separated bands, called the *compression zone*. Thus, use of the shortest pulse time which permits separation of the desired upper size range will provide optimal resolution. The use of progressive increases in pulse time, called pulse time ramping or switch time ramping, can increase the linearity of size-dependent velocity of DNA fragments, particularly by using an exponential time ramp. A typical ramp from 5 s to 40 s enables optimal separation of DNA fragments from 50 kb to 600 kb, while increasing the pulse time to 75 s extends the separation to 1 Mb with standard CHEF conditions (see below; Birren *et al.*, 1988).

The DNA migration rate will increase with voltage. At a constant pulse time, an increase in voltage will mobilise larger molecules, but with a decrease in band sharpness. A voltage of 6V/ cm is used for the separation of molecules from 50 kb to 1.5 Mb in size, but should be reduced to separate larger molecules (Birren *et al.*, 1988). The reorientation angle taken by alternating electric fields can be decreased from the normal 120° to 96–106° to faster separate molecules of >1 Mb in size, such as yeast chromosomes (Birren *et al.*, 1988).

The temperature and ionic strength of the electrophoresis buffer affects DNA velocity and resolution. At increased temperature, DNA migrates faster, but band resolution decreases. PFGE gels are normally run at temperatures between 12 and 15°C. This temperature is maintained with the use of a buffer cooling and re-circulating system to prevent the generation of temperature gradients within the gel during the run.

The agarose type and concentration, as well as the buffer strength, affect DNA migration in PFGE in a similar way to conventional electrophoresis. Special PFGE-grade agarose with a high physical strength and low electro-endosmosis is recommended. An agarose concentration of 1% is normally used for the separation of DNA fragments ranging from 50 kb to 1 Mb, but this concentration should be decreased for larger size separation. Use of more concentrated agarose (1.2–1.6%) increases band sharpness and resolution, but at the cost of longer running times and a lower range of DNA size separation (Kim *et al.*, 1999). Low ionic strength buffers (e.g., 0.25–0.5 × TAE or TBE) are preferred to reduce heat generation and shorten the run times (Birren *et al.*, 1988).

7.3 PRINCIPLES OF GENOMIC MACRORESTRICTION ANALYSIS

Macrorestriction analysis consists of PFGE separation of a limited number (<30) of large (>50 kb) fragments of a bacterial chromosome digested with infrequently cleaving restriction endonucleases, also called "rare cutters". These enzymes are selected on the basis of the rarity of occurrence of their recognition sequence in the target genome. For example, the tetra nucleotide CTAG is counter-selected and is therefore very rare in most bacterial genomes with high (>45%) G+C base composition. Endonucleases that include CTAG in their recognition sequence cleave such genomes less than once per 1 kb: e.g., *Xba*I (TCTAGA), *Spe*I (ACTAGT) and *Nhe*I (GCTAGC). Hexamer endonucleases with G+C recognition sequences are rare cutters of A+T rich prokaryotic genomes, and vice-versa (McClelland *et al.*, 1987). Likewise, enzymes with eight and 10 nucleotide-long recognition sites are also used (e.g., *Not*I, *Sfi*I). As a general rule, the following enzymes are useful macrorestriction cutters based on genomic G+C content:

(i) < 40% G+C: *Sma*I (CCCGGG), *Sac*II (CCGCGG), *Rsr*II (CGGWCGG), *Eag*I (CGGCCG), *Nar*I (GGCGCC), *Nae*I (GCCGGC);
(ii) 40–50% G+C: *Avr*II (CCTAGG), *Nhe*I (GCTAGC), *Sfi*I (GGCCN$_5$GGCC);
(iii) 50–65% G+C: *Xba*I (TCTAGA), *Spe*I (ACTAGT);
(iv) > 65% G+C: *Xba*I, *Spe*I, *Dra*I (TTTAAA), *Ssp*I (AATATT).

Table 7.1. Infrequently cleaving endonucleases suitable for macrorestriction analysis of selected bacteria

Organism	Enzymes[a]
Staphylococcus aureus	*Sma*I, *Sst*II, *Csp*I
Staphylococcus epidermidis	*Sma*I, *Sst*II, *Kpn*I
Enterococcus spp.	*Sma*I, *Apa*I
Streptococcus pneumoniae	*Sma*I, *Apa*I
Clostridium difficile	*Sma*I, *Sst*II, *Nru*I
Escherichia coli	*Xba*I, *Not*I, *Sfi*I
Klebsiella pneumoniae	*Xba*I, *Asn*I
Enterobacter spp.	*Xba*I, *Spe*I
Serratia marcescens	*Xba*I
Citrobacter spp.	*Xba*I
Haemophilus influenzae	*Sma*I, *Rsr*II
Bordetella pertussis	*Xba*I, *Asn*I, *Dra*I
Neisseria meningitidis	*Not*I, *Bgl*II
Pseudomonas aeruginosa	*Spe*I, *Dra*I, *Ssp*I, *Xba*I
Stenotrophomonas maltophilia	*Xba*I
Acinetobacter baumannii	*Sma*I, *Apa*I
Burkholderia spp.	*Spe*I
Campylobacter spp.	*Sma*I
Bacteroides spp.	*Not*I
Mycobacterium spp.	*Ase*I, *Xba*I, *Dra*I, *Asn*I
Legionella pneumophila	*Sfi*I, *Not*I

[a] These enzymes generate between five and 50 chromosomal fragments.

Examples of restriction endonucleases commonly used for macrorestriction and PFGE typing of important human pathogens are shown in Table 7.1.

7.4 METHODOLOGY

A. DNA preparation

Conventional DNA extraction methods in solution are unsuitable for macrorestriction analysis because mechanical sharing will degrade genomic DNA into fragments of several hundred kb that are inadequate for megabase-size restriction digestions. To obtain intact chromosomal DNA, it is necessary to protect the DNA by incorporating bacteria into ultra-pure, nuclease-free, low-melting temperature agarose blocks, plugs or beads. All DNA preparative steps are then conducted by exposing agarose-embedded cells to lytic solutions and enzymes.

It is very important to standardise the DNA concentration in the agarose plugs used for restriction and PFGE analysis to obtain comparable and informative fragment patterns. Indeed, insufficient DNA will produce faint or incomplete patterns, whereas excess DNA will generate smears and result in retarded mobility of fragments during PFGE. The DNA concentration can be estimated from the bacterial cell density, which should be measured with a densitometer or nephelometer device and adjusted to a standard value for a given species, usually to between 1–5 × 10^9 cfu/ml. The cell preparation will depend on the particular organism. Non-fastidious organisms are normally grown overnight in appropriate broth medium with orbital shaking, whereas fastidious organisms are preferably harvested from colonies grown on an appropriate solid medium. Bacteria are suspended in EET buffer (100 mM EDTA, 10 mM EGTA, 10 mM Tris-HCl, pH7.8), recovered at 4°C by centrifugation, and then resuspended in SE buffer (75 mM NaCl, 25 mM EDTA, pH7.8) to a density of 1–5 × 10^9 cfu/ml, depending on the organism. A simpler method, recommended by the manufacturer of a ready-to-use PFGE typing kit (GenePath; BioRad), is to prepare the bacterial suspension, centrifuge it, and then visually compare the size of the pellet obtained in a micro-tube with a sketch of an optimal pellet.

The bacterial suspension is briefly heated to 50°C, mixed with an equal volume of 2% low-melting temperature agarose maintained at 56°C, and then dispensed into plug molds. Molds are provided with PFGE equipment or can be custom-made. As the quality of the macrorestriction PFGE patterns depends on plug thickness and regularity of shape, thin plugs are preferable. A mold chamber of 6 × 10 × 0.5 mm produces a block which can be cut with a scalpel blade after DNA preparation and restriction into smaller 4 × 2 × 0.5 mm blocks that can be inserted into the slots in a PFGE gel. When set, the agarose plugs are removed, placed into tubes with 0.5 ml of lysis solution, incubated at 56°C for a minimum period of 2 h (preferably overnight) with gentle shaking. The lysis solution used for Gram-negative bacteria contains 1% sodium lauroyl sarcosine (or sodium docecyl sulphate), 1 mg/ml proteinase K (or pronase), and 0.5 M EDTA, pH9.5. The detergent and proteolytic enzyme act together to remove cellular constituents, while the high EDTA concentration inhibits nuclease activity. Other microorganisms require additional cell-wall digesting enzyme treatments, including lysostaphin for staphylococci, lysozyme and mutanolysin for enterococci and streptococci, and zymolase for yeasts. To reduce the incubation time, these cell-wall lytic enzymes can be mixed with the cell suspension immediately prior to incorporation in the agarose (Goering & Winters, 1992)

After cellular and protein lysis, the agarose blocks containing high molecular size DNA are extensively washed (at least four times for 30 min each) in TE buffer (10 mM Tris, 1 mM EDTA, pH7.5) to remove proteolytic activity, detergent and excess EDTA. Inactivation of proteinase K by the addition of phenylmethyl sulfonyl fluoride (which is very toxic) is sometimes recommended in the literature, but is not normally required if the washing procedure outlined above is followed. Agarose blocks containing intact DNA can be stored for several months in TE buffer at 4°C, preferably with monthly changes of buffer.

B. DNA macrorestriction digestion

Prior to digestion, plugs with intact DNA are cut into smaller portions (e.g., 2 × 4 × 0.5 mm) by placing them in a plastic sterile Petri dish placed over millimeter grid paper. Plugs must be equilibrated into a small volume (120 μl) of the appropriate restriction enzyme reaction buffer for 30 min at the optimal temperature for enzyme activity. A sufficient amount (20 U) of enzyme in buffer is added, mixed and the plug is incubated for 4 h at the appropriate reaction temperature recommended by the manufacturer. Prior to loading in the gel slots, the agarose blocks containing digested DNA are washed in TE buffer and pre-cooled to 4°C together with the PFGE gel. Plugs are carefully inserted in the gel slots and sealed with 1% low-melting temperature agarose.

C. PFGE

The optimal running parameters used for separation of macrorestriction fragments will depend on the organism, enzyme and PFGE apparatus used. To determine these conditions, it is possible to either adapt previously published protocols or select parameters based empirically on the predicted size of the restriction fragments. Some systems (e.g., the CHEF-Mapper) provide algorithms that define optimal separation conditions based on the molecular size range of interest.

Using CHEF electrophoresis, the following conditions are a good starting point for the analysis of bacterial macrorestriction fragments from 50–800 kb in size: i.e., pulse time ramp from 5 s to 50 s, at 6 V/cm, in a 1% gel, run at 13°C for 24 h. Based on the results obtained, modifications in pulse time and/or run time can be made to adjust the resolution for smaller or larger DNA fragments.

DNA fragments are normally visualised after staining the gel for 30 min in ethidium bromide solution (0.5 μg/ml) and destaining in distilled water, followed by photography or digital image capture by a CCD camera under UV light. Alternatively, Southern blotting of DNA fragments on a nylon membrane and hybridisation of membranes with labelled DNA probes can be performed with PFGE gels (Römling *et al.*, 1994). Because DNA transfer is less efficient than with conventional agarose gels, it is best to depurinate the DNA in 0.25 M HCl or to expose the DNA to UV-nicking before denaturation. Transfer is preferably performed with vacuum-blotting equipment.

D. Molecular size standards

The major limitation of PFGE, like any electrophoretic typing system, is the level of reproducibility achievable between patterns obtained within and between gels. To ensure adequate normalisation of patterns and accurate fragment size estimates, suitable molecular size markers should be included in at least every fifth lane. Suitable size markers should span the entire size range of chromosomal fragments at regular intervals to enable accurate standardisation of migration distance and correction of intra-gel distortions or "smiling". The usual size range required for macrorestriction analysis of bacterial genomes is from 20 kb to 1 Mb.

Commercially available ladders available for PFGE include polymers of λ phage DNA (monomer size of 48.5 kb) or plasmid pBR328, which can be used to cover the scale up to 200–300 kb, and *Saccharomyces cerevisiae* chromosomes which extend from 200 kb to >1.6 Mb. However, these markers tend to produce broader bands than bacterial chromosomal fragments, and are not easily standardised to contain equivalent DNA concentrations.

A more advantageous molecular size standard can be prepared in-house by performing a chromosomal digest of reference bacterial strains that have been accurately mapped. As an example, a *Sma*I digest of *Staphylococcus aureus* strain NCTC 8325/3 DNA, which yields 16 evenly distributed fragments of 10–673 kb, can be used. Alternatively, *Escherichia coli* strain MG1655 DNA digested with *Not*I, or *Pseudomonas aeruginosa* DNA digested with *Spe*I (Römling *et al.*, 1994) can be used as size standards.

E. Troubleshooting

Faint bands or a broad smear may indicate DNA degradation, which occurs during bacterial cell harvesting by the activity of endogenous nucleases. This occurs frequently with strains of *Clostridium* spp., and occasionally with strains of *Campylobacter jejuni* or *P. aeruginosa*. Treatment of the bacterial suspension in 10% formalin for 1 h, followed by three washes in saline, inhibits DNAse activity (Gibson *et al.*, 1994). Another recently observed cause of this DNA degradation with *P. aeruginosa* and *Streptomyces* spp. (Römling & Tümmler, 2000) is cleavage by reactive Tris radicals in the electrophoresis buffer, which can be neutralised by adding 50 μM thiourea to the buffer.

Broad bands may indicate incomplete digestion of DNA (if one heavy broad band is apparent on top of the lane) or DNA overload in the plugs. Bands can become distorted if the plugs are unevenly cut, only partially inserted into the gel, or damaged during insertion, or if the bacterial suspension is not homogeneous prior to agarose embedding. Care should be taken to avoid bacterial clumps being trapped in the agarose.

Lanes can become distorted if the gel is not fastened securely during electrophoresis, if air bubbles are trapped underneath and generate localised heating in the gel, or if one electrode burns out and distorts the geometry of the electric field. To avoid gel displacement, a special adhesive support plate is available for casting gels from some PFGE equipment manufacturers (e.g., BioRad).

7.5 ANALYSIS AND INTERPRETATION OF DATA

A. *Visual analysis*

When PFGE typing is applied to the comparison of a limited number of isolates from a presumptive outbreak, the analysis can be performed on one gel and patterns are then compared directly with each other. Strains of the same species share the same genome size, G+C content and codon usage. Therefore, macrorestriction patterns of unrelated isolates exhibit a similar number of chromosomal macrorestriction fragments in the same size range, but only a small minority of fragments will be of the same size. In contrast, clonally-related isolates that arise from the same contamination source, or from the same infection chain, and which share a recent common ancestor, produce either indistinguishable or closely related patterns.

The question arises as to how minor pattern differences should be interpreted in terms of genetic relatedness, and this is related to the confidence with which an epidemiological association between isolates can be inferred (Tenover *et al.*, 1995; Struelens *et al.*, 1996). Minor differences arise naturally in a clonal lineage over the course of infection in individual patients or during epidemic spread. These differences are due to mutations or larger genetic rearrangements, such as insertion or deletion of mobile genetic elements, including insertion sequences, transposons, plasmids and bacteriophages. These changes occur at frequencies that vary according to the genomic plasticity of each bacterial species (or even lineage), and which also depend on the environment and the number of replication cycles undergone over time (Struelens *et al.*, 1996). For example, the following empirical observations were made in two long-term monitoring studies of *P. aeruginosa* isolates colonising unrelated cystic fibrosis patients (Grothues & Tümmler, 1991; Struelens *et al.*, 1993c). Genomic macrorestriction analysis revealed that clonally related isolates from individual patients or siblings exhibited up to four *Xba*I and six *Dra*I fragment differences, equivalent to 80% fragment matching. In contrast, isolates from unrelated patients differed by more than 10 *Xba*I or 20 *Dra*I fragments, and patterns were less than 70% similar. Likewise, epidemiologically related strains of methicillin-resistant *S. aureus* were found to exhibit one to three fragment differences in their *Sac*II profiles, at >85% similarity, over the course of a hospital outbreak lasting 3 years (Struelens *et al.*, 1992a).

In keeping with these empirical findings and theoretical biological considerations, Tenover and colleagues (Tenover *et al*, 1997) have proposed interpretative criteria for macrorestriction patterns from small sets of isolates related to putative outbreaks of bacterial infection (Table 7.2). These criteria, which are widely used, are also consistent with typing guidelines proposed by the European Study Group of Epidemiological Markers (Struelens *et al* 1996). Fig. 7.2 illustrates the typical differences arising from a single mutation or rearrangement in one fragment of a "prototype" epidemic strain. This *central reference pattern* is defined by its modal frequency among the patterns exhibited by a set of outbreak-related isolates. The

Table 7.2. Proposed criteria for interpreting PFGE patterns of small sets of isolates ($n<30$) related to a putative outbreak (adapted from Tenover *et al.*, 1997)

Number of fragment differences compared with the epidemic strain	Number of genetic differences	Category of genetic relatedness	Epidemiological interpretation
None	0	Indistinguishable	Part of the outbreak
2–3	1	Closely related	Probably part of the outbreak
4–6	2	Possibly related	Possibly part of the outbreak
≥7	≥3	Different	Not part of the outbreak

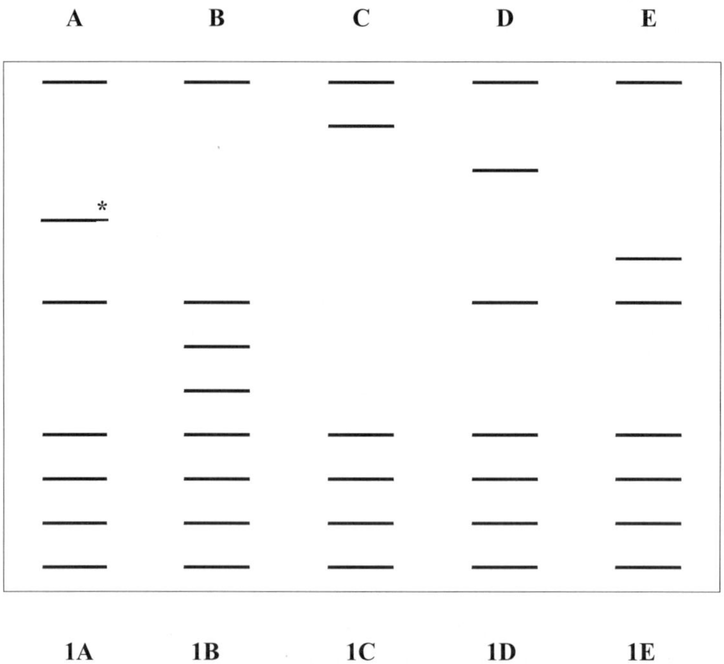

Fig. 7.2. Schematic diagram of typical variations in PFGE patterns of clonally-related isolates following the occurrence of single genetic events in the epidemic strain (at position of fragment labelled with asterisk). Lane **A**, epidemic strain pattern; **B**, mutation with gain of restriction site (three fragment differences); **C**, mutation with loss of restriction site (three fragment differences; **D**, insertion (two fragment differences); **E**, deletion (two fragment differences). All patterns are considered subtypes (A-E) of PFGE epidemic type 1 and would be labelled 1A to 1E (adapted from Tenover *et al.*, 1995).

situation becomes more complicated during a prolonged outbreak, where the "epidemic type" may progressively shift in frequency due to genetic drift and successive waves of spread. These criteria also need to be applied with caution in settings where only a limited number of widespread endemic clones is circulating. In such a setting, even genotypically closely related isolates may not be part of the same outbreak.

The classification of PFGE patterns into types that are closely or possibly related is usually achieved by attributing a letter or numeral to indicate a PFGE type, with a letter or numeral suffix to indicate a subtype, e.g., type A1, A2 etc. (Fig. 7.2; Struelens *et al.*, 1992a; Tenover *et al.*, 1995).

B. Computer-assisted data analysis

Numerisation of PFGE fingerprints, together with computer-assisted normalisation and introduction of data into a relational database, is necessary to analyse patterns from large series of isolates compared on multiple gels. Computer-assisted analysis allows systematic quality control of patterns and rapid identification of identical or closely related patterns in a large database of several hundred patterns. Quantitative analysis of pattern similarity in large databases, assembled following multicentre surveys or long term surveillance programmes, can then be performed.

The first step involves the digital acquisition of the gel image. This can be achieved by direct gel image capture with a CCD (charge coupled device) camera or by scanning a gel photograph with a flatbed document scanner. Although both systems offer similar resolution and dynamic range (100–300 dpi with 256 grey scale levels is sufficient), the CCD camera is more convenient and approximately 10-fold less costly in terms of consumables. For example, our laboratory uses a 30-well PFGE gel with a migration distance of 10 cm for separating *S. aureus* NCTC 8325/3 *Sma*I fragments spanning 10–674 kb. A CCD camera with a 587 × 768 pixel resolution will capture the informative band track with a vertical resolution of 380 data points.

Digitised images are saved in TIFF files (tagged image file format) which can then be imported into fingerprint analysis software for personal computers. Several specialised software packages are commercially available, including GelCompar and BioNumerics (Applied Maths, Kortrijk, Belgium), Taxotron (Taxolab, Institut Pasteur, Paris, France), and the BioImage system (BioImage, Ann Arbor, MI, USA). These programs use different approaches for comparing electrophoretic patterns (Gerner-Smidt *et al.*, 1998). Some programs allow processing of a crude gel image, including removal of artefacts due to non-specific staining by background subtraction algorithms, optimisation of signal contrast over 0–256 grey levels, and rectification of distorted bands/tracks by interpolation within each lane and each gel.

Pattern comparison can proceed in two different ways: either by normalising patterns between gels and performing pattern comparison, or by identifying each band fragment and performing band matching analysis, based either on their normalised migration distance or on their calculated molecular size. In the GelCompar and BioNumerics programs, a standard reference pattern in one gel is used as a reference system to normalise all subsequent gels by interpolation of the migration distances in tracks with the same reference molecular size standards in every gel. In the BioImage and Taxotron programs, the molecular size of each fragment is calculated from the size standards included in each gel. As described in detail elsewhere in this book, the comparison of patterns can be based on two types of similarity coefficient, i.e., densitometric curve comparisons based on the Pearson correlation coefficient, or fragment matching coefficients of similarity, such as the Dice coefficient.

When using the Pearson coefficient of similarity, a bias is introduced in the comparison of PFGE patterns by factors such as the intensity of band staining (giving a higher weight to large DNA fragments), band shape and gel background, all of which are taken into account in the coefficient calculation. This coefficient is, however, the simplest to use. It is also more forgiving than band matching coefficients when analysing PFGE patterns containing a very large number of similar size fragments, or many faintly staining low molecular size fragments.

However, the Dice coefficient, calculated as the number of matching size fragments multiplied by two and divided by the total number of fragments in a pair of patterns, is the preferred coefficient for comparing RFLP or PFGE patterns (Struelens *et al.*, 1996). It is also known as the coefficient of Nei and Li when it is applied to RFLP data. Depending on the software, it will be calculated based on either normalised patterns with band identification, or on a table of fragments of calculated size. Although all software programs provide automatic band identification, the user must always verify that faint bands and double bands are detected and that artefacts are eliminated. Calculation of the molecular size of fragments by interpolation is preferably performed by using the spline regression method because the relationship between fragment size and migration distance is neither linear nor log-linear for PFGE gels. The precision of fragment size determination depends on the quality of the gels and on the molecular size standards used for normalisation. The percent size error can be limited to 5–7% of the fragment size or 0.5–1% of the total track length. These limits can be used for quality control of gels and for setting the tolerance limit in fragment matching comparisons.

It is useful to include an internal control strain in every gel in addition to the molecular size standards that should be loaded on each side and in every fifth well of the gel. This internal control strain allows quality control of the reproducibility of the gel. Its pattern should be ≥96% similar between gels by the Pearson coefficient and 100% similar by the Dice coefficient with a 1% track length tolerance level. Any gel with lower reproducibility values should not be included in the database.

Phenetic clustering of PFGE profiles can be performed by constructing similarity trees from the triangular matrix of pairwise similarity coefficients between isolates of interest in the database (Fig. 7.3). The most neutral and commonly used algorithm is the unweighted pair group method using arithmetic averages (UPGMA; Struelens *et al.*, 1996). Another method, which aggregates similar profiles based on minimising inter-group branch length, is the neighbour-joining method. In these trees, the horizontal branch length measures the distance between isolates or groups of isolates (Fig. 7.4). Although such dendrograms facilitate visualisation of hierarchical groups of isolates based on macrorestriction polymor-

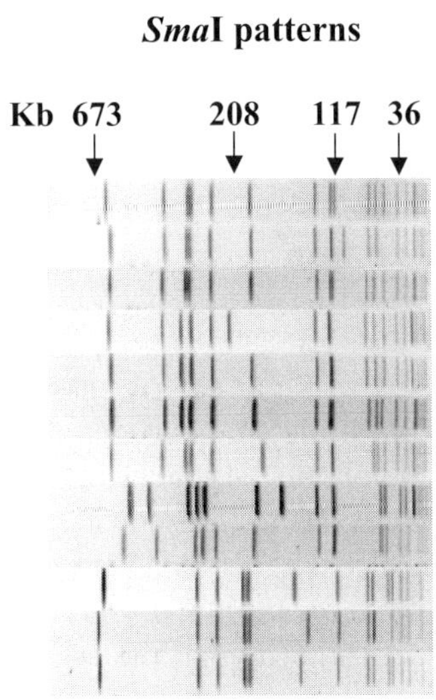

Type	Matrix of similarity											
1e	100											
1d	80.0	100										
1b	92.3	88.0	100									
1c	84.6	72.0	84.6	100								
1a	84.6	80.0	92.3	92.3	100							
1a	84.6	88.0	92.3	92.3	100	100						
1f	69.2	72.0	84.6	76.9	76.9	76.9	100					
3	46.2	24.0	38.5	38.5	38.5	23.1	38.5	100				
4	48.0	33.3	40.0	32.0	40.0	24.0	32.0	80.0	100			
2b	53.8	32.0	46.2	30.8	46.2	15.4	30.8	84.6	88.0	100		
2c	46.2	48.0	46.2	30.8	38.5	53.8	53.8	38.5	24.0	38.5	100	
2a	56.0	41.7	56.0	48.0	64.0	56.0	56.0	48.0	33.3	48.0	64.0	100

Fig. 7.3. CHEF gel and matrix of Dice similarity coefficients (determined with BioNumerics software using a band matching tolerance level set at 0.8% of lane length) for *Sma*I patterns of methicillin-resistant *S. aureus* strains. Groups of closely-related subtypes (1a–f; 2a–c) and unrelated types (3 and 4) are shown.

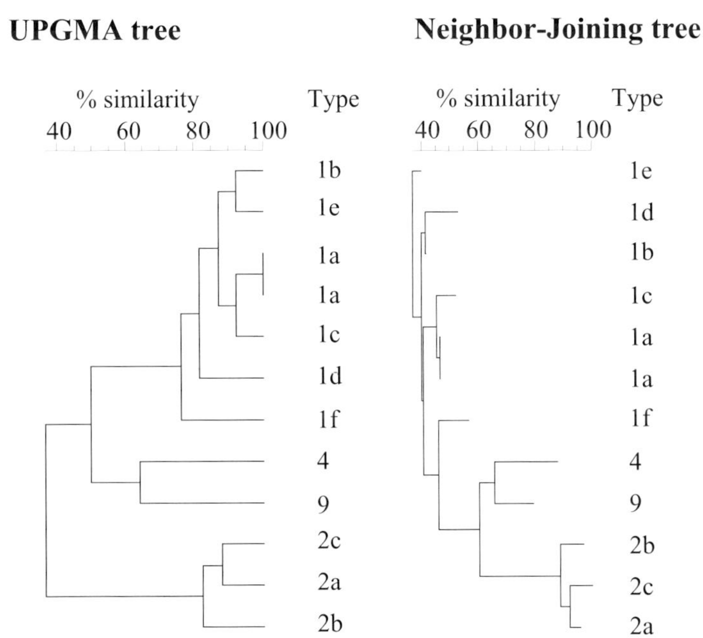

Fig. 7.4. Dendrograms of *Sma*I pattern similarity calculated with the Dice coefficient for the methicillin-resistant *S. aureus* strains typed in Fig. 7.3, constructed with the UPGMA (left) and neighbor-joining (right) clustering methods.

phisms, they should not be inferred to depict phylogenetic structure. In general, the correlation of PFGE pattern similarity with genomic relatedness decreases rapidly below 70% similarity values although, under certain conditions (e.g., use of multiple rare cutters), PFGE can provide valuable taxonomic depth beyond clonal lineages to the genus and species levels (Grothues & Tümmler, 1991).

7.6 APPLICATIONS

PFGE analysis of microbial genomes has been applied in a variety of studies, including the physical mapping of bacterial chromosomes, the study of genomic evolution of bacterial clones in a specific habitat, and taxonomic studies from the genus to pathovar level. In molecular epidemiology, PFGE has been used to compare macrorestriction patterns of >120 bacterial species, belonging to 45 genera, and >10 species of yeasts. Moreover, PFGE has been used for typing yeasts, filamentous fungi and protozoa based on their electrophoretic karyotype (Voss *et al.*, 1995). A Medline search in mid-2000 yielded 1350 publications on "PGFE and bacterial typing", thereby illustrating the popularity of this technique.

In addition to its broad range applicability with minor technical modifications, PFGE analysis of macrorestriction patterns has been found to perform with optimal discriminatory power (95–100%) for most bacterial species so far evaluated with large collections of unrelated isolates (Tenover *et al.*, 1994; Grundmann *et al.*, 1995; Struelens *et al.*, 1996). This high resolving power is related to the fact that macrorestriction patterns scan >90% of the chromosome for large rearrangements (affecting >5% of fragment length) and approximately 0.05% of the genome for point mutations affecting restriction sites. In comparison with other phenotypic and genotypic markers, PFGE typing shows equal or superior epidemiological concordance and discrimination for many pathogens, including *S. aureus* (Struelens *et al.*, 1993a; Kumari *et al.*, 1997), Enterobacteriaceae (Arbeit *et al.*, 1990; Gori *et al.*, 1996), *P. aeruginosa* (Grundmann *et al.*, 1995; Kersulyte *et al.*, 1995), *Acinetobacter baumannii* (Seifert & Gerner-Smidt., 1995) and *Legionella pneumophila* (Struelens *et al.*, 1992b; Fry *et al.*, 1999). These characteristics explain why PFGE is one of the current methods of choice for typing most nosocomial and many community-acquired bacterial pathogens (Tenover *et al.*, 1997; Struelens, 1998).

The commonest application of PFGE typing is its use for comparative typing in local epidemiological investigations of suspect outbreaks of infection (De Gheldre *et al.*, 1997; Tenover *et al.*, 1997; Struelens, 1998; Struelens *et al.*, 1998). This high resolution method can also provide useful insights in the study of clonality of acute and chronic infections (Arbeit *et al.*, 1990; 1993; Struelens *et al.*, 1993b, c; Hakim *et al.,* 1998). It can also assist in differential diagnosis of relapse and re-infection, and in ascertaining the portal of entry of systemic infection by clonal delineation of isolates recovered from multiple samples in individual patients (Arbeit *et al.*, 1993; Wendt *et al.*, 1999).

Macrorestriction analysis performed with standardised protocols and computer-assisted normalised databases has been applied successfully to both local surveillance of nosocomial infection (Harstein *et al.*, 1997; Lemaitre *et al.*, 1998) and national and international surveys of the emergence, molecular evolution and geographical spread of epidemic clones of methicillin-resistant *S. aureus* (Roman *et al.*, 1997; Deplano *et al.*, 2000), *Streptococcus pneumoniae* (Lefevre *et al.*, 1993; Ip *et al.,* 1999), *Salmonella enterica* Typhimurium (Tassios *et al.*, 1999), *Neisseria meningitidis* (Van Looveren *et al.*, 1998) and *P. aeruginosa* (Tassios *et al.*, 1998). However, efforts aimed at providing an international consensus on standard PFGE protocols for library typing of major epidemic pathogens such as MRSA have been hampered by the difficulties that many laboratories have in obtaining reproducible results, even when using ready-made reagents from a single manufacturer (Van Belkum *et al.*, 1998). However, recent progress has been achieved in the USA, where 'PulseNet', coordinated by the Centers for Disease Control (CDC), enables standardised PFGE typing of Shiga-toxin-producing *E.coli* and salmonellae to be used by a network of public health laboratories (*www.cdc.gov/ncidod/dbmd/pulsenet/pulsenet.htm*). Data can be compared directly in the central database after transfer on the Internet. Similar procedures are now being piloted in Europe, e.g., the HARMONY network (*www.phls.co.uk/international/harmony.htm*).

7.7 CONCLUSION AND PERSPECTIVES

Genomic macrorestriction analysis by PFGE (PFGE typing) has emerged over the past decade as the typing method of choice for most bacteria causing nosocomial infections, and many community-acquired pathogens as well. Although technical protocols have been simplified over recent years and the duration of some of the incubation periods has been shortened, the technique remains technically demanding and still requires at least a 72 h turnaround time. Further automation, possibly by means of development of PFCE, would enhance the practicability, throughput and speed of the analysis. Likewise, further international agreement needs to be achieved to develop standard protocols for data generation, quality control procedures, database structures and analytical criteria to enable the full use of PFGE typing for the international surveillance of global pathogens.

REFERENCES

Arbeit, R.D., Arthur, M., Dunn, R., Kim, C., Selander, R.K. & Golstein, R. (1990). Resolution of recent evolutionary divergence among *Escherichia coli* from related lineages: the application of pulsed field electrophoresis to molecular epidemiology. *Journal of Infectious Diseases* **161**, 230–235.

Arbeit, R.D., Slutsky, A., Barber, T.W., Maslow, J.N., Niemczyk, S., Falkinham, J.O., O'Connor, G.T. & von Reyn, C.F. (1993). Genetic diversity among strains of *Mycobacterium avium* causing monoclonal and polyclonal bacteremia in patients with AIDS. *Journal of Infectious Diseases* **167**, 1384–1390.

Birren, B.B., Lai, E., Clark, S.M., Leroy, H. & Simon, M.I. (1988). Optimised conditions for pulsed field gel electrophoretic separations of DNA. *Nucleic Acids Research* **16**, 7563–7583.

De Gheldre, Y., Maes, N., Rost, F., De Ryck, R., Clevenbergh, P., Vincent, J.L. & Struelens, M.J. (1997). Molecular epidemiology of an outbreak of multidrug-resistant *Enterobacter aerogenes* infections and in vivo emergence of imipenem resistance. *Journal of Clinical Microbiology* **35**, 152–160.

Deplano, A., Witte, W., Van Leewen, W.J., Brun, Y. & Struelens, M.J. (2000). Clonal dissemination of epidemic methicillin-resistant *Staphylococcus aureus* in Belgium and neighbouring countries. *Clinical Microbiology and Infection* **6**, 1–7.

Fry, N.K., Alexiou-Daniel, S., Bangsborg, J.M., Sverker, B., Pastoris, M.C., Etienne, J., Forsblom, B., Gaia, V., Helbig, J.H., Lindsay, D, Lück, C., Pelaz, C., Uldum, S.A. & Harrison, T.G. (1999). A multicenter evaluation of genotypic methods for epidemiologic typing of *Legionella pneumophila* serogroup 1: results of a pan-European study. *Clinical Microbiology and Infection* **5**, 462–477.

Gerner-Smidt, P., Graves, L.M., Hunter, S. & Swaminathan, B. (1998). Computerized analysis of restriction fragment length polymorphism patterns: comparative evaluation of two commercial software packages. *Journal of Clinical Microbiology* **36**, 1318–1323.

Gibson, J.R., Sutherland, K. & Owen, R.J. (1994). Inhibition of DNAse activity in PFGE analysis of DNA from *Campylobacter jejuni. Letters in Applied Microbiology* **19**, 357–358.

Goering, R.V. & Winters, M.A. (1992). Rapid method for epidemiological evaluation of Gram-positive cocci by field inversion gel electrophoresis. *Journal of Clinical Microbiology* **30**, 577–580.

Gori, A., Espinasse, F., Deplano, A., Nonhoff, C., Nicolas, M.H. & Struelens, M.J. (1996) Comparison of pulsed-field gel eletrophoresis and randomly amplified DNA polymorphism analysis for typing extended spectrum-β-lactamase-producing *Klebsiella pneumoniae. Journal of Clinical Microbiology* **34**, 2448–2453.

Grothues, D. & Tümmler, B. (1991). New approaches in genome analysis by pulsed-field gel electrophoresis: application to the analysis of *Pseudomonas* species. *Molecular Microbiology* **5**, 2763–2776.

Grothues, D., Koopmann, U., Von der Hardt, H. & Tümmler, B. (1998) Genome fingerprinting of *Pseudomonas aeruginosa* indicates colonization of cystic fibrosis siblings with closely related strains. *Journal of Clinical Microbiology* **26**, 1973–1977.

Grundmann, H., Schneider, C., Hartung, D., Daschner, F.D. & Pitt, T.L. (1995). Discriminatory power of three DNA-based typing techniques for *Pseudomonas aeruginosa. Journal of Clinical Microbiology* **33**, 528–534.

Gurrieri, S., Rizzarelli, E., Beach, D. & Bustamante, C. (1990) Imaging of kinked configurations of DNA molecules undergoing orthogonal field alternating gel electrophoresis by fluorescence microscopy. *Biochemistry* **29**, 3396–3401.

Hakim, A., Deplano, A., Maes, N., Kentos, A., Rossi, C. & Struelens, M.J. (1998). Polyclonal coagulase-negative staphylococcal catheter-related bacteremia documented by molecular identification and typing. *Clinical Microbiology and Infection* **5**, 224–227.

Hartstein, A.I., LeMonte, A.M. & Iwamoto, P.K.L. (1997). DNA typing and control of methicillin-resistant *Staphylococcus aureus* at two affiliated hospitals. *Infection Control and Hospital Epidemiology* **18**, 42–48.

Ip, M., Lyon, D.J., Yung, R.W.H., Chan, C. & Cheng, F.B. (1999). Evidence of clonal dissemination of multidrug-resistant *Streptococcus pneumoniae* in Hong Kong. *Journal of Clinical Microbiology* **37**, 2834–2839.

Kersulyte, D., Struelens, M., Deplano, A. & Berg, D.E. (1995). Comparison of arbitrarily primed PCR macrorestriction (pulsed-field gel electrophoresis) typing *of Pseudomonas aeruginosa* strains from cystic fibrosis patients. *Journal of Clinical Microbiology* **33**, 2216–2219.

Kim, Y. & Morris, M.D. (1995). Rapid pulsed field capillary electrophoretic separation of megabase nucleic acids. *Analytical Chemistry* **67**, 784–786.

Kim, Y., Jett, J.H., Larson, E.J., Penttila, J.R., Marrone, B.L. & Keller, R.A. (1999). Bacterial fingerprinting by flow cytometry: bacterial species discrimination. *Cytometry* **36**, 324–332.

Kozulik, B. (1995). Models of gel electrophoresis. *Analytical Biochemistry* **231**, 1–12.

Kumari, D.N.P., Keer, V., Hawkey, P.M., Parnell, P., Joseph, N., Richardson, J.F. & Cookson, B. (1997). Comparison and application of ribosome spacer DNA amplicon polymorphisms and pulsed-field gel eletrophoresis for differentiation of methicillin-resistant *Staphylococcus aureus* strains. *Journal of Clinical Microbiology* **35**, 881–885.

Lefevre, J.C., Faucon, G., Sicard, A.M. & Gasc, M. (1993). DNA fingerprinting of *Streptococcus pneumoniae* strains by pulsed field gel electrophoresis. *Journal of Clinical Microbiology* **31**, 2724–2728.

Lemaître, N., Sougakoff, W., Masmoudi, A., Fievet, M.H., Bismuth, R. & Jarlier, V. (1998). Characterization of gentamicin-susceptible strains of methicillin-resistant *Staphylococcus aureus* involved in nosocomial spread. *Journal of Clinical Microbiology* **36**, 81–85.

McClelland, M., Jones, R, Patel, Y. & Nelson, M. (1987). Restriction endonucleases for pulsed field mapping of bacterial genomes. *Nucleic Acids Research* **15**, 5985–6005.

Moissac, Y.R., Sheryl, L.R. & Peppler, M.S. (1994). Use of pulsed-field gel electrophoresis for epidemiological study of *Bordetella pertussis* in a whooping cough outbreak. *Journal of Clinical Microbiology* **32**, 398–402.

Roman, R.S., Smith, J., Walker, M., Byrne, S., Ramotar, K., Dyck, B., Kabani, A. & Nicolle, L.E. (1997). Rapid geographic spread of a methicillin-resistant *Staphylococcus aureus* strain. *Clinical Infectious Diseases* **25**, 698–705.

Römling, U. & Tümmler, B. (2000). Achieving 100% typeability of *Pseudomonas aeruginosa* by pulsed-field gel electrophoresis. *Journal of Clinical Microbiology* **38**, 464–465.

Römling, U., Wingender, J., Müller, H. & Tümmler, B. (1994). A major *Pseudomonas aeruginosa* clone common to patients and aquatic habitats. *Applied and Environmental Microbiology* **60**, 1734–1738.

Schwartz, D.C. & Cantor, C.R. (1984). Separation of yeast chromosome-sized DNA by pulsed field gradient gel electrophoresis. *Cell* **37**, 67–75.

Seifert, H. & Gerner-Smidt, P. (1995). Comparison of ribotyping and pulsed-field gel electrophoresis for molecular typing of *Acinetobacter* isolates. *Journal of Clinical Microbiology* **33**, 1402–1407.

Struelens, M.J. (1998). Molecular epidemiologic typing systems of bacterial pathogens: current issues and perspectives. *Mémorias do Instituto Oswaldo Cruz* **93**, 581–585.

Struelens, M.J., Deplano, A., Godard, C., Maes, N. & Serruys, E. (1992a). Epidemiologic typing and delineation of genetic relatedness of methicillin-resistant *Staphylococcus aureus* by macrorestriction analysis of genomic DNA by using pulsed-field gel electrophoresis. *Journal of Clinical Microbiology* **30**, 2599–2605.

Struelens, M.J., Maes, N., Rost, F., Deplano, A., Jacobs, F., Liesnard, C., Bornstein, N., Grimont, F., Lauwers, S., McIntyre, M.P. & Serruys, E. (1992b). Genotypic and phenotypic methods for the investigation of a nosocomial *Legionella pneumophila* outbreak and efficacy of control measures. *Journal of Infectious Diseases* **166**, 22–30.

Struelens, M.J., Bax, R., Deplano, A, Quint, W.G.V. & Van Belkum, A. (1993a). Concordant clonal delineation of methicillin-resistant *Staphylococcus aureus* by macrorestriction analysis and polymerase chain reaction genome fingerprinting. *Journal of Clinical Microbiology* **31**, 1964–1970.

Struelens, M.J., Rost, F., Deplano, A., Maas, A., Schwam, V., Serruys, E. & Cremer, M. (1993b). *Pseudomonas aeruginosa* and *Enterobacteriaceae* bacteremia after biliary endoscopy: an outbreak investigation using DNA macrorestriction analysis. *American Journal of Medicine* **95**, 489–498.

Struelens, M.J., Schwam, V., Deplano, A. & Baran, D. (1993c). Genome macrorestriction analysis of diversity and variability of *Pseudomonas aeruginosa* strains infecting cystic fibrosis patients. *Journal of Clinical Microbiology* **31**, 2320–2326.

Struelens, M.J. & Members of the European Study Group on Epidemiological Markers (ESGEM) of the European Society for Clinical Microbiology and Infectious Diseases (ESCMID) (1996). Consensus guidelines for appropriate use and evaluation of microbial epidemiologic typing sys-

8.1 INTRODUCTION

AFLP was originally developed at KeyGene International (Wageningen, The Netherlands) as a universal DNA fingerprinting method with applications in a large variety of fields, including the monitoring of traits in plant and animal breeding, the diagnosis of disease in humans, animals and plants, parentage analysis, forensic analytical studies, and the genotypical characterisation of fungi, yeast or bacteria (Zabeau & Vos, 1993). Looking back over the past 7 years during which the technique has been put to use, it is clear that any kind of DNA, regardless of its source, composition or complexity, can be analysed by this method. During this period, AFLP has proven to be an important alternative in genotypic analysis and its concept has formed the basis for new applications, including genetic mapping and expression profiling. The aim of this chapter is to give the reader a thorough understanding of the method's principles, with attention to possible technical pitfalls and advances, and to provide on overview of what has been accomplished so far with AFLP in regard to the typing and characterisation of bacteria.

8.2 METHODOLOGY

The name AFLP chosen by its inventors reflects the resemblance of the method to Restriction Fragment Length Polymorphism (RFLP) analysis and has been filed as a trademark by KeyGene International, Wageningen. Unfortunately, in the literature there seems to be some confusion on this point, and often AFLP is used as an acronym for 'Amplified Fragment Length Polymorphism'. It is important to make this distinction as the latter definition has been used on several occasions in reference to genotypic methods that produce banding patterns comprising amplicons of variable length, including randomly amplified polymorphic DNA (RAPD) analysis (Williams *et al.*, 1990), arbitrarily primed PCR (AP-PCR) analysis (Welsh & McClelland, 1990), and DNA amplification fingerprinting (DAF; Caetano-Annolés *et al.*, 1991). Strictly speaking, and based on the technical layout, AFLP should be classified under the category of Selective Restriction Fragment Amplification (SRFA) techniques employing short double stranded 'adaptors' or 'indexers' that are ligated to genomic restriction fragments, and which serve as primer binding sites for amplification (Vaneechoutte, 1996).

The AFLP concept can be largely divided into three steps (Fig. 8.1): (i) digestion of total cellular DNA with one or more restriction enzymes and ligation of restriction half-site-specific adaptors to all restriction fragments; (ii) selective amplification of some of these fragments with two PCR primers that have corresponding adaptor and restriction site-specific sequences; and (iii) electrophoretic separation of amplicons on a gel matrix, followed by visualisation of the banding pattern.

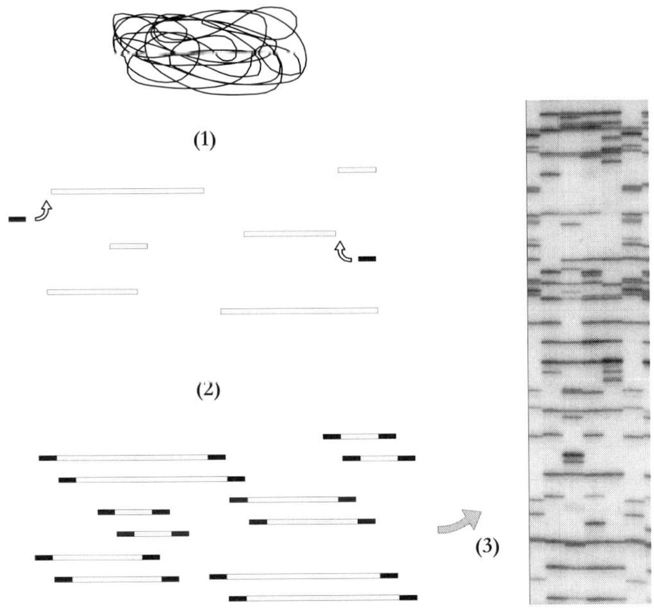

Fig. 8.1. Main principles of the AFLP method: (**1**) digestion of total genomic DNA; (**2**) ligation of adaptors; (**3**) amplification with adaptor-specific primers and electrophoretic separation of amplicons (see text for details).

A. *Digestion of genomic DNA*

(i) Choice of restriction enzymes

It is well known that microorganisms display a wide range in G+C content of 26 to 72 mol%. Because an adequate high number of restriction fragments must be obtained to examine the entire genome, a good choice of restriction enzymes is important. For instance, when choosing *Eco*RI (G/AATTC) for a high GC genome (>60 mol%), too few representative fragments will be obtained and large segments of the genome will remain unexplored. In the original AFLP procedures (Vos *et al.*, 1995), two restriction enzymes were used: one digesting the DNA infrequently (called the '6-base cutter') and one that digests the DNA frequently (known as the '4-base cutter').

The simultaneous use of two restriction enzymes is beneficial. First, heterologous ends are produced, allowing ligation of adaptors with different sequences so that many primer pairs can be used on the same template. This facilitates the analysis of highly monomorphic genomes to a great extent (Keim *et al.*, 1997; see later). Second, by using a mix of enzymes, only a limited number of fragments (due to the 6-base cutter) are obtained that are relatively small (due to the 4-base cutter) and amplification is thus straightforward. In addition, the resulting amplicons are in an optimal size range (50–1000 bp) for high resolution separation on denaturing 5–6% (w/v) polyacrylamide gels (i.e., sequencing gels).

In spite of these advantages, several papers describe the use of a single restriction enzyme for AFLP in which amplicons are separated on conventional 1.5–2% (w/v) agarose gels stained with ethidium bromide (Valsangiacomo *et al.*, 1995; Picardeau *et al.*, 1997; Clerc *et al.*, 1998; Gibson *et al.*, 1998). This obviously saves time because the DNA does not require labelling and there is no need for pouring ultra-thin slab gels involving meticulous plate cleaning. It also saves costs as no special equipment is required. However, it is unclear whether these technical simplifications have an effect on the usefulness of the method with respect to bacterial typing and characterisation.

Clerc *et al.* (1998) have described the use of a single 4-base cutter *Msp*I (C/CGG) and AFLP primers that contain three selective nucleotides. In this case, the rationale of using +3 primers is that a 4-base cutter will produce many more fragments. Assuming random base distribution, the cutting frequency F of a restriction enzyme can be estimated from the probability p that a defined string of n bases will occur in any given sequence of DNA as follows:

$$F = p(N_1N_2N_3 \ldots N_n) = \frac{1}{4} \times \frac{1}{4} \times \frac{1}{4} \times \ldots = \frac{1}{4^n}. \tag{1}$$

The number of restriction sites in a genome with size S is then given by S × F. Hence, for a genome size of 5 million bases, roughly 20,000 fragments are expected when using a 4-base cutter. Because simplified banding patterns are easier to interpret and analyse, this large number of fragments must be reduced substantially during the selective PCR. This can be cumbersome because the use of +3 primers may affect reproducibility due to mismatches at the first or second position of the selective extension (see later).

In the study of Clerc *et al.* (1998), *Msp*I-generated banding patterns contained up to 37 amplicons. This is about seven times more than would be expected theoretically (e.g., 20,000 divided by 64 × 64 equals 4.77), a good indication that mismatching does indeed occur. Of course, the exact size and base composition of the genomes under study and the composition of the 3'-extensions of the AFLP primers should be taken into consideration. Consequently, Clerc *et al.* (1998) tested more than 40 +3 primers, 12 of which gave suitable patterns. Unfortunately, data on reproducibility experiments were not included, and no information was given on the average size of the amplicons, so that the quality of the banding patterns obtained remains uncertain.

Valsangiacomo *et al.* (1995) and Picardeau *et al.* (1997) used *Pst*I (CAGCT/G) as the sole restriction enzyme for AFLP. After selective amplification, banding patterns contained 5–10 and 3–8 amplicons, respectively, in the size range of 0.5–2.5 kb. Whereas Valsangiacomo *et al.* (1995) used AFLP primers with one or two selective bases, Picardeau *et al.* (1997) resorted to the combined use of +2 and +5 primers to limit the number of amplicons. In the latter study, deploying a total of 4–10 selective bases casts serious doubts on the specificity of the PCR since, theoretically (e.g., for a *Pst*I digest on bacterial DNA), only a few thousand restriction fragments are expected as starting material. Mismatching in this case is thus very likely. In fact, the banding patterns generated by Picardeau *et al.* (1997) had a clearly visible background of what appears to be aberrantly amplified fragments. It would be interesting to know which PCR profile for AFLP was used, but this information was not included.

Table 8.1 lists some of the restriction enzymes and selective primers that have been used so far for AFLP with microbial DNA. As a thumb of rule, using the original AFLP procedures (Vos *et al.*, 1995), the combined digests of *Eco*RI / *Mse*I or *Apa*I / *Taq*I seem to be most suited for microbial DNAs with low (<40 mol%) and high (>55 mol%) G+C content, respectively, and the combination *Hin*dIII / *Taq*I appears to be useful for genomes with intermediate G+C content (40–55 mol%) (Janssen *et al.*, 1996). The enzymes *Apa*I (GGGCC/C), *Hin*dIII (A/AGCTT), *Eco*RI (G/AATTC), *Taq*I (T/CGA), and *Mse*I (/TTAA) are reliable and, with the exception of *Mse*I, inexpensive.

(ii) DNA quality

There are numerous methods available for the preparation of microbial genomic DNA. The most commonly used methods are based on cell lysis through addition of cell wall degrading enzymes (e.g., lysozyme, lysostaphin, proteinase K, etc.) and ionic agents such as SDS or sarcosyl, followed by phenol and chloroform extractions. Methods that aim to obtain high molecular size DNA from organisms known to release high levels of DNAses often make use of nucleic acid stabilising agents, such as cetyltrimethylammonium bromide (CTAB) or guanidinium thiocyanate (GTC). For AFLP it is strongly recommended to use DNA of adequate purity and high molecular size. The simple reason lies in the fact that the activity of restriction enzymes may be sensitive to co-purified impurities, leading to inspecific cleavage ('star-activity') or incomplete digestion. If this occurs, partial fragments of aberrant size will be amplified with serious consequences for data interpretation and reproducibility. Nonetheless, Clerc *et al.* (1998) reported the use of a quick boiling method to produce DNA from *Pseudomonas syringae* grown on a specific medium. Cell suspensions were boiled for 10 min and the cleared lysate was used directly for combined restriction-ligation reactions and subsequent selective PCR. As no reproducibility experiments were conducted, it remains unclear if there were any significant differences between AFLP patterns obtained with purified DNA compared to those obtained with crude lysate DNA. Presumably, because Clerc *et al.* (1998) used a frequent cutter (*Msp*I) as the sole restriction enzyme, low level degradation was not detected.

Recently, a quick DNA method for AFLP analysis of pneumococci and staphylococci was tested (P. Janssen & J. Van Eldere, unpublished results). Cells were lysed and genomic DNA was prepared with the InstaGene kit (Bio-Rad, Richmond, VA), involving a boiling step and intensive vortexing. The DNA was digested with *Hin*dIII and *Taq*I without further purification. Fragments were tagged with corresponding adaptors, selectively amplified, and separated on an automated sequencing apparatus. In the range of 0–500 bp, banding patterns obtained with crude lysate preparations were very similar to AFLP patterns obtained with purified

Table 8.1. Restriction enzymes and selective primers used for AFLP analysis of bacterial genomes

Organism	G + C[a]	Size[a]	Enzymes[b]	Primers[c]	Visualization[d]	No. bands[e]	Reference
Clostridium beijerinckii	28	5.8	*Eco*RI / *Mse*I	C / C (C / A)[f]	DP, [γ-^{32}P]	26 (50–60)[f]	Janssen *et al.* (1996)
Bacillus anthracis	33	4.5	*Eco*RI / *Mse*I	N / N	DP, [γ-^{32}P]	50–90	Keim *et al.* (1997)
Campylobacter spp.	30–38	1.7–2.0	*Hin*dIII / *HhaI*	A / A	DP, fAFLP	40–50	Duim *et al.* (1999)
Campylobacter spp.	30–38	1.7–2.0	*Bgl*II / *Csp*6I	– / A	DP, FAM	50	Kokotovic & On. (1999)
Mycoplasma spp.	30–35	0.6–1.0	*Bgl*II / *MfeI*	– / –	DP, FAM	60–80	Kokotovic *et al.* (1999)
Staphylococcus aureus	34	2.8	E–M and A–T	– / C resp. – / G	DP, Cy5	70–80	Grady *et al.* (1999)
Staphylococcus aureus	34	2.8	*Eco*RI / *Mse*I	– / AT	DP, FAM	40–75	Hookey *et al.* (1999)
Staphylococcus epidermidis	30–37	2.4	*Eco*RI / *Mse*I	– / A	DP, [γ-^{32}P]	40–70	Sloos *et al.* (1998)
Acinetobacter spp.	38–47	±2.6	*Hin*dIII / *Taq*I	A / AA	DP, [γ-^{32}P]	20–50	Janssen *et al.* (1997)
Acinetobacter spp.	38–47	±2.6	*Eco*RI / *Mse*I	A / C	DP, Tex, [γ-^{32}P]	30–55	Koeleman *et al.* (1998)
Vibrio spp.	38–51	2.8–3.5	*Hin*dIII / *Taq*I	A / A	DP, Tex, [γ-^{33}P]	50–60	Pedersen *et al.* (1998)
Helicobacter pylori	39	1.7	*Hin*dIII	A, G, C, T	Aga, EtBr	15–20	Gibson *et al.* (1998)
Legionella pneumophila	39	4.1	*Pst*I	G, GC, A, AT	Aga, EtBr	5–10	Valsangiacomo *et al.* (1995)
Paenibacillus larvae	42	NA	*Eco*RI / *Mse*I	C / A	DP, [γ-^{32}P]	60–70	Heyndrickx *et al.* (1996)
Chlamydia spp.	45	1.0–1.2	*Eco*RI / *Mse*I	– / C	DP, Tex	45–50	Meijer *et al.* (1999)
Escherichia coli	51	4.6	*Eco*RI / *Mse*I	– / AA	DP, FAM	50	Arnold *et al.* (1999b)
Salmonella spp.	50–53	4.4–4.7	*Eco*RI / *Mse*I	AA / –	DP, [γ-^{33}P]	45–55	Aarts *et al.* (1998)
Mycobacterium kansasii	NA	6.2	*Pst*I	GC / ATTAG[g]	Aga, EtBr	3–8	Picardeau *et al.* (1997)
Streptococcus pneumoniae	43	2.2	*Hin*dIII / *Taq*I	A / A	DP, Cy5	60–70	Van Eldere *et al.* (1999)
Streptococcus pyogenes	NA	2.0	*Eco*RI / *Mse*I	– / T	DP, FAM	100–120	Desai *et al.* (1998)
Aeromonas spp.	57–63	4.5	*Apa*I / *Taq*I	A / A	DP, TAMRA	25–35	Huys & Swings (1999)
Aeromonas spp.	57–63	4.5	*Apa*I / *Taq*I	A / A	DP, [γ-^{32}P]	40–50	Huys *et al.* (1996b)
Stenotrophomonas spp.	NA	NA	*Apa*I / *Taq*I	G / G	DP, [γ-^{32}P]	45–60	Hauben *et al.* (1999)
Pseudomonas syringae	60	5.6	*Msp*I	NNN	Aga, EtBr	24–37	Clerc *et al.* (1998)

(Continued.)

Table 8.1. Continued.

Organism	G + C[a]	Size[a]	Enzymes[b]	Primers[c]	Visualization[d]	No. bands[e]	Reference
Pseudomonas aeruginosa	63	5.9	*Eco*RI / *Mse*I	– / C	DP, Cy5	50	Speijer *et al.* (1999)
Xanthomonas axonopodis	64	4.5	E–M–T–P[h]	N / N	DP, [γ-^{32}P]	21–64	Restrepo *et al.* (1999)
Pseudomonas spp.	58–68	±6.0	*Eco*RI / *Mse*I	A / C	DP, $AgNO_3$ staining	15–25	Geornaras *et al.* (1999)
Burkholderia spp.	67–69	4.1–8.1	*Apa*I / *Taq*I	A / A (– / –)	DP, FAM, ([γ-^{32}P])	40 (25)	Coenye *et al.* (1999)

[a] The GC-content is expressed in mol%, approximate genome size is in million bases (Mbp); NA, not available.
[b] Abbreviations: E, *Eco*RI; P, *Pst*I; A, *Apa*I; M, *Mse*I; T, *Taq*I.
[c] Only 3'-selective bases are given. Primers separated by a comma were used separately. N, A or G or C or T.
[d] Abbreviations: DP, denaturing PAGE; Aga, agarose gel electrophoresis; EtBr, ethidium bromide; fAFLP, fluorescent label not specified; Cy5, 5-carboxyfluorescein; TAMRA, carboxytetramethyrhodamine; FAM, 6-carboxyfluorescein; Tex, Texas Red ® (sulphonyl chloride derivative of sulphorhodamine 101).
[e] Only for primer pairs that were used in further analyses.
[f] In total, 17 *C. beijerinckii* strains were investigated (unpublished).
[g] Presumably used as a primer pair.

DNA from the same strains, although one- or two-band differences were occasionally observed. In addition, the level of reproducibility was higher when DNA of high integrity and molecular size was used as a template. Nevertheless, such short protocols may become useful for the rapid automated typing of microbial isolates, particularly in preliminary screening schemes.

(iii) Genome complexity, compositional bias and DNA modification

According to genomic data available on the Web (i.e., The Institute for Genomic Research – TIGR – Microbial Database; *http://www.tigr.org/*), the size of prokaryotic genomes falls in a 20-fold range between 0.5 and 10 million basepairs (Mb), with most genomes measuring between 2 and 5 Mb in length. Although the final number of restriction fragments is determined primarily by the size of the genome, it is also very important to consider the base composition and distribution of the genome, and to bear in mind that a DNA modification system may be present. The formula of Nei & Li (1979) takes into account the variation in mol% G+C content and calculates the restriction frequency F as:

$$F = \left(\frac{g}{2}\right)^{m} + \left(1 - \frac{g}{2}\right)^{n}, \qquad (2)$$

where g is the fractional G+C content of the genome DNA, and m and n are the numbers of G+C base pairs and A+T base-pairs in the recognition site, respectively. However, when calculated values were compared to the outcome of chromosomal digests, it was found that this formula was actually a poor predictor of restriction frequency (Phillips *et al.*, 1987; Owen, 1989) because genomes often exhibit compositional variance in di-and trinucleotide sequences (Karlin *et al.*, 1997). These variations are thought to be introduced by several factors, including mutation pressure, base stacking stability, DNA repair and codon usage.

To account for such variations, Phillips *et al.* (1987) used Markov chain analysis to determine the mono- through hexanucleotide composition of the *Escherichia coli* genome. A few years later, Forbes *et al.* (1991) adopted this idea to calculate the mean fragment length for a given combination of restriction enzyme and genome, and compared the computed values with electrophoretic results for a range of restriction enzymes digesting a number of bacterial genomes. Given that restriction frequency and mean fragment length are reciprocal, the restriction frequency (F) for a 4-base cutter with recognition sequence ($N_1N_2N_3N_4$) can be calculated from:

$$F = \frac{f(N_1N_2N_3).f(N_2N_3N_4)}{f(N_2N_3)}, \qquad (3a)$$

with f being the frequency of di- and trinucleotides for a given genome in percentages. For a 6-base cutter with restriction site ($N_1N_2N_3N_4N_5N_6$), this would translate into:

$$F = \frac{f(N_1N_2N_3).f(N_2N_3N_4).f(N_3N_4N_5).f(N_4N_5N_6)}{f(N_2N_3).f(N_3N_4).f(N_4N_5)}. \qquad (3b)$$

Nowadays, entire microbial genomes are being sequenced at an astonishingly high rate, and it has become a rather simple matter to obtain reliable estimates of di- and trinucleotide frequencies for a genome of interest. In fact, some Web-based applications provide this kind of information for any set of genes, for a given number of input sequences, or even for entire genomes, bringing precise calculation of expected restriction frequencies within easy reach. Still, computations such as those outlined above do not necessarily agree with the real situation. Detailed comparison of the expected and observed oligonucleotide frequencies in various bacterial species actually suggest that palindromic restriction sites (e.g., recognition sites for class II restriction enzymes), are under-represented in bacterial genomes (Rocha *et al.*, 1998). Nonetheless, for a rough estimate of the expected number of fragments, formula (1) is quite useful to remember, while Markov chain analysis, as illustrated by formula (3), appears to be better suited for accurate estimates.

An additional factor that may influence the cutting frequency of restriction enzymes is DNA modification, particularly in the form of methyltransferases that 'protect' recognition sites from restriction by base methylation. Methyltransferases have been observed in many bacteria, although detailed information on their occurrence and functional diversity is scarce.

As a final remark, it is generally a good idea to gain further knowledge of cleavage frequencies by empirical determination, and to test out a whole range of restriction enzymes on a number of strains in order to find out which enzyme combinations (and, therefore, which primer combinations) will ultimately produce the most suitable banding patterns.

B. Template preparation

(i) Ligation of adaptors

After digestion of the genomic DNA, fragments are tagged with short double stranded DNA molecules that bind in a complementary fashion to the restriction half-sites. In the presence of T4 ligase and ATP, a phosphodiester link is created between the 3'-OH and 5'-PO_4 groups at the end of each adaptor and fragment (Fig. 8.2, step 1). Note that, because unphosphorylated oligonucleotides are used for the adaptors, there is only one phosphodiester link for each adaptor-to-fragment ligation (Fig. 8.2, step 2), meaning that if DNA was denatured prior to the addition of *Taq* DNA polymerase and dNTPs, no amplification would occur (Vos *et al.*, 1995). The PCR in AFLP works because *Taq* polymerase fills in the 3'-recessed ends or displaces the non-ligated strand at ambient temperatures (Vos *et al.*, 1995), thereby ensuring the presence of the originally intended primer binding site (Fig. 8.2, step 3).

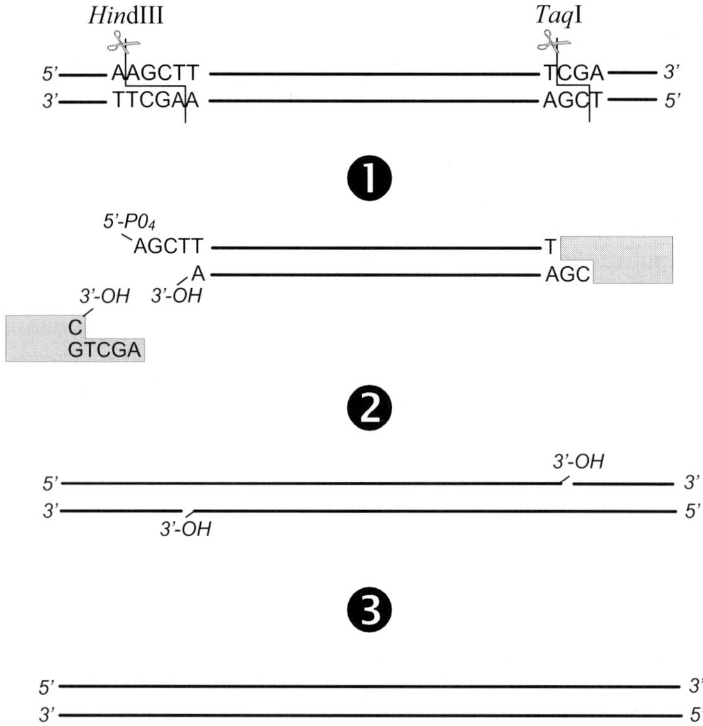

Fig. 8.2. Preparation of template DNA for selective amplification (see text for details).

(ii) Adaptor features

When two restriction enzymes are used, adaptors have different sequences allowing a combination of PCR primers. Obviously, the core sequences of the adaptors are dictated by proper primer design, i.e., inverted repeats and long nucleotide runs should be avoided. Also, although the sequences must be sufficiently different to ensure specificity, the T_m of both primers should be in the same range. Once the primer sequences are chosen, adaptors can be designed. An essential feature of AFLP adaptors is that they contain a base change so that the original restriction site is not reconstructed during ligation. In the example of Fig. 8.2, a *Hin*dIII-adaptor with a 3'-cytosine is used for this purpose. This has the advantage that the ligation can be done in the presence of restriction enzymes, thereby preventing fragment-to-fragment ligation and simultaneously stimulating adaptor-to-fragment ligation. In addition, because adaptors are made of unphosphorylated oligonucleotides, they are unable to ligate to each other. The combination of these features ensures that virtually all restriction fragments are tagged with the appropriate adaptor sequence.

In the original AFLP procedure (Zabeau & Vos, 1993), the adaptor corresponding to the 6-base cutter half-site was labelled with a biotin group. Prior to amplification, fragments created by the rare cutter (i.e., *Hin*dIII-*Taq*I templates) were purified with streptavidine-coated magnetic beads. By so doing, essentially all

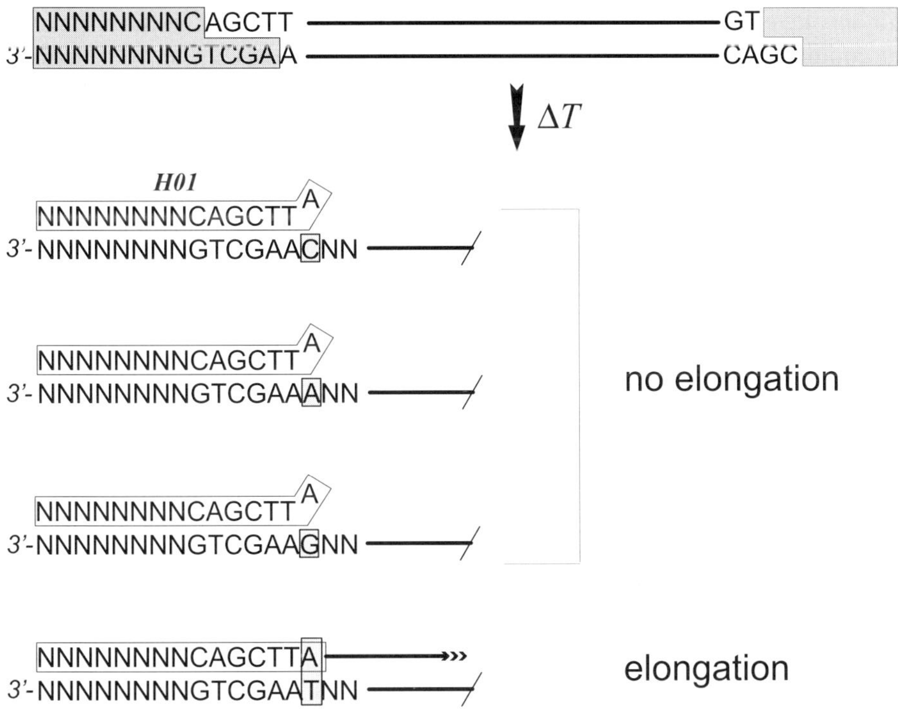

Fig. 8.3. Selective amplification with AFLP primers; only a perfect match at the 3'-end between primer and template results in DNA chain elongation and subsequent PCR amplification.

fragments with 4-base cutter half-sites located at both ends, otherwise accounting for up to 90% of the complete template population, were successfully removed. This renders the reaction mixture less complex, while achieving an enrichment for fragments with heterologous ends so that all reaction components are used optimally. This purification step often improves the quality of the banding pattern, especially when using eukaryotic DNAs (Zabeau & Vos, 1993). However, for AFLP analysis of bacterial genomes, template mixtures contain many fewer fragments, and there is no need to eliminate the templates produced by the frequent cutter (Vos *et al.*, 1995; Janssen *et al.*, 1996). In addition, due to careful primer design and extensive optimisation of PCR conditions, purification with coated beads has now become redundant (see next section).

C. *Selective PCR*

(i) Principles

AFLP primers consist of the adaptor-derived core sequence, including the 3'-part of the restriction half-site, and an extension sequence made up by a number of so-called selective bases. These selective bases are complementary to nucleotides flanking the restriction sites. For a given template, elongation of the DNA chain will only take place if the primer binds with high specificity, e.g., if the corresponding complementary nucleotide is present in the fragment (Fig. 8.3). In addition, exponential amplification is only achieved when DNA synthesis occurs from both ends. The stringency of primer binding is essential for the success of reproducible AFLP. According to Vos *et al.* (1995), primers with mispaired 3'-ends don't participate in the amplification process, especially if only one or two selective bases are used per primer.

In addition, a typical AFLP amplification profile starts with a high annealing temperature (63–65°C), which is gradually lowered (usually by 0.7–1.0°C per cycle) until the optimal annealing temperature is reached. This part of the PCR profile, often referred to as the 'touch down', usually consists of 10–12 cycles and ensures dissociation of mismatched primers, thus improving specificity. The final annealing temperature, usually about 52–56°C, is subsequently maintained for a further 18–24 cycles, depending on the application.

Another characteristic of selective PCR in AFLP with two enzymes is that small fragments with homologous ends (i.e., fragments produced by the 4-base cutter only) appear to be amplified far less efficiently. Possible explanations are a difference in annealing temperatures between the two selective primers and the formation of 'panhandle' stem-loop structures (Vos *et al.*, 1995). The fact that these fragments do not actually participate in the amplification process has made the purification step over magnetic beads (see previous section) superfluous.

(ii) Primer structure

Obviously, proper primer design is very important. Hairpin structures should be avoided and the formation of self-dimers and cross-dimers should be kept to a minimum as much as possible to prevent the loss of reaction components. Also, it is preferred that a primer sequence does not contain long stretches of As or Ts, because this may cause local instabilities in the primer-template hybrid, and the G+C-content should range between 40 and 60%, allowing highly specific annealing between primer and template at 56–60°C. For AFLP primers, another feature is the obligatory presence of a 5'-guanine residue. This prevents the generation of so-called 'double bands' that were observed occasionally in older AFLP gels due to incomplete addition of an extra nucleotide to the synthesised strands. This terminal transferase activity, also known as 'extendase' activity, of the DNA polymerase is 3'-base dependent and has been studied in detail by Hu (1993). Apparently, if the 3'-base is a cytosine, extendase activity is quite strong and mainly adenines are added. Using AFLP primers with a 5'-guanine thus guarantees that most amplicons will be identical and have the same electrophoretic properties, leading to the formation of discrete bands and improving comparative analysis of banding patterns.

(iii) Selection criteria

Assuming random base distribution, one out of four fragments will be amplified for each selective base used. However, in reality, the final complexity of the AFLP pattern is determined not only by the number of selective bases used, but also by the choice of the selective bases and the base composition of the genome. Table 8.1 provides a summary of selectivity according to the primers used and the genome being studied. A tendency towards matched base composition between primer extension and genome can be clearly seen from these data. For instance, a C/C combination of AFLP primers to analyse the *Clostridium beijerinckii* genome (28 mol% G+C) generates an average of 26 amplicons, whereas a C/A combination of primers gives roughly two times that number, simply because more Ts are available for base pairing.

In general, when properly executed, AFLP methodology deploying +1 or +2 primers (i.e., primers with one or two selective bases, respectively) ensures a proportional reduction in the number of amplicons for each selective base used, and mismatching – if it occurs at all – does not reach a detectable level because of the high-stringency conditions of the PCR. However, low-level mismatching does occur for +3 primers, and selectivity is lost when +4 primers are used (Vos *et al.*, 1995).

To circumvent the problem of mismatching, a two-step amplification strategy was developed for the AFLP analysis of complex genomes (Vos *et al.*, 1995). First, fragments are amplified with +1 primers in a pre-amplification reaction. After completion, the reaction mixture is diluted and used for the main amplification reaction with +3 primers. This reduces background 'smears' to a large extent, and has the additional advantage that large quantities of template DNA can be produced. For most bacterial genomes, it will suffice to use one or two selective bases and tune the final number of amplicons by testing out a few primer pairs.

D. Fragment separation and pattern visualisation

(i) Use of denaturing polyacrylamide gels

The original AFLP procedures described by Zabeau & Vos (1993) and Vos *et al.* (1995) made use of conventional sequencing gels containing 7–8 M urea and 4.5–6% cross-linked polyacrylamide. Typical run conditions for electrophoresis were a constant power of 50 W and 40–50 V/cm, a running buffer of 0.6 × TBE (TBE: 100 mM Tris, 100 mM boric acid, 2 mM EDTA, pH 8.3) and a run time of about 2.5 h. With these conditions, the loading front migrated through approximately 90% of the gel length, with an excellent separation of fragments ranging in size between 50 and 600 nucleotides. Typically, detection of AFLP amplicons required the labelling of one of the primers with either P-32 or P-33, followed by visualisation of the banding patterns by autoradiography or by the use of a phosphorimager (Fig. 8.1).

However, many research institutes prefer to limit the use of radioactivity, and methods for non-radioactive detection of nucleic acids have been put forward. Chalhoub *et al.* (1997) evaluated the use of silver-staining for the visualisation of AFLP products, and concluded that autoradiography and silver-staining displayed a similar resolution with equal sensitivities. Unfortunately, silver-staining detects both strands of the AFLP products, thereby giving rise to double-band patterns,

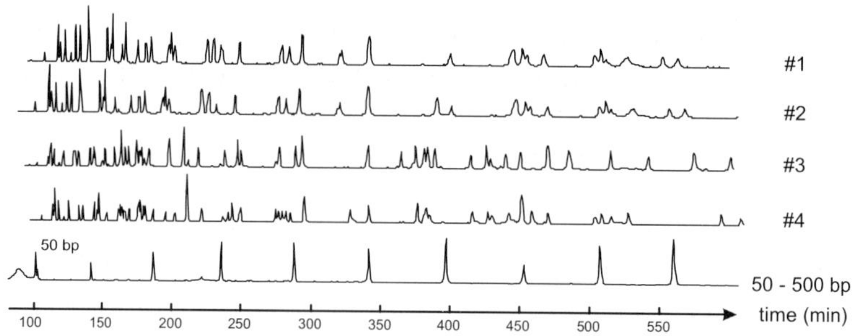

Fig. 8.4. Electropherogram obtained for four strains of *S. pneumoniae* via the use of fluorescently labelled primers and an automated sequencing apparatus (see text).

and complicating the interpretation of the AFLP results. A recent paper by Lin *et al.* (1999) described the chemiluminescent detection of AFLP products. Detection of the amplicons was achieved by blotting the electrophoresed DNA on to a nylon membrane, followed by hybridisation of the transferred DNA with an alkaline phosphotase-labelled probe. Patterns were subsequently visualised by exposure to a light-sensitive film and appeared to be similar to the autoradiographic patterns when primers labelled with P-32 were used. The drawback of this approach is that blotting and hybridisation of large gels is technically very difficult, and Lin *et al.* (1999) resorted to the use of small 15 × 17 cm gels in which only fragments in the size range 50–330 nucleotides could be examined.

A much better, but also much more expensive, approach involves the use of Cy5-labelled primers in combination with on-line laser detection of fluorescent amplicons while they pass through the gel (Fig. 8.4). This has been done for AFLP analysis of a number of bacterial species (Table 8.1). If an automated apparatus is equipped with a multiple fluorescence detection system (e.g., the ABI 377 system; Applied Biosystems), a separate fluorescent label (e.g., 6-carboxy-x-rhodamine; ROX) can be used as an internal standard for addition to each sample, thus greatly improving the normalisation of the gels. Improved resolution is also achieved because the fragments are detected at a fixed distance from the origin, leading to a more uniform spacing between fragments. General advantages of 'automated' fluorescent AFLP (fAFLP) include the large simultaneous throughput of samples, a rapid turnaround, and the direct processing of raw data and subsequent error-free storage of results through linkage with a computer. An interesting alternative to online detection of fluorescently-labeled AFLP products has been reported by Roman *et al.* (1999) who describe the scanning of wet or dry AFLP gels with a fluorescence imager in as little as 11 min.

(ii) Agarose gel electrophoresis

Several groups have reported on a simplified AFLP procedure in which only one restriction enzyme is used (Valsangiacomo *et al.*, 1995; Picardeau *et al.*, 1997; Clerc *et al.*, 1998; Gibson *et al.*, 1998; Table 8.1). Because the average size range of amplicons appeared to be much higher (0.4–2.5 kb) as compared to conventional AFLP using two restriction enzymes (50–550 bp), separation of AFLP products was routinely done on agarose gels. Depending on the application, agarose concentrations varied from 1.5–3% and nucleic acids were stained with ethidium bromide. In spite of the much lower resolution inherent to agarose gel electrophoresis, AFLP results obtained in these studies generally agreed well with other typing data available. Obvious advantages of agarose-based AFLP are the ease of use and the low cost. On the other hand, large templates (> 2 kb) may be amplified with a lower efficiency, so that PCR conditions must be adjusted accordingly, and ethidium bromide staining is notoriously unreliable for very small DNA fragments. Taken together, AFLP analysis on agarose represents a cost-effective means to produce preliminary typing data, but appears to be less suited for detailed taxonomic studies of homogeneous species or for epidemiological investigations of highly related strains (i.e., in outbreak situations).

8.3 DATA ANALYSIS

A. *Digitisation of AFLP data*

A common theme in the non-automated processing of AFLP data is the generation of image files from the electrophoresis patterns (e.g., those obtained by autoradiography, by phosphor imaging, etc.). These image files are usually saved in 'tagged image file format' (TIFF), which is a generally accepted computer file format and supported by most image processing tools for both workstations and personal computers. A TIFF file consists of a number of labels (tags) which describe certain properties of the file (such as gray levels, colour table, byte format, compression size, etc.). After the inital tags comes the data, which may be interrupted by more descriptive tags. The size of each TIFF file is predetermined by the resolution in which the digital image was created (i.e., by the specifications of the scanner). A typical resolution of an AFLP image obtained from a 35 × 43 cm polyacrylamide slab gel is 1200 by 2000 datapoints. The image is used for defining 'tracks' for each lane of interest. Within the boundaries of these tracks, data points are resampled and average densitometric values are lined out as a function of the run-time, resulting in electrophoresis 'peak profiles'. These profiles are then used for comparative analysis (cf. the Pearson product-moment correlation coefficient). Images can also be used to define the bands for each lane that will participate in the comparative analysis. Similarities are then calculated from the presence and absence of bands (cf. the band-based similarity coefficient of Dice). However, it may sometimes be difficult to assign bands correctly because of the high complexity of the AFLP banding patterns, in which case the Pearson correlation coefficient is often preferred.

With 'automated' AFLP technology, fluorescently labelled amplicons are detected with a laser while they migrate downwards through the gel. The detection signals are collected, digitised and sent to the computer for storage and processing. These raw data are converted to the TIFF format with software that is normally included in the automatic sequencing apparatus package. The actual size of the image file depends on the conversion settings used, with a typical 7- to 10-fold reduction of the original vertical resolution (as defined by the run-time and signal detection interval). Although TIFF images obtained by classical or fluorescent AFLP are processed in the same way, the actual banding patterns obtained by the two different approaches differ in one important aspect. Classically obtained patterns (i.e., via a scan of an autoradiogram) contain large bands that are compressed together and smaller bands that are relatively diffuse as compared to the larger bands. In contrast, fAFLP patterns contain many small discrete bands while the larger bands are diffuse, simply because the smaller fragments migrate much faster through the gel and are detected very early, with little time in between each detection. A possible improvement to overcome the 'stacking' of small bands in fAFLP patterns would be the use of a signal detection interval that is initially very small (e.g., 500 ms) and gradually increases during the runtime; i.e., more data points are created at the start of the run so that the pattern is somewhat stretched and small bands can be assigned more accurately. Alternatively, Stoffel fragment of DNA polymerase (which has a low rate of processivity) may be added to facilitate the discrimination of small (50–100 bp) PCR products (Hookey *et al.*, 1999).

B. Basic requirements for comparative analysis of AFLP patterns

A general prerequisite for a reliable and accurate intra- and inter-gel comparison of banding patterns is the use of reference patterns that are included at regular intervals in each gel. Usually, in the classical approach of AFLP or in fAFLP with only one fluorophore, a reference pattern is included every 4–6 lanes, and one reference pattern is chosen as the standard. Gels are 'normalised' by aligning all reference strains to this standard so that all patterns positioned between these references are spatially adjusted for optimal comparison. However, minute differences in the gel matrix may cause slightly different migration of the amplicons, influencing the analysis to some extent. Nonetheless, the high resolution of large polyacrylamide slab gels in AFLP methodology ensures that external reference patterns on the same gel usually correlate well (95–100%), and inter-gel correlations of 90% or higher are routinely achieved. A slightly higher reliability and consistency of results may be reached by fAFLP that makes use of two or more fluorophores, with the added advantage of a significantly increased throughput, since sample patterns can be run together with an internal standard that is labelled differently.

Another point that needs attention is the signal strength. Peaks that are topped off, either by the use of a scanner with an insufficient optical density range, or by software that does not permit proper conversion of the raw data, should be avoided as they will affect analysis quite badly. Similarly, some patterns may contain one or two very heavy bands due to highly efficient amplification and it is advisable to omit these amplicons from the analysis altogether (e.g., by deselecting the corresponding zones in the digitised gel).

Finally, the authenticity of banding patterns must also be considered. Random spots on the autoradiogram, or background noise in fAFLP, should be removed prior to the generation of raw data files. In the latter case, special precautions that are outlined in the apparatus manual (such as never cleaning the glass plates with soap), the use of a standardised gel matrix, and additional features that allow automatic background subtraction, have virtually eliminated all false signals. However, for those that still work with autoradiograms, the random specks and spots may pose a real problem, as they may become registered as actual bands during the subsequent scan. These false signals can be easily removed from the primary TIFF image by graphics software (e.g., PhotoStyler or similar), but the operation can be very tedious and time-consuming. The source of the random signals is not always clear, but regular cleaning of the intensifying screen and the film cassette used for autoradiography may help, and special care should be taken in preparing and pouring the gels. For instance, small dust particles, impurities, or tiny pieces of paper towel may contribute to anomalous signal generation, requiring filtering and/or deionizing of the acrylamide solution and thorough cleaning of the glassplates. As a consequence, many researchers resort to the use of highly purified and standardised acrylamides purchased as a premixed solution, or even buy prefabricated polymerised gels. In addition, the use of UV-crosslinking to ensure uniform polymerisation of the gel matrix has become very popular because it reduces the risk for local differences in gel properties and it is relatively fast.

C. *Reproducibility*

The reproducibility of AFLP has been tested in various studies and involved the inclusion of duplicate strains and/or the generation of multiple sub-cultures on which the same DNA extraction and AFLP procedures were performed (whether or not simultaneously). In general, the intra- and inter-gel similarity levels were at least 95% and 90%, respectively. However, much depends on the quality of the DNA because impurities that are being co-purified with the DNA may have a detrimental effect on the activity of restriction nucleases or other enzymes such as the T4 ligase or the DNA polymerase. In addition, the integrity of the DNA obviously plays a role as degraded DNA would lead to a large decrease or total loss of target fragment.

Substantial differences in banding patterns may also be observed when massive amounts of RNA are present in the DNA preparation (consequently, a routine RNAse step is strongly recommended) and some brands of DNA polymerase give slightly different banding patterns for the same template DNA (P. Janssen & R. Coopman, unpublished data). It has been further suggested (Gibson *et al.*, 1998) that repeated thawing and freezing may damage template DNA – resulting in decreased amounts of product – presumably due to the higher fragility of the single-strand linkage between adapter and fragment (Fig. 8.2). Although variations in signal strength (i.e., peak height) have been noted in almost all of the fAFLP studies (where signal strength is more readily visible), this only poses a problem in extreme instances (local differences in PCR efficiency may be caused by secondary DNA structures, long stretches of As or Ts, or GC-rich regions).

Although fAFLP is rapidly becoming the standard (automated) technique in diagnostic laboratories, especially in those that undertake frequent epidemiological studies (making the use of a DNA sequencer cost-effective), multicentre standardisation studies (e.g., for different types of PCR equipment, different operators, minimal variations in the buffer conditions, etc.) have yet to be undertaken for AFLP. Perhaps the ultimate test for the accuracy of the AFLP method is to compare AFLP fragments derived experimentally from a given bacterial strain with those predicted by analysis of its published sequence. This has been done for an *E. coli* K12 strain, for which 92% of the *in silico* predicted fragments could be produced experimentally by AFLP (Arnold *et al.*, 1999a). To date, 28 bacterial genomes are fully sequenced, with sequence analysis of another 80 bacteria well underway, many of them health-related. Without doubt, sequence-derived (i.e., 'virtual') AFLP data will be useful to fine-tune experimental AFLP conditions and may form the basis of future standardisation schemes.

8.4 APPLICATIONS

Over the past few years, many new genotypic typing methods have been developed, including PFGE, AP-PCR, ARDRA, RAPD, mixed-linker PCR, restriction fragment end-labelling, and sampled sequencing (Van Belkum, 1994; Vaneechoutte, 1996). These methods, some of which are discussed in this book, are either based on the use of restriction enzymes, PCR amplification, or both, and have all been used on bacterial genomes with variable success. With the advent of AFLP in 1993 for the detection of molecular markers in plants (Zabeau & Vos, 1993), and the general publishing of the AFLP concept two years later (Vos *et al.*, 1995), microbiologists were quick to apply the AFLP technique to address specific questions in microbial taxonomy and epidemiology. Below are some examples on the use and applications of AFLP in microbial systems. More generalised reviews on AFLP are available in the literature (Blears *et al.*, 1998; Koeleman *et al.*, 1998).

A. AFLP as a taxonomic tool

(i) The genus Aeromonas

The first report on the use of AFLP for bacterial typing and characterisation was presented in 1993 at the Fourth International Symposium on *Aeromonas* and *Plesiomonas* (Janssen *et al.*, 1993) as part of an ecological study on the presence of aeromonads in drinking water production plants (Huys *et al.*, 1993). A total of 38 reference and type strains belonging to the known hybridisation groups (HGs) of *A. hydrophila* and *A. caviae* were subjected to AFLP fingerprinting. All six HGs (three of *A. hydrophila* and three *of A. caviae*) were clearly separated and AFLP-based clustering fully corroborated genotypic and chemotaxonomic data. Strains of the same HG were easily distinguished from each other according to their AFLP profile, and newly isolated aeromonads were allocated to the correct HG as confirmed by fatty acid analysis.

In early 1996, the potential of AFLP to characterise bacterial strains at the sub generic level was evaluated (Janssen *et al.*, 1996). In this study, 90 *Aeromonas* strains were included, representing all 14 HGs known within the genus at that time. The AFLP clusters obtained were in perfect concordance with the grouping of these strains according to hybridisation data, and 85 strains were assigned to the correct HG. Apparently, the five remaining strains belonging to *A. sobria* (two strains), *A. veronii* biogroup *veronii* (one strain) and HG11 (two strains), could not be assigned due to statistical variance as these species were represented by only two to three strains each.

In a follow-up study by Huys *et al.* (1996b; Table 8.1), the use of AFLP allowed the clarification of some unresolved issues in the taxonomic stucture of the genus *Aeromonas*. This elaborate study included the phenospecies *A. allosacharophila*, *A. encheleia*, *A. enteropelogenes* and *A. ichthiosmia*, none of which were previously associated with any of the existing HGs. According to the newly obtained AFLP data, the former two species were related genotypically to subgroups of *A. veronii* and *A. eucrenophila*, respectively, and the latter two species were found to be identical to *A. trota* and another subgroup of *A. veronii*, respectively. The same AFLP data also showed a significant genotypic heterogeneity among A. *eucrenophila*, suggesting a subdivision of this species, with an affiliation of some strains with the unnamed species HG11. This was confirmed in a separate paper that investigated the genotypic relationship between eight *A. eucrenophila* strains, six *A. encheleia* strains, the two HG11 strains, and 14 mainly aquatic *Aeromonas* isolates using AFLP, 16S rRNA RFLP analysis, SDS-PAGE of whole-cell proteins, and fatty acid analysis (Huys *et al.*, 1996a), eventually leading to an emended description of the species *A. eucrenophila* and *A. encheleia* (Huys *et al.*, 1997a).

In the same period, a library of AFLP profiles for *Aeromonas* species was established. This database (AERO94), consisting mainly of 90 reference and previously reported type strains (Huys *et al.*, 1996b), supplemented with 17 new strains, was used to determine the genotypic diversity among 168 *Aeromonas* isolates originating from five drinking water production plants in Flanders, Belgium (Huys *et al.*, 1996c). Of these, 144 strains (86%) could be allocated by AFLP to one of the currently known *Aeromonas* taxa. The remaining 24 unidentified isolates formed a homogeneous AFLP cluster and were found to be closely related to, but clearly distinct from, *A. hydrophila* HG2 (now classified as *A. bestiarum*). A subsequent study, whereby some of these strains were analysed in detail using a polyphasic approach, eventually resulted in the proposal of a new *Aeromonas* species, *A. popoffii* (Huys *et al.*, 1997b). A recent publication describes the use of automated AFLP to fingerprint *Aeromonas* strains (Huys & Swings, 1999).

(ii) The genus Acinetobacter

In an extensive study, 151 classified strains of the genus *Acinetobacter* (representing 18 genomic species, including type, reference and field strains) and eight previously unclassified acinetobacters were analysed by AFLP (Janssen *et al.*, 1997).

Using a single set of restriction enzymes and one combination of corresponding AFLP primers, all classified strains could be allocated to the correct genomic species, and all groups were separated properly, with minimal intraspecific similarity levels (S_D) ranging from 29% to 74%. Even the closely related DNA groups 1 (*A. calcoaceticus*), 2 (*A. baumannii*), 3 and 13TU (*sensu* Tjernberg & Ursing, 1989), together forming the so-called *A. calcoaceticus – A. baumannii* (Acb) complex, were clearly distinguishable by AFLP, with intra-specific linkage levels above 50%. AFLP analysis also successfully highlighted some incongruities within the 13BJ-14TU group and placed four unclassified strains, obtained from diverse sources and origins, firmly together in one group at a minimal similarity level of 50%.

(iii) AFLP genotyping in other taxonomic investigations

AFLP fingerprinting was used in a polyphasic taxonomic study to support reclassification and emended discriptions of *Paenibacillus larvae*, the causative agent of disease in honeybee larvae (Heyndrickx *et al.*, 1996). Two subgroups, *P. larvae* subsp. *larvae* and *P. larvae* subsp. *pulvifaciens*, could be separated by AFLP in separate clusters in spite of the high linkage level between them (>80%) and the respective internal similarities of 88% and 95% (using the Dice coefficient). These results were confirmed by the other taxonomic data obtained in this study by fatty acid analysis, pyrolysis mass spectrometry, DNA-DNA binding studies and whole-cell protein profiling.

Recently, 94 *Vibrio* strains that are closely related to *V. harveyi*, an ubiquitous marine bacterium, were analysed with AFLP (Pedersen *et al.*, 1998). One single enzyme combination, *Hin*dIII / *Taq*I, and one particular pair of primers were used to group 77 strains in nine AFLP clusters with a cut-off similarity level of 50% (using the Dice similarity coefficient; S_D). Twelve of the remaining 17 strains were type strains of separate *Vibrio* species, and five strains could not be grouped. The AFLP data obtained were supported in the same study by DNA-DNA hybridisation, ribotyping and plasmid profile results.

AFLP methodology has also helped to unravel some of the taxonomic fine-structure in *Xanthomonas*, a genus that mainly comprises phytopathogenic and plant-associated bacteria. In one study, 68 *X. translucens* strains were investigated by AFLP (Bragard *et al.*, 1997). This species encompasses the so-called 'translucens group' of closely related pathovars within the former species *X. campestris*, and also includes five other pathovars isolated from various *Poaceae*. The classification of the *X. translucens* pathovars and their strains is very intricate. Nonetheless, AFLP-based clustering was consistent with the current pathogenicity grouping, and recent reclassifications within these taxa were fully supported by the AFLP data obtained.

In another study, a particular pathovar of *X. axonopodis*, pv. *manihotis* (Xam), the causative agent for cassava bacterial blight, was studied in detail by AFLP (Restrepo *et al.*, 1999). To identify the best enzyme combination and selective primers, 64 different primer pairs, comprising all 16 possible +1/+1 primer combinations for each enzyme combination (Table 8.1), were tested on six strains. Eight enzyme/primer pairs that generated informative banding patterns were selected for all further analyses. From the literature, it appears that this is one of the most comprehensive studies on the effect of enzyme and primer use on AFLP banding pattern composition for a given bacterial species, and the results obtained confirmed the bias described earlier for restriction sites and selective bases towards the G+C content of the genome analysed (Janssen *et al.*, 1996). In total, 47 *X. axonopodis* strains, originating from three different edaphoclimatic zones (ECZs), were subjected to AFLP with eight pre-selected primer combinations, giving rise to a total of 173 polymorphic bands that were considered for multiple component analysis (MCA) and cluster analyses. For nine of the 10 AFLP clusters that were discerned at a 70% similarity cut-off point (using Jaccard's coefficient), an agreement with the ECZ allocation (i.e., the region from which the strains originated) was observed. In addition, the AFLP data closely matched the RFLP results obtained for the pathogenicity gene *pthB*. In fact, AFLP analysis appeared to have a higher discriminative power in differentiating highly related strains as it revealed the existence of sub-groups in the ECZ5 Xam population and strengthened the hypothesis that ECZ5 ("from high-altitude tropic regions") actually forms a genetically and evolutionary separate group.

Another example of efficient assessment of intra-pathovar diversity by AFLP was presented by Clerc *et al.* (1998), who studied the genetic relatedness of 23 strains belonging to various pathovars of *Pseudomonas syringae* with AFLP and RAPD. Both techniques were equally able to discriminate the otherwise indistinguishable pathovars *P. syringae* pv. *tomata* and *P. syringae* pv. *maculicola*, but in addition, AFLP was more efficient in addressing intra- and inter-specific distances.

B. *Epidemiological typing of bacteria*

(i) Genotyping in the hospital environment

Valsangiacomo *et al.* (1995) reported on the application of AFLP to type *Legionella pneumophilia* strains from different regions of Switzerland. A simplified procedure was used, employing a single restriction enzyme, *Pst*I, and agarose gel electrophoresis (Table 8.1). AFLP analysis of 10 clinical isolates (from three cases of legionellosis) and 18 environmental strains (mainly obtained from the local water systems) enabled the authors to assess the origin of infection and the likely route of transmission of the pathogen. These results were corroborated by ribotyping and RFLP analysis.

An epidemiological relationship between environmental and clinical isolates of *Mycobacterium kansasii* was established in a polyphasic study using RFLP and PFGE analyses, PCR restriction analysis of the *hsp65* gene, and AFLP (Picardeau *et al.*, 1997). Thirty-eight strains of clinical origin (including 14 isolates from AIDS patients) and 24 isolates from local tapwater were investigated. For AFLP analysis, only one restriction enzyme, *Pst*I, was used, and separation of the PCR products was done in 2% agarose gels (Table 8.1). Although the AFLP patterns consisted of only 3–8 bands, they were found to be as informative as the PFGE patterns, since both methodologies displayed polymorphisms within each of the five clusters defined by this study.

A single enzyme approach was also developed for AFLP-based typing of *Helicobacter pylori* (Gibson *et al.*, 1998). In this study, the genomic DNA of 46 isolates obtained from 28 patients (some of which belonged to the same family group) was digested by *Hin*dIII, tagged with appropriate adaptors, and selectively amplified with +1 primers (Table 8.1). Amplicons were separated on a 1.5% agarose gel and the resulting AFLP patterns were used for pairwise analysis of strains. All strains were characterised by at least one other genotypic method (ribotyping and/or urease genotyping). AFLP profiles differentiated strains of unrelated individuals and confirmed the common origin of strains in some of the family groups.

To investigate whether certain pathogenic *Aeromonas* strains were circulating among patients with diarrhoea in Bangladesh and to gain a better understanding of the virulence of *Aeromonas* spp., Kuhn *et al.* (1997) set out to investigate 120 isolates, including 80 faecal isolates from patients (69 isolates from patients with diarrhoea, and 11 from healthy controls) and 40 environmental isolates from surface water. All isolates were phenotyped with the Phene-Plate (PhP) method, a high-resolution biochemical fingerprinting system. Most (106) were assigned to hybridisation groups (HGs) by fatty acid analysis, while some (29) required further characterisation using the AERO94 database of *Aeromonas* AFLP profiles (Huys *et al.*, 1996b, c). The predominant PhP type in the human isolates group was type BD-2 (all of which were identified as *A. hydrophila* HG1), which produced relatively high levels of virulence factors, including haemolysin and cytotoxin. This PhP type occurred only once in the environmental group, suggesting that the HG1/BD-2 type represents a true human pathogen, and confirming other reports associating *A. hydrophila* HG1 with human intestinal disease.

AFLP has also been used in conjunction with other typing methods for the epidemiological typing of *Acinetobacter baumannii* strains (Dijkshoorn *et al.*, 1996). The AFLP analysis in this study followed the conventional approach of applying two restriction enzymes, *Hin*dIII and *Taq*I, and using +1 and +2 selective primers (Table 8.1). Thirty-one strains, comprising 14 strains from 14 outbreaks in different European cities and 17 sporadic strains from different origins and sources were investigated. Twelve of the 14 outbreak strains grouped in one AFLP cluster (at 92.8% using the Pearson correlation coefficient), while a second cluster comprised three outbreak strains and one sporadic strain that linked at 89.9%. Strains in these clusters had patterns with differences in one or two band positions only. At a delineation level of 89.0%, only one other cluster containing three sporadic strains could be discerned. All other strains, including the two remaining outbreak strains, were left ungrouped. The other methods used in this study (biotyping, antiobiogram analysis, cell envelope protein profiling and ribotyping) fully corroborated the AFLP data, indicating that AFLP was a reliable and accurate typing method. A subsequent study on 25 *Acinetobacter* strains isolated from five hospital outbreaks in three countries was undertaken (Janssen & Dijkshoorn, 1996). Isolates from the same outbreak displayed identical banding patterns and each set of outbreak strains was found in one particular AFLP cluster with a minimum of 94% similarity (using the Pearson correlation coefficient), confirming previously published typing data on these strains. In a recent study, automated AFLP with fluorescently labelled primers was used to type *A. baumannii* isolates, including type and reference strains and clinical isolates (Koeleman *et al.*, 1998).

Conventional AFLP with two restriction enzymes and one radioactively labelled primer has also been used for typing *Staphylococcus epidermidis* isolates in multiple blood cultures (Sloos *et al.*, 1998). Fifteen unrelated strains from various cities across The Netherlands and 44 strains from 11 patients were investigated (Table 8.1). These patients were from four different departments of the Leiden University Medical Center. With AFLP, the former group was heterogeneous with a similarity level ranging between 78 and 93% (using the Pearson correlation coefficient). Isolates from nine patients were found in nine separate clusters, each containing the four representative blood culture strains originally isolated from each patient. Within each set of four strains, the AFLP patterns were highly similar (>94%), indicating identity (the identity level was set at 94% by reproducibility tests). For the sets of strains isolated from the two remaining patients, dissimilar AFLP patterns were found with relatively low linkage levels of 83 and 73%. This may be explained by the fact that contamination with skin inhabitants during blood-taking procedures is not uncommon.

A number of research groups have used fluorescent AFLP (fAFLP) for typing clinical isolates (Table 8.1). Desai *et al.* (1998) investigated two possibly related outbreaks of group A streptococcal (GAS) invasive disease that took place in a nursing home and a district general hospital in North London, and 16 other GAS isolates collected from various hospitals across England. All but four strains were of serotype M77. The two outbreaks contained two and eight strains, respectively, and were all of serotype T13 M77. Within each outbreak, only one AFLP genotype was observed, with virtually no pattern variance, whereas the non-outbreak M77 serotype strains displayed unique AFLP profiles. In addition, AFLP readily distinguished the two clones from each other. Through comparison of the AFLP data with macro-restriction analysis of the same strains, it was concluded that the discriminatory power of fAFLP for typing GAS was much higher, and that AFLP methodology is suited for subtyping within a single serotype. In a follow-up study, Desai *et al.* (1999) analysed 35 *S. pyogenes* M1 isolates by fAFLP. These isolates were from a different clinical background and formed a clonal group displaying identical ribotyping and macro restriction profiles. Nonetheless, fAFLP readily subtyped them, grouping 25 isolates in seven (multi-isolate) profiles and assigning further individual profiles to the remaining ten isolates.

Fluorescent AFLP analysis and macro-restriction analysis were also compared by Van Eldere *et al.* (1999) who genotyped 48 pneumococci isolated from blood and cerebrospinal fluid. All isolates originated from hospitals in the northern part of France or southern regions of Belgium bordering France and represented five serotypes (types 6, 9, 14, 19 and 23F). Of these strains, 42% showed intermediate or full penicillin resistance, with the majority of penicillin-resistant isolates (78%) located in serotypes 9V and 23F. These groups could be easily differentiated with both DNA typing methods. This was also the case for the serotype 14 and 19 penicillin-susceptible strains which formed well-separated clusters following PFGE and AFLP analysis. However, the susceptible type 23F strains were only found as a distinct cluster when AFLP was used. Overall, data analysis showed that AFLP and macro restriction were equally efficient in assessing intraserotype diversity.

A recent paper by Hookey *et al.* (1999) has described the use of fAFLP for the genotypic analysis of methicillin-resistant *Staphylococcus aureus* (MRSA). A collection of 34 isolates from 22 hospitals in the south of England, together with a single isolate of each of the current predominant UK phage types, and one reference strain of MRSA, were subjected to AFLP using *Eco*RI and *Mse*I enzymes and a single primer set (Table 8.1). The resulting data were compared with data obtained by standard phenotypic methods (including phage-typing, protein-A production and antibiogram analysis) and data generated by genotypic methods such as RFLP analysis of the coagulase (*coa*) gene and macro-restriction analysis. Based on replicate studies, the level of identity for AFLP was set at 93.7% (using the Dice coefficient). Using this criterion, all but two strains could be distinguished from each other. Thirty-one of the 34 strains fell into four major fAFLP clusters with an internal linkage level of at least 80%. The two largest clusters, containing 10 and 13 strains, could be further subdivided into two subgroups, roughly confirming the grouping of these strains according to the phenotypic and other genotypic characteristics determined in this study.

An extension of this work was reported by Grady *et al.* (1999) who took 24 MRSA isolates of phage-type 15 (EMRSA-15) and subjected them to separate fAFLP analyses using the restriction enzymes *Apa*I + *Taq*I and *Eco*RI + *Mse*I. Both template DNAs were selectively amplified with a combination of +0 (for *Apa*I-, resp. *Eco*RI-adaptors) and +1 (for *Taq*I-, resp. *Mse*I-adaptors) primers, resulting in AFLP patterns containing *c*.75 bands in the size range of 50–800 and 50–300 bp, respectively. In each of the two separate analyses, EMRSA-15 isolates could be differentiated from other MRSA isolates included in the study. By combining both data sets, fAFLP divided the 24 EMRSA-15 isolates into 11 profiles. In contrast, RFLP analysis of the coagulase gene of these isolates failed to discriminate between any of these isolates, and macro-restriction of the 24 isolates, although discriminative, was not as reproducible as fAFLP.

(ii) AFLP analysis of food-borne pathogens

To investigate the possible source and route of food contamination, sensitive and reliable typing methods are needed. Although classical methods such as biotyping, phage-typing and serotyping may still be used, they usually give incomplete information, and diagnostic laboratories nowadays prefer to integrate one or more DNA-based typing methods in their identification and typing schemes. The usefulness of AFLP for the typing and identification of food pathogens is well documented. In the study by Aarts *et al.* (1998), 78 different *Salmonella* strains, comprising 62 serotypes, were analysed by AFLP. Choosing *Eco*RI and *Mse*I as enzymes and one particular primer combination, reproducible and informative AFLP profiles with up to 50 bands were obtained. All serotypes displayed a unique profile, and AFLP appeared to group all strains with identical bacteriophage specificity in the same cluster. However, a phylogenetic analysis based on AFLP data was not performed.

In another study, 50 type, reference and field strains of *Campylobacter jejuni* and *C. coli* were subjected to fAFLP analysis (Kokotovic & On, 1999; Table 8.1). Of the 27 *C. jejuni* and 23 *C. coli* strains studied, 19 and 18 different fAFLP profiles, respectively, were recognised. In general, outbreak strains could be readily discriminated from the sporadic isolates within the same *Campylobacter* species. However, a numerical analysis of AFLP data was not included.

Campylobacters were also subjected to fAFLP by Duim *et al.* (1999) who investigated 45 strains of *C. jejuni* and *C. coli*, including 31 isolates from poultry, 10 human isolates and four reference strains. Some of these strains had been genotypically characterised previously. Informative patterns were obtained for all strains using one enzyme combination and one set of +1 selective primers (Table 8.1). Isogenic mutants of *Campylobacter*, or highly related strains that produced identical PFGE patterns, showed highly similar or identical AFLP patterns. Twenty-five randomly chosen poultry isolates grouped in two AFLP clusters representing the two species *C. jejuni* and *C. coli* (as confirmed by species-specific multiplex PCR) with average linkage levels of 32% and 58%, respectively. The 10 *C. jejuni* strains of human origin displayed heterogeneous AFLP patterns. When the results of all the *C. jejuni* strains were combined, human isolates scattered throughout the dendrogram and an epidemiological link between the strains was not apparent. Nonetheless, the AFLP data showed clearly that some human strains were highly related to poultry strains (with a Pearson correlation coefficient of 90% or higher), supporting the thesis that poultry products are a source of human infection.

In a recent study by Arnold *et al.* (1999b), 87 strains of *E. coli*, including 72 strains from a EcoR reference collection and 15 strains of the clinically important serogroup O157, were subjected to fAFLP analysis (Table 8.1). The composition of the EcoR group had been defined previously by multilocus enzyme electrophoresis (MLEE). Sixty-three of these EcoR strains were grouped by AFLP in the correct MLEE subdivision, while 11 of the O157 serotype strains were found in a separate AFLP cluster.

C. AFLP for studying the molecular evolution of microbes

In general terms, nucleotide substitutions that are introduced inadvertently, i.e., due to replication infidelities, or are brought forward under selective pressure, constitute a major source for naturally occuring DNA polymorphisms. However, genetic diversity in prokaryotes appears to be driven largely by a number of dynamic processes that enable them to react swiftly to changes in their environ-

and the abundant presence of rRNA species do exist. For instance, Amara & Vijaya (1997) reported on the specific addition of a poly(A) tail to *E. coli* transcripts with a yeast poly(A) polymerase. Apparently, in the presence of manganese and magnesium, ribosomal RNAs remain part of the polysome and, as such, are not available for polyadenylation, while adenines are added to the free 3'-ends of the transcripts. Another report has described how rRNA can be removed succesfully from total RNA by subtractive hybridisation with antisense rRNA produced *in vitro* (Robinson *et al.*, 1994). This involved cloning the 16S-23S operon (in reverse direction) in a transcription vector and incorporation of biotin in the antisense rRNA. With these and similar approaches, the AFLP method should become a valuable tool for analysis of differential expression in bacteria. Recently, a commercial kit from Display Systems Biotech (Vista, CA) has become available that allows differential display of prokaryotic genes without the need of poly(A)-primed PCR by making use of random octamers for cDNA production.

8.5 COMPARISON WITH OTHER METHODS

This chapter (sections 8.3 and 8.4) provides ample evidence for the versatility and reliable performance of AFLP in microbial taxonomy and epidemiology. However, there is a lack of congruity in the methodology used, i.e., the use of restriction enzymes, gel systems and visualisation technologies may all differ from laboratory to laboratory, and a standardised scheme for AFLP does not yet exist. Nonetheless, the technique is robust and reproducible when executed within basic guidelines, much more so than other PCR-based methods such as RAPD or REP-PCR, which are intrinsically prone to variations in amplification efficiency (Towner & Cockayne, 1993; van Belkum, 1994). These basic guidelines are inspired by common sense. For instance, high resolution AFLP with two restriction enzymes, high quality DNA and automated laser detection can be expected to give better and more reproducible results than, say, AFLP on crude DNA with one restriction enzyme and separation of amplicons through agarose. In addition, possible sources of variation in AFLP analyses are minimalised by using preset electrophoresis conditions and a standardised PCR profile at high stringency.

AFLP does not require knowledge of genomic sequences (unlike REP-PCR) and covers the entire genome (unlike locus-specific RFLP; e.g. ribotyping or ARDRA). The technique can also be used with any DNA, regardless of its origin or complexity. Even the smallest bacterial genome (that of *Mycoplasma*) produces complex AFLP patterns (Table 8.1), while AFLP can also be used to analyse larger genomes (e.g., from eukaryotes). However, the variable base composition of bacterial genomes requires the use of suitable restriction enzymes and the applicability of AFLP remains limited to the sub-generic level (Janssen *et al.*, 1996).

The turn-around time of AFLP is relatively good, with less than three days needed for template preparation, selective amplification, electrophoresis, data acquisition and analysis (34 strains, starting from chromosomal DNA). For fAFLP, even two days suffice and more strains can be processed simultaneously (depending on the apparatus). The downside is that the cost of a DNA sequencer (US $40,000–125,000) may be prohibitive for some laboratories. Recently, Olive & Bean (1999) have compared the characteristics of various molecular typing methods, including PFGE and RAPD.

The concordance of AFLP-generated data with existing taxonomic data or data simultaneously generated by other taxonomic or epidemiological methods is very good (Dijkshoorn *et al.*, 1996; Huys *et al.*, 1996b; Janssen *et al.*, 1996; 1997; Koeleman *et al.*, 1998; Desai *et al.*, 1999; Kokotovic *et al.*, 1999; Speijer *et al.*, 1999; Van Eldere *et al.*, 1999). In particular, some of these reports provide evidence that AFLP data are in very good agreement with the data obtained by DNA-DNA hybridisation, which is still regarded as the 'gold' standard method in microbial systematics. Recently, Hauben *et al.* (1999) and Rademaker *et al.* (2000) established correlation plots of DNA homology data versus AFLP correlation values for large sets of bacterial strains, and found an overall high correlation between AFLP fingerprinting and DNA-DNA pairing data. However, AFLP is unlikely to replace this technique for bacterial species delineation, as outlined in a cautionary note by Esteve (1997). Rather, because high resolution AFLP data can be easily stored and exchanged by computer, AFLP should be considered as an ideal preliminary screening method, while any significant taxonomic findings should be confirmed by DNA hybridisation studies.

8.6 FUTURE PROSPECTS AND CONCLUSIONS

The range of microbial genomes that have been sequenced to completion has grown rapidly over the past few years and will certainly expand steadily over the next year or two, with the sequencing of more than 100 microbial genomes finished by the year 2002. The resulting biological information has enabled scientists to predict possible protein functions and to identify potential target genes for antimicrobial drug discovery. Databases are now accessible to the average researcher in the laboratory via Web browser interfaces (reviewed by Moir *et al.*, 1999) and many of these permit downloading of sequence data for customised analysis. For some human pathogens, such as *Streptococcus pneumoniae* and *Helicobacter pylori*, multiple strains have already been completely sequenced, allowing complete genome comparison with specially developed software (MUMmer; Delcher *et al.*, 1999). However, such an approach of 'armchair genomics' needs to link up with reality since it is an impossible task to sequence all microorganisms in the environment or to analyse the entire genome of every clinical isolate of interest. High resolution DNA fingerprinting techniques such as AFLP should thus remain very useful, especially for the analysis of microbial communities or for detailed epidemiological studies. In addition, genetic markers that can be located to a particular amplicon can be easily isolated by excision of the AFLP fragment, reamplified and directly sequenced. In this context, AFLP-based probe development and the use of microarray technology should be seen as logical continuations in microbial genome analysis.

REFERENCES

Aarts, H.J. , van Lith, L.A. & Keijer, J. (1998). High-resolution genotyping of *Salmonella* strains by AFLP-fingerprinting. *Letters in Applied Microbiology* **26**, 131–135.

Amara, R.R. & Vijaya, S. (1997). Specific polyadenylation and purification of total messenger RNA from *Escherichia coli*. *Nucleic Acids Research* **25**, 3465–3470.

Arnold, C., Metherell, L., Clewley, J. & Stanley, J. (1999a). Predictive modelling of fluorescent AFLP: a new approach to the molecular epidemiology of *E. coli*. *Research in Microbiology* **150**, 33–44.

Arnold, C., Metherell, L., Willshaw, G., Maggs, A. & Stanley, J. (1999b). Predictive fluorescent amplified-fragment length polymorphism analysis of *Escherichia coli*: high-resolution typing method with phylogenetic significance. *Journal of Clinical Microbiology* **37**, 1274–1279.

Bachem, C.W., van der Hoeven, R., de Bruijn, S., Vreugdenhil, D., Zabeau, M. & Visser, R.G. (1996). Visualization of differential gene expression using a novel method of RNA fingerprinting based on AFLP: analysis of gene expression during potato tuber development. *Plant Journal* **9**, 745–753.

Blears, M.J., De Grandis, S.A., Lee, H. & Trevors, J.T. (1998). Amplified fragment length polymorphism (AFLP): a review of the procedure and its applications. *Journal of Industrial Microbiology and Biotechnology* **21**, 99–114.

Bragard, C., Singer, E., Alizadeh, A., Vauterin, L., Maraite, H. & Swings, J. (1997). *Xanthomonas translucens* from small grains: diversity and phytopathological relevance. *Phytopathology* **87**, 1111–1117.

Brikun, I., Suziedelis, K. & Berg, D.E. (1994). DNA sequence divergence among derivatives of *Escherichia coli* K-12 detected by arbitrary PCR (random amplified polymorphic DNA) fingerprinting. *Journal of Bacteriology* **176**, 1673–1682.

Caetano-Annolés, G., Bassam, B. & Gresshoff, P.M. (1991). DNA amplification using very short arbitrary primers. *BioTechnology* **9**, 553–557.

Chalhouk, B.A., Thibault, S., Laucou, V., Rameau, C., Höfte, H. & Cousin, R. (1997). Silver staining and recovery of AFLP amplification products on large denaturing polyacrylamide gels. *BioTechniques* **22**, 216–220.

Clerc, A., Manceau, C. & Nesme, X. (1998). Comparison of randomly amplified polymorphic DNA with amplified fragment length polymorphism to assess genetic diversity and genetic relatedness within genospecies III of *Pseudomonas syringae*. *Applied and Environmental Microbiology* **64**, 1180–1187.

Coenye, T., Schouls, M., Govan, J.R., Kersters, K. & Vandamme, P. (1999). Identification of *Burkholderia* species and genomovars from cystic fibrosis patients by AFLP fingerprinting. *International Journal of Systematic Bacteriology* **49**, 1657–1666.

Delcher, A.L., Kasif, S., Fleishmann, R.D., Peterson, J., White, O. & Salzberg, S.L. (1999). Alignment of whole genomes. *Nucleic Acids Research* **27**, 2369–2376.

Desai, M., Tanna, A., Wall, R., Efstratiou, A., George, R. & Stanley, J. (1998). Fluorescent amplified-fragment length polymorphism analysis of an outbreak of group A streptococcal invasive disease. *Journal of Clinical Microbiology* **36**, 3133–3137.

Desai, M., Efstratiou, A., George, R. & Stanley, J. (1999). High-resolution genotyping of *Streptococcus pyogenes* serotype M1 isolates by fluorescent amplified-fragment length polymorphism analysis. *Journal of Clinical Microbiology* **37**, 1948–1952.

Dijkshoorn, L., Aucken, H., Gerner-Smidt, P., Janssen, P., Kaufmann, M.E., Garaizar, J., Ursing, J. & Pitt, T.L. (1996). Comparison of outbreak and non-outbreak *Acinetobacter baumannii* strains by genotypic and phenotypic methods. *Journal of Clinical Microbiology* **34**, 1519–1525.

Duim, B., Wassenaar, T.M., Tigter, A. & Wagenaar, J. (1999). High-resolution genotyping of *Campylobacter* strains isolated from poultry and humans with amplified fragment length polymorphism fingerprinting. *Applied and Environmental Microbiology* **65**, 2369–2375.

Esteve, C. (1997). Is AFLP fingerprinting a true alternative to the DNA-DNA pairing method to assess genospecies in the genus *Aeromonas*? *International Journal of Systematic Bacteriology* **47**, 245–246.

Forbes, K.J., Bruce, K.D., Jordens, J.Z., Ball, A. & Pennington, T.H. (1991). Rapid methods in bacterial fingerprinting. *Journal of General Microbiology* **137**, 2051–2058.

Geornaras, I., Kunene, N.F., von Holy, A. & Hastings, J.W. (1999). Amplified fragment length polymorphism fingerprinting of *Pseudomonas* strains from a poultry processing plant. *Applied and Environmental Microbiology* **65**, 3828–3833.

Gibson, J.R., Slater, E., Xerry, J., Tompkins, D.S. & Owen, R.J. (1998). Use of an amplified-fragment length polymorphism technique to fingerprint and differentiate isolates of *Helicobacter*. *Journal of Clinical Microbiology* **36**, 2580–2585.

Grady, R., Desai, M., O'Neill, G., Cookson, B. & Stanley, J. (1999). Genotyping epidemic methicillin-resistant *Staphylococcus aureus* phage-type 15 by fluorescent amplified-fragment length polymorphism. *Journal of Clinical Microbiology* **37**, 3189–3203.

Hauben, L., Vauterin, L., Moore, E.R., Hoste, B. & Swings, J. (1999). Genomic diversity of the genus *Stenotrophomonas*. *International Journal of Systematic Bacteriology* **49**, 1749–1760.

Heyndrickx, M., Vandemeulebroucke, K., Hoste, B., Janssen, P., Kersters, K., De Vos, P., Logan, N.A. & Berkeley, C.W. (1996). Reclassification of *Paenibacillus* (formerly *Bacillus*) *pulvifaciens* (Nakamura 1984) Ash *et al.* 1194, a later subjective synonym of *Paenibacillus* (formerly *Bacillus*) *larvae* (White 1960) Ash *et al.* 1994, as a subspecies of *P. larvae*. Emended description of *P. larvae* with *P. larvae* subsp. *larvae* and *P. larvae* subsp. *pulvifaciens*. *International Journal of Systematic Bacteriology* **46**, 270–279.

Hookey, J.V., Edwards, V., Patel, S., Richardson, J.F. & Cookson, B.D. (1999). Use of fluorescent amplified fragment length polymorphism (fAFLP) to characterise methicillin-resistant *Staphylococcus aureus*. *Journal of Microbiological Methods* **37**, 7–15.

Hu, G. (1993). DNA polymerase-catalyzed addition of non-templatal extra nucleotides to the 3' end of a DNA fragment. *DNA Cell Biology* **12**, 763–770.

Huys, G. & Swings, J. (1999). Evaluation of a fluorescent amplified fragment length polymorphism (FAFLP) methodology for the genotypic discrimination of *Aeromonas* taxa. *FEMS Microbiology Letters* **177**, 83–92.

Huys, G., Coopman, R., Vancanneyt, M., Kersters, I., Verstraete, W., Kersters, K. & Janssen, P. (1993). High resolution differentiation of aeromonads. *Medical Microbiology Letters* **2**, 248–255.

Huys, G., Altwegg, M., Hänninen, M.-L., Vancanneyt, M., Vauterin, L., Coopman, R., Torck, U., Lüthy-Hottenstein, J., Janssen, P. & Kersters, K. (1996a). Genotypic and chemotaxonomic description of two subgroups in the species *Aeromonas eucrenophila* and their affiliation to *A. encheleia* and *Aeromonas* DNA hybridization group 11. *Systematic and Applied Microbiology* **19**, 616–623.

Huys, G., Coopman, R., Janssen, P. & Kersters, K. (1996b). High-resolution genotypic analysis of the genus *Aeromonas* by AFLP fingerprinting. *International Journal of Systematic Bacteriology* **46**, 572–580.

Huys, G., Kersters, I., Coopman, R., Janssen, P. & Kersters, K. (1996c). Genotypic diversity among *Aeromonas* isolates recovered from drinking water production plants as revealed by AFLP™ analysis. *Systematic and Applied Microbiology* **19**, 428–435.

Huys, G., Kämpfer, P., Altwegg, M., Coopman, R., Janssen, P., Gillis, M. & Kersters, K. (1997a). Inclusion of *Aeromonas* DNA hybridization group 11 in *Aeromonas encheleia* and extended descriptions of the species *Aeromonas eucrenophila* and *A. encheleia*. *International Journal of Systematic Bacteriology* **47**, 1157–1164.

Huys, G., Kämpfer, P., Altwegg, M., Kersters, I., Lamb, A., Coopman, R. Lüthy-Hottenstein, J., Vancanneyt, M., Janssen, P. & Kersters, K. (1997b). *Aeromonas popoffii* sp. nov., a mesophilic bacterium isolated from drinking water production plants and reservoirs. *International Journal of Systematic Bacteriology* **47**, 1165–1171.

Janssen, P. (1993). The application of a novel PCR-based genomic fingerprinting method for the high-resolution differentiation of aeromonads. In *Abstracts of the fourth international symposium on Aeromonas and Plesiomonas*, p.17. ASM Press, Herndon, VA.

Janssen, P. & Dijkshoorn, L. (1996). High resolution fingerprinting of *Acinetobacter* outbreak strains. *FEMS Microbiology Letters* **142**, 191–194.

Janssen, P., Coopman, R., Huys, G., Swings, J., Bleeker, M., Vos, P., Zabeau, M. & Kersters, K. (1996). Evaluation of the DNA fingerprinting method AFLP as a new tool in bacterial taxonomy. *Microbiology* **142**, 1881–1893.

Janssen, P., Maquelin, K., Coopman, R., Tjernberg, I., Bouvet, P., Kersters, K. & Dijkshoorn, L. (1997). Discrimination of *Acinetobacter* genomic species by AFLP fingerprinting. *International Journal of Systematic Bacteriology* **47**, 1179–1187.

Karlin, S., Mrazek, J. & Campbell A.M. (1997). Compositional biases of bacterial genomes and evolutionary implications. *Journal of Bacteriology* **179**, 3899–3913.

Keim, P., Kalif, A., Schupp, J., Hill, K., Travis, S.E., Richmond, K., Adair, M., Hugh-Jones, M., Kuske, C.R. & Jackson, P. (1997). Molecular evolution and diversity in *Bacillus anthracis* as detected by amplified fragment length polymorphism markers. *Journal of Bacteriology* **179**, 818–824.

Kersters, I., Huys, G., Van Duffel, H., Vancanneyt, M., Kersters, K. & Verstraete, W. (1996). Survival potential of *Aeromonas hydrophila* in freshwaters and nutrient-poor waters in comparison with other bacteria. *Journal of Applied Bacteriology* **80**, 266–276.

Knowles, D.J.C. (1997). New strategies for antibacterial drug design. *Trends in Microbiology* **5**, 379–383.

Koeleman, J.G., Stoof, J., Biesmans, D.J., Savelkoul, P.H. & Vandebroucke-Grauls, C.M. (1998). Comparison of amplified ribosomal DNA restriction analysis, random polymorphic DNA analysis, and amplified fragment length polymorphism fingerprinting for identification of *Acinetobacter* genomic species and typing of *Acinetobacter baumannii*. *Journal of Clinical Microbiology* **36**, 2522–2529.

Kokotovic, B. & On, S.L.W. (1999). High-resolution genomic fingerprinting of *Campylobacter jejuni* and *Campylobacter coli* by analysis of amplified fragment length polymorphisms. *FEMS Microbiology Letters* **173**, 77–84.

Kokotovic, B., Friis, N.F., Jensen, J.S. & Ahrens, P. (1999). Amplified-fragment length polymorphism fingerprinting of *Mycoplasma* species. *Journal of Clinical Microbiology* **37**, 3300–3307.

Kozian, D.H. & Kirschbaum, B.J. (1999). Comparative gene-expression analysis. *Trends in Biotechnology* **17**, 73–78.

Kühn, I., Albert, M.J., Ansaruzzaman, M., Bhuiyan, N.A., Alabi, S.A., Islam, M.S., Neogi, P.K., Huys, G., Janssen, P., Kersters, K. & Möllby, R. (1997). Characterization of *Aeromonas* spp. isolated from humans with diarrhea, from healthy controls, and from surface water in Bangladesh. *Journal of Clinical Microbiology* **35**, 369–373.

Lin, J.-J., Ma, J. & Kuo, J. (1999). Chemiluminescent detection of AFLP markers. *BioTechniques* **26**, 344–348.

Matz, M.V. & Lukyanov, S.A. (1998). Different strategies of differential display: areas of application. *Nucleic Acids Research* **26**, 5537–5543.

Meijer, A., Morré, S., van den Brule, A., Savelkoul, P. & Ossewaarde, J. (1999). Genomic relatedness of *Chlamydia* isolates determined by amplified fragment length polymorphism analysis. *Journal of Bacteriology* **181**, 4469–4475.

Moir, D.T., Shaw, K.J., Hare, R.S. & Vovis, G.F. (1999). Genomics and antimicrobial drug discovery. *Antimicrobial Agents and Chemotherapy* **43**, 439–446.

Money, T., Reader, S., Qu, L.J., Dunford, R.P. & Moore, G. (1996). AFLP-based mRNA fingerprinting. *Nucleic Acids Research* **24**, 2616–2617.

Naas, T., Blot, M., Fitch, W.M. & Arber, W. (1995). Dynamics of IS-related genetic rearrangements in resting *Escherichia coli* K-12. *Molecular Biology and Evolution* **12**, 198–207.

Nei, M. & Li, W.-H. (1979). Mathematical model for studying genetic variations in terms of restriction endonucleases. *Proceedings of the National Academy of Sciences of the United States*

of America **76**, 5269–5273.
Olive, D.M. & Bean, P. (1999). Principles and applications of methods for DNA-based typing of microbial organisms. *Journal of Clinical Microbiology* **37**, 1661–1669.
Owen, R.J. (1989). Chromosomal DNA fingerprinting – a new method of species and strain identification applicable to microbial pathogens. *Journal of Medical Microbiology* **30**, 89–99.
Quinn, F.D., Newman, G.W. & King, C.H. (1997). In search of virulence factors of human bacterial disease. *Trends in Microbiology* **5**, 20–26.
Pedersen, K., Verdonck, L., Austin, B., Austin, D.A., Blanch, A.R., Grimont, P.A.D., Jofre, J., Koblavi, S., Larsen, J.L., Tiainen, T., Vigneulle, M. & Swings, J. (1998). Taxonomic evidence that *Vibrio carchariae* Grimes *et al.* 1985 is a junior synonym of *Vibrio harveyi* (Johnson and Shunk 1936) Baumann *et al.* 1981. *International Journal of Systematic Bacteriology* **48**, 749–758.
Phillips, G.J., Arnold, J. & Ivarie, R. (1987). Mono- through hexanucleotide composition of the *Escherichia coli* genome: a Markov chain analysis. *Nucleic Acids Research* **15**, 2611–2626.
Picardeau, M., Prod'hom, G., Raskine, L., LePennec, M.P. & Vincent, V. (1997). Genotypic characterization of five subspecies of *Mycobacterium kansasii*. *Journal of Clinical Microbiology* **35**, 25–32.
Rademaker, J.LW., Hoste, B., Louws, F.J., Kersters, K., Swings, J., Vauterin, L., Vauterin, P., & de Bruijn, F.J. (2000). Comparison of AFLP and rep-PCR genomic fingerprinting with DNA-DNA homology studies: *Xanthomonas* as a model system. *International Journal of Systematic and Evolutionary Microbiology* **50**, 665–677.
Restrepo, S., Duque, M., Tohme, J. & Verdier, V. (1999). AFLP fingerprinting: an efficient technique for detecting variation of *Xanthomonas axonopodis* pv. *Manihotis*. *Microbiology* **145**, 107–114.
Robinson, K.A., Robb, F.T. & Schreier, H.J. (1994). Isolation of maltose-regulated genes from the hyperthermophilic archaeum, *Pyrococcus furiosus*, by subtractive hybridization. *Gene* **148**, 137–141.
Rocha, E.P., Viari, A. & Danchin, A. (1998). Oligonucleotide bias in *Bacillus subtilis*: general trends and taxonomic comparisons. *Nucleic Acids Research* **26**, 2971–2980.
Roman, B.L., Pham, V.N., Bennett, P.E. & Weinstein, B.M. (1999). Non-radioisotopic AFLP method using PCR primers fluorescently labeled with Cy5™. *BioTechniques* **26**, 236–238.
Sarkar, N. (1997) Polyadenylation of mRNA in prokaryotes. *Annual Review of Biochemistry* **66**, 173–197.
Sloos, J.H., Janssen, P., van Boven, C.P.A. & Dijkshoorn, L. (1998). AFLP™ typing of *Staphylococcus epidermidis* in multiple sequential blood cultures. *Research in Microbiology* **149**, 221–228.
Speijer, H., Savelkoul, P., Bonten, M., Stobberingh, E. & Tjhie, J. (1999). Application of different genotyping methods for *Pseudomonas aeruginosa* in a setting of endemicity in an intensive care unit. *Journal of Clinical Microbiology* **37**, 3654–3661.
Tjernberg, I. & Ursing, J. (1989). Clinical strains of *Acinetobacter* classified by DNA-DNA hybridization. *Acta Pathologica Microbiologica Scandinavica* **97**, 595–605.
Towner, K.J. & Cockayne, A. (1993). *Molecular methods for microbial identification and typing.* Chapman & Hall, London.
Valsangiacomo, C., Baggi, F., Gaia, V., Balmelli, T., Peduzzi, R. & Pifferetti, J.-C. (1995). Use of amplified fragment length polymorphism in molecular typing of *Legionella pneumophila* and application to epidemiological studies. *Journal of Clinical Microbiology* **33**, 1716–1719.
Van Belkum, A. (1994). DNA fingerprinting of medically important microorganisms by use of PCR. *Clinical Microbiology Reviews* **7**, 174–184.
Van Eldere, J., Janssen, P., Hoefnagels-Schuermans, A., van Lierde, S. & Peetermans, W. (1999). Amplified-fragment length polymorphism analysis versus macro-restriction fragment analysis for molecular typing of *Streptococcus pneumoniae* isolates. *Journal of Clinical Microbiology* **37**, 2053–2057.
Vaneechoutte, M. (1996). DNA fingerprinting techniques for microorganisms. A proposal for

classification and nomenclature. *Molecular Biotechnology* **6**, 115–142.

Vos, P., Hogers, R., Bleeker, M., Reijans, M., van de Lee, T., Hornes, M., Frijters, A., Pot, J., Peleman, J., Kuiper, M. & Zabeau, M. (1995). AFLP: a new technique for DNA fingerprinting. *Nucleic Acids Research* **23**, 4407–4414.

Welsh, J. & McClelland, M. (1990). Fingerprinting genomes using PCR with arbitrary primers. *Nucleic Acids Research* **18**, 7213–7218.

Williams, J.G.K., Kubelik, A.R., Livak, K.J., Rafalski, J.A. & Collins, M.D. (1990). DNA polymorphisms amplified by arbitrary primers are useful as genetic markers. *Nucleic Acids Research* **18**, 6531–6535.

Zabeau, M. & Vos, P. (1993). *Selective restriction fragment amplification: a general method for DNA fingerprinting*. European Patent Office 0 534 858 A1.

9 Application and Analysis of ARDRA Patterns in Bacterial Identification, Taxonomy and Phylogeny

Mario Vaneechoutte[1] and Marc Heyndrickx[2]
[1]Department of Clinical Chemistry, Microbiology & Immunology, Ghent University Hospital, Belgium; [2]Department for Animal Product Quality, Center for Agricultural Research, Melle, Belgium

CONTENTS

9.1 INTRODUCTION

A. *General remarks on PCR-RFLP analysis*

(i) Principle

PCR-RFLP analysis consists of polymerase chain reaction (PCR)-based amplification of a stretch of DNA – usually a gene or a part of a gene – combined with subsequent restriction digestion of the PCR product and electrophoretic analysis of restriction fragment length polymorphism (RFLP analysis). It can therefore be used to differentiate between species and strains of living organisms as a shortcut to sequence determination. The PCR step enables enrichment and purification of a certain part of the genome, followed by restriction digestion of the amplified DNA to reveal sequence polymorphism in a rapid, technologically simple and highly reproducible manner. PCR-RFLP analysis compares well with other techniques which combine amplification with single strand conformation polymorphism (SSCP) analysis (Widjojoatmodjo *et al.*, 1995) or cleavase fragment length polymorphism (CFLP) analysis (Brow *et al.*,1996; Lyamicheva *et al.*, 1996; Sreevatsan *et al.*, 1998), or with denaturing gradient (DGGE) (Muyzer *et al.*, 1993; Marsh *et al.*, 1998) or temperature gradient gel electrophoresis (TGGE) (Nübel *et al.*, 1996; Smit *et al.*, 1999) analysis, all of which also reveal sequence polymorphism among strains without the need for full sequence determination (Vaneechoutte, 1996).

(ii) Monitoring the discriminatory power of PCR-RFLP analysis

The discriminatory power of PCR-RFLP analysis can be monitored by the choice of more or less polymorphic (i.e., variable) regions to be amplified, by amplifying stretches of different length, and by performing restriction with enzymes which digest more or less frequently. This flexibility enables differences to be studied between strains or between species.

(iii) PCR-RFLP analysis for the differentiation of species

This chapter focuses on the application of PCR-RFLP analysis to differentiate bacterial (and eukaryotic) species. The genes which can be used for this purpose have to be well-conserved in order to minimise intra-specific variability which would obscure the species differentiation possibilities. Protein-encoding genes have been used successfully to differentiate bacterial species: e.g., *recA* for *Acinetobacter* spp. (Nowak & Kur 1996; Jawad *et al.*, 1998), *hsp65* for *Mycobacterium* and *Nocardia* spp. (Telenti *et al.*, 1993; Steingrübe *et al.*, 1995a, b) and the histidine operon for *Azospirillum* spp. (Grifoni *et al.*, 1995). Intra-specific polymorphism in protein-encoding genes is generally high, and amplification of short regions is often necessary to avoid too much intra-specific variability in the RFLP patterns. For example, Plikaytis *et al.* (1992) found 10 different restriction types for 31 *Mycobacterium gordonae* strains when amplifying a 1380-bp fragment of the *hsp65* gene, while Telenti *et al.* (1993) found only five different types among 24 strains when amplifying only 439 bp.

Most applications for bacterial species differentiation have been based on the rRNA cistron, which contains the 16S (also named small subunit rRNA or ssu rRNA), 23S and 5S rRNA genes, ordered in the same manner in most bacteria and with the genes separated by spacer regions of variable length. For eukaryotes, mitochondrial DNA (Boudry *et al.*, 1998; Orui, 1998) or rRNA genes (Vilgalys & Hester, 1990; Clark, 1997) have been used.

Several regions of RNA genes are highly conserved, such that primers complementary to these regions will enable amplification of these genes for most or all bacteria or eukaryotes (Greisen *et al.,* 1994; Clark & Diamond, 1997; Liu *et al.*, 1997; Smit *et al.,* 1997). Besides highly conserved regions, more variable regions are present within the rRNA operon, which allow for differentiation between most species. The intra-specific variability of these regions is limited, leading to relatively homogeneous restriction fingerprints for organisms of the same species. For example, only one 16S rRNA gene restriction fingerprint is observed with *M. gordonae* (Vaneechoutte *et al.,* 1993; personal unpublished data). Different regions of the rRNA cistron have been used for PCR-RFLP analysis, including the 16S rRNA gene promotor region (Dobner *et al.*, 1996), the 16S rRNA gene (Gürtler *et al.*, 1991; Jayarao *et al.*, 1992; Ralph *et al.*, 1993; Vaneechoutte *et al.*, 1993; Carlotti & Funke 1994), the 16S-23S spacer region (Harasawa *et al.*; 1993; Dolzani *et al.*, 1995; Nowak *et al.*, 1995; Liveris *et al.*, 1999), the 16S + spacer + part of the 23S rRNA gene (Vaneechoutte *et al.*, 1992; Salzano *et al.*, 1994), the 23S-5S spacer region (Wittenbrink *et al.*, 1998), and the (almost) complete rRNA cistron (Smith-Vaughan *et al.*, 1995; Ibrahim *et al.*, 1996). A variation that has been described involves combined amplification of the 16S rRNA and *recA* genes, followed by simultaneous electrophoretic analysis of restriction digests of both products (Nowak & Kur, 1996).

This chapter focuses on ARDRA – i.e., amplified rDNA restriction analysis, a short name for PCR-RFLP analysis of the rRNA gene(s) – aimed at differentiating between bacterial and eukaryotic species.

9.4 ARDRA AS A SCREENING METHOD FOR THE STUDY OF MICROBIAL ECOLOGY, EPIDEMIOLOGY AND BIODIVERSITY

A. *Screening collections of cultured isolates*

Rapid genotypic techniques such as ARDRA enable large collections of cultured isolates to be screened for the presence of different species, and are therefore well-suited for the screening of bacterial (and eukaryotic) communities and for epidemiological studies. While large-scale screening is capable of providing an impression of the biological diversity present in a population, representatives of the different ARDRA types observed can then be further characterised, e.g., by full sequence determination.

The following few examples serve to illustrate use of the technique. Thus, using ARDRA, it was shown that the genotypic diversity among the bacterial genera *Xenorhabdus* and *Photorhabdus*, which are symbionts of entomopathogenic nematodes, reflected the genotypic diversity of their hosts (Fischer-Le Saux *et al.*, 1998). In another study combining phenotypic identification and ARDRA to examine the commensal occurrence of *Acinetobacter* spp. on human skin, it was demonstrated that *A. baumannii* and *Acinetobacter* sp. 13TU, the clinically most important nosocomial *Acinetobacter* species, occur only rarely on skin, which leaves their natural habitat as yet still unknown (Seifert *et al.,* 1997). Other examples include the use of ARDRA by Andrighetto *et al.* (1998) to study the occurrence of homofermentative thermophilic lactobacilli in dairy products, by Navarro *et al.* (1992) to study *Nitrobacter* populations, and by Becker *et al.* (1998) to study lignite carbonisation wastewater populations.

ARDRA of cultured bacteria from different communities can also be used to estimate bacterial diversity. A study by Fulthorpe *et al.* (1998) showed that most (91%) of the 3-chlorobenzoate mineralising soil organisms isolated from six regions on five continents were endemic (see also Staley, 1999). Barberio & Fani (1998) applied ARDRA to study that part of the microbial community of a sewage treatment plant which could be cultured on a selective medium containing two nonylphenol ethoxylates as the sole carbon source. In this way, the cultivable bacteria with bioremediating capacities could be readily isolated and characterised. Accordingly, the cultivable part of an oil-degrading bacterial community from the Venice lagoon was first studied by means of ARDRA (Di Cello *et al.*, 1997). Characterisation of large collections of cultured strains by means of ARDRA has been carried out for agricultural soils (Ovre & Torsvik, 1998).

B. *Screening population structure and diversity, starting from cloned amplified rRNA genes*

One possible way to study bacterial communities is by direct amplification of the 16S rRNA genes present in a sample, using universal primers, whereafter the amplified mixture of 16S rRNA genes is cloned and the cloned genes are differentiated by ARDRA. Using DNA reassociation analysis, Torsvik *et al.* (1998) demonstrated that bacterial communities in pristine soil and sediments may contain more than 10,000 different bacterial types. The diversity of the total soil community was at least 200-fold higher than the diversity estimated on the basis of cultivable bacterial isolates. This indicated that the culture conditions selected for only a distinct sub-population of all the bacterial species present in the environment. LaMontagne *et al.* (1998) showed that the more abundant species are not amplified any more efficiently than the less abundant species, and thus that direct amplification combined with cloning offers a reliable picture of the true composition of such a community. Other studies have also reported little amplification bias with mixed cultures (Smit *et al.*, 1997; Dojka *et al.*, 1998; Wood *et al.*, 1998). However, Nusslein & Tiedje (1998) found that preceding fractionation of DNA on the basis of G+C content enabled the detection of less dominant organisms which would have been overlooked by direct amplification of eubacterial rRNA genes.

Dojka *et al.* (1998) cloned 812 amplified ssu rRNA genes from hydrocarbon- and chlorinated solvent-contaminated aquifers under bioremediation. More than 50% of the clones had a unique ARDRA fingerprint. Sequencing of the amplified gene of each ARDRA type which occurred more than once revealed 94 bacterial sequence types, of which 10 were found to have no phylogenetic association with known taxonomic divisions. Similarly, Chandler *et al.* (1998) amplified and cloned bacterial and archaeal 16S rRNA genes from a low biomass paleosol community. Of 746 bacterial and 190 archaeal clones that were characterised by ARDRA, 242 bacterial and 16 archaeal clones were partially sequenced and compared against the ssu rRNA gene database (RDP) and GenBank. Six novel eubacterial sequences, clustering with or near the Chloroflexaceae, and 16 unique archaeal ARDRA groups were recognised.

Characterisation by ARDRA of more than 300 clones from the amplified rRNA gene mixture of a hot spring at Yellowstone National Park, followed by sequence determination of 122 clones with representative ARDRA patterns, indicated that 30% of the sequence types were unaffiliated with 14 previously recognised bacterial divisions, and that they comprised 12 novel candidate divisions (Hugenholtz *et al.*, 1998). Accordingly, the epibiotic bacterial flora of the hydrothermal vent polychaete *Alvinella pompejana* has been characterised in this manner (Haddad *et al.*, 1995). Other studies include the characterisation of bacterial communities in marine sediments (Gray & Herwig, 1996; Rath *et al.*, 1998), in tundra soil (Zhou *et al.*, 1997), in PCB-dechlorinating anaerobic enrichments (LaMontagne *et al.*, 1988), and the microorganisms associated with the seagrass *Halophila stipulacea* (Weidner *et al.*, 1996).

Instead of starting from cloned amplified rRNA genes, microbial diversity can also be assessed by means of sequence-dependent electrophoretic separation of the mixture of amplified rRNA genes (Muyzer *et al.*, 1993; Nübel *et al.*, 1996; Smit *et al.*, 1999), followed by reamplification of excised bands with the same primers and sequence analysis, or by restriction digestion of the amplified products (see Chapter 11).

C. *Studying profiles of whole communities*

Instead of performing ARDRA on cultured isolates or on individually cloned rRNA genes, it is possible to create profiles of total communities in a single step by performing restriction digestion on the mixture of amplified rRNA genes. This approach enables a quick estimate of bacterial diversity in a community and investigations of the population dynamics (Liu *et al.*, 1997; Princic *et al.*, 1998). Thus, ARDRA (with rRNA gene primers specific for ammonium-oxidising bacteria) has been used to study the influence of varying ammonium levels on the composition of a bacterial community (Princic *et al.*, 1998). Reversal towards the original composition was observed when the initial ammonium levels were restored. Marsh *et al.* (1998) found that ARDRA of fluorescently-labelled end restriction fragments was more sensitive then DGGE for studying changes in the eukaryotic community of activated sludge. Smit *et al.* (1997) studied shifts in microbial community structure and diversity caused by copper contamination of soil, while ARDRA of reactor community DNA has demonstrated how fixed-film reactor communities with different starting compositions can converge to a community of the same composition which is then stable for several months (Massol-Deya *et al.*, 1997). Alternative community fingerprinting methods make use of sequence dependent electrophoresis techniques (Muyzer *et al.*, 1993; Nübel *et al.*, 1996; Smit *et al.*, 1999).

9.5 ARDRA AS A TOOL IN BACTERIAL PHYLOGENY AND TAXONOMY

A. *Overview*

Bacterial taxonomy consists of classification, nomenclature and identification of microorganisms. It is now generally accepted that the phylogenetic relationships between microorganisms should be used as a framework for modern bacterial taxonomy . The most generally applied method for determining phylogenetic relationships between microorganisms is based on comparative analysis of the 16S rRNA gene sequences (Neefs *et al.*, 1990). PCR-based amplification of the rRNA gene, combined with cycle sequencing, has led to an explosion of available sequences deposited in several publically accessible international libraries. More than 41,000 entries were listed in July 2000 when searching for "16S" in the GenBank nucleotide sequence database (*http://www.ncbi.nlm.nih.gov/htbin-post/Entrez/Query?db=n*).

The 16S rRNA gene is normally chosen for sequencing because it contains conserved as well as variable regions – as dictated by the structural and functional constraints of the rRNA molecule – thereby enabling study of phylogenetic relationships between all bacterial taxa (Woese, 1987). Instead of full sequence determination, cluster analysis of the combination of ARDRA patterns obtained with different restriction enzymes has been used successfully for phylogenetic analysis and/or taxonomic classification of microorganisms (Table 9.2). However, before

Table 9.2. Examples of the application of ARDRA for phylogenetic and/or taxonomic studies

Alcaligenes	Vandamme *et al.* (1996)
Bacillus sensu lato	Heyndrickx *et al.* (1995; 1996a, b; 1997; 1998)
Thermophilic soil bacilli	Mora *et al.* (1998)
Bordetella	Vandamme *et al.* (1996)
Borrelia	Wang *et al.* (1997a); Liveris *et al.* (1999)
Capnocytophaga	Wilson *et al.* (1995)
Dairy lactobacilli	Andrighetto *et al.* (1998)
Facultative hydrogenotrophs	Brim *et al.* (1999)
Phytoplasma	Gundersen *et al.* (1996); Davis *et al.* (1998); Lee *et al.* (1998)
Photorabdus	Fischer-Le Saux *et al.* (1998)
Prevotella	Haraldson & Holbrook (1998)
Pseudomonas	Laguerre *et al.* (1994a); Keel *et al.* (1996); Manceau & Horvais (1997)
Rhizobia from root nodules	Rome *et al.* (1996); Khbaya *et al.* (1998)
Uncultured microorganisms from a seagrass	Weidner *et al.* (1996)
Xenorhabdus	Fischer-Le Saux *et al.* (1998)
Xanthomonas	Nesme *et al.* (1995)

focusing on the application of ARDRA for phylogeny and taxonomy, several remarks should be made with regard to the general applicability of rRNA genes and of restriction digestion for these purposes.

B. *Theoretical considerations*

(i) *Limitations of rRNA genes as tools for studying phylogenetic relationships and bacterial taxonomy*

(a) Discriminatory power. It is generally accepted that organisms sharing >97% rRNA gene similarity may belong to a single species, but that the resolution of 16S rRNA gene sequence analysis between closely related species is usually low, so that DNA-DNA hybridisation or other techniques remain necessary for precise species delineation (Stackebrandt & Goebel, 1994).

Furthermore, it is well known that some species which are clearly different and valid, as shown by DNA-DNA hybridisation, may have identical or nearly identical (>99% similarity) 16S rRNA gene sequences (Fox *et al.*, 1992). For example, the species *Acinetobacter haemolyticus* and *A. johnsonii* are only related distantly according to DNA-DNA homology, but have almost identical ARDRA patterns (Vaneechoutte *et al.*, 1995a).

It should also be mentioned that there are some specific cases in which microorganisms have a separate species status for other than mere taxonomic reasons, e.g., because of different levels of clinical importance or pathogenicity. Thus, *Bor-*

detella pertussis, *B. parapertussis* and *B. bronchiseptica* share >80% DNA-DNA homology and 99.7% 16S rRNA gene similarity, and would normally be regarded as a single species on mere taxonomic grounds (Vandamme *et al.*, 1997).

(b) Mosaic rRNA genes. The existence of mosaic rRNA genes (Sneath, 1993; Gürtler, 1999) indicates the occurrence of horizontal transfer and recombination of (parts of) rRNA cistrons between strains and species, possibly limiting the usefulness of the rRNA genes for phylogenetic purposes in some bacterial groups. Horizontal exchange might be an explanation for the observation with some genera that there is an apparent discongruence between 16S rRNA gene similarity and total DNA homology values. For example, the four described genomic species of the *Acinetobacter calcoaceticus – A. baumannii* complex have high DNA homology and are also biochemically closely related. However, restriction digestion of the 16S rRNA gene readily reveals several differences between these species. These findings for *Acinetobacter* are largely confirmed by 16S rRNA gene sequencing (Ibrahim *et al.*, 1997).

(c) Introns or intervening sequences. Introns or intervening sequences (IVSs) have been identified in some species. A 235-bp IVS was found at the same location in the ssu rRNA gene of different strains of *Helicobacter canis* (Linton *et al.*, 1994), and multiple introns of different length were detected at different positions in the ssu rRNA gene of *Thermoproteus* strains (Itoh *et al.*, 1998). The archaeon *Pyrobaculum aerophilum* contains a 713-bp intron, while the closely related species *P. islandicum* contains no intron (Burggraf *et al.*, 1993). In most of the multiple 16S rRNA genes of *Clostridium paradoxum,* heterogeneous IVSs were found (Rainey *et al.*, 1996). It is clear that the occurrence of introns will influence estimates of phylogenetic relatedness.

(d) Intra-specific rRNA gene variability. A high level of intra-specific variability of the 16S rRNA gene has been observed in some species (Clayton *et al.*, 1995; Graf, 1999). As mentioned in section 9.1.A.iii, this variability is usually much more limited than for protein-encoding genes, and may sometimes indicate the existence of overlooked species or subspecies.

(e) Micro-heterogeneity. Another problem encountered when using rRNA gene sequences for phylogeny and taxonomy is the occurrence of inter-operon sequence variability within an individual genome (also called micro-heterogeneity). In most species, the genome of individual cells contains multiple RNA operons or alleles (Gürtler *et al.*, 1991; Cole & St. Girons, 1994; Gürtler & Stanisch, 1996; Schmidt, 1998). Apparently, the different alleles of the 16S rRNA gene on the bacterial genome can also show allelic sequence differences, at least in some species, with different degrees of micro-heterogeneity between strains within the species. Inter-operon variability has been well-documented for *Bacillus* (Stewart *et al.*, 1982), *Clostridium* (Gürtler *et al.*, 1991; Gürtler & Stanisch, 1996; Rainey *et al.*, 1996),

Paenibacillus polymyxa, with 10 variant nucleotide positions in the 16S rRNA genes (Nübel *et al.*, 1996), *Bacillus sporothemodurans,* with three different copies (Pettersson *et al.*, 1996), and in *Haloarcula marismortui* (Mylvaganam & Dennis, 1992) and *Thermobispora bispora* (Wang *et al.*, 1997b), each harbouring two 16S rRNA genes with sequence variation of >5%. Bascunana *et al.* (1994) showed micro-heterogeneity for the two 16S rRNA operons of *Mycoplasma* sp. strain F38.

Sequence differences between alleles of multi-copy genes (like the rRNA cistron in most bacteria) will result in sequencing ambiguities at these positions when direct sequencing (sequencing starting from amplified rRNA genes without prior cloning) is performed, since the amplification mixture will contain alleles with different sequences. These ambiguities will interfere with the phylogenetic analysis.

Micro-heterogeneity can also cause interpretation problems for ARDRA, since it may give rise to additional bands with lower intensity that result from low copy number alleles having specific sequence differences which influence the number of restriction sites. This phenomenon is observable in the *Hae*III restriction patterns of the type strain of *Paenibacillus polymyxa* (Fig. 9.1a), which has two additional bands – 685 bp and 225 bp long, respectively – compared to other *P. polymyxa* strains, indicating that an initial 910-bp fragment has an additional *Hae*III restriction site in some 16S rRNA alleles of the type strain. Both the variable intensity and presence of these fragments in the *Hae*III restriction patterns of different strains (Fig. 9.1a) indicate that the number of alleles carrying this additional restriction site is strain-dependent, which corresponds well with the results obtained by sequence-dependent separation with TGGE of the 16S rRNA gene amplicon (Nübel *et al.*, 1996). In a similarity calculation based on coding the presence or absence of bands, these additional bands will be assigned equal value as the other major bands, although they only represent a fraction of the 16S rRNA alleles carrying the restriction site difference. A solution could be to ignore the (very) weak additional bands, but then the problem arises of how to define the cut-off level of band intensity for scoring. In spite of these problems, ARDRA is more informative than sequencing in cases of micro-heterogeneity because it actually shows the nature and degree of micro-heterogeneity within a strain or species.

(f) A high level of intra-specific rRNA variability. This may also be due to the existence of poorly delineated species which need taxonomic revision. A good example of this is *Bacillus circulans*, which actually consists of several DNA homology groups (Nakamura & Swezey, 1983), with ARDRA confirming that this 'species' consists of very diverse phylogenetic groups (De Vos *et al.*, 1997).

From the above remarks it follows that phylogenetic studies based on ribosomal genes have to be considered with caution, since normally one or only a few strains per species have been sequenced. Again, rapid techniques like ARDRA can allow a larger number of strains per species to be studied, thus providing a better estimate of possible intra-specific heterogeneity.

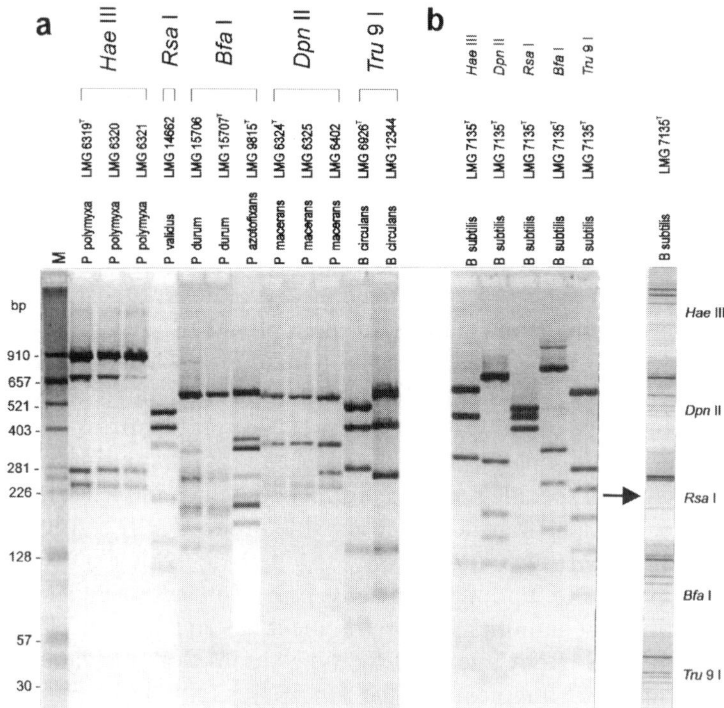

Fig. 9.1. (**a**) Normalised computer profiles of the ARDRA patterns of some representatives of the genera *Paenibacillus* (P) and *Bacillus* (B) obtained with different restriction enzymes. Lane M represents the molecular size marker; (**b**) Example of the combination by GelCompar of five ARDRA patterns obtained with the enzymes *Hae*III, *Dpn*II, *Rsa*I, *Bfa*I and *Tru*9I for the *B. subtilis* subsp. *subtilis* type strain ATCC 6051. The resulting combined pattern shown to the right of the arrow is used for the numerical analysis and clustering (reprinted from Heyndrickx *et al.*, 1996c).

(ii) Problems with the cluster analysis of restriction patterns

Counting genetic events as differences in the number of restriction fragments poses several problems. For phylogenetic purposes, restriction analysis of genes is an inferential technique compared to sequence determination, and some limitations must be taken into account. The most important pitfall with regard to restriction digestion is that the presence or absence of every band will contribute equally to the homology or similarity calculation between any two fingerprints. However, this is not always a correct assessment of the number of genetic transformations that has occurred between the two corresponding taxa, as is illustrated below.

A mutation, which introduces an additional restriction site in the 16S rRNA gene of taxon A represents only one genetic event or transformation, but it will cause a difference of three restriction fragments compared with the corresponding fingerprint obtained for taxon B in which this mutation did not occur. This follows from the fact that one fragment present in the fingerprint of taxon B disappears in the fingerprint of taxon A, to be replaced by two additional smaller fragments. In

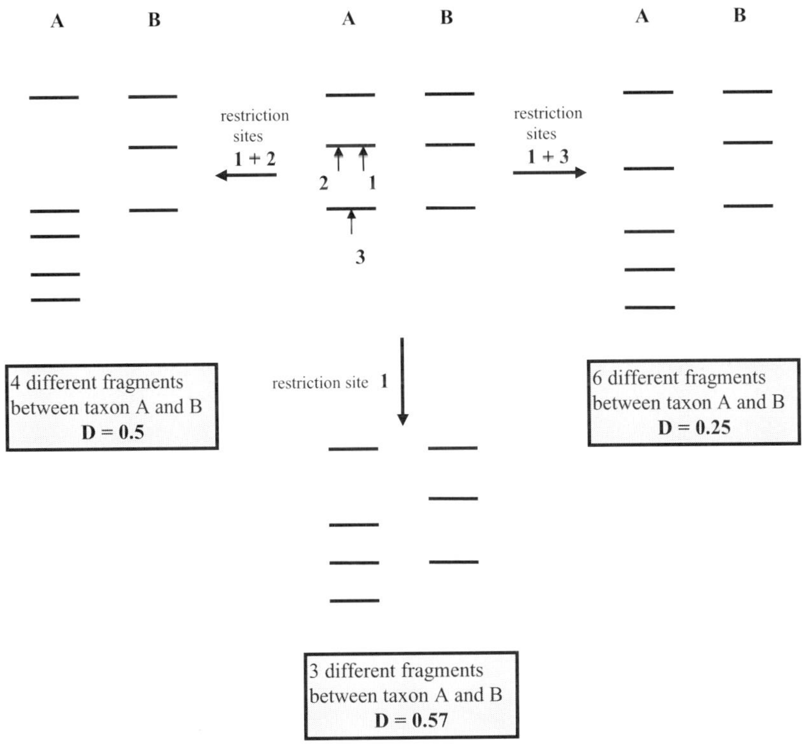

Fig. 9.2. Schematic representation of the influence of the number and relative position of restriction sites on the theoretical ARDRA patterns of taxa A and B, illustrating the influence of the relative physical distance between restriction sites introduced by independent genetic events on the number of different restriction fragments generated between the two taxa. The top middle part of the figure presents the original situation in which there is no difference between the two taxa. A first new restriction site in taxon A, indicated by arrow 1, results in a three fragment difference between taxa A and B (lower part of the figure). A second new restriction site in taxon A results in either a four fragment difference between taxa A and B (left part of the figure) when this happens close to the first restriction site, i.e., on the same fragment indicated by arrow 2, or a six fragment difference between the two taxa (right part of the figure) when this happens further away from the first restriction site, i.e., on another fragment indicated by arrow 3. In each case, the effect is shown on the Dice coefficient D ($D =. 2 \times n_{AB}/(n_A + n_B)$, in which n_{AB} is the number of fragments common to pattern A and B, n_A is the total number of fragments in pattern A, and n_B is the total number of fragments in pattern B.

the similarity calculation, this single event is counted three times using the coding strategy of presence or absence of bands. This would not present a problem if all independent genetic events were counted in a similar manner. However, this is not always the case (Fig. 9.2). If two independent genetic events or transformations in taxon B introduce two restriction sites which are distantly located from each other in the 16S rRNA gene of taxon A, this will result in a difference of six fragments between the two corresponding fingerprints, because a second larger fragment of taxon B is replaced by two additional smaller fragments in taxon A. In the similarity calculation, this double event is counted six times, which is still in accordance with the similarity calculation based on triple counting of a single event.

In contrast, if two independent genetic events introduce two closely located restriction sites in the 16S rRNA gene of taxon A, this will cause a difference of four fragments with the fingerprint of taxon B, because a single large fragment of taxon B is replaced by three smaller fragments in taxon A. In the similarity calculation, this double event is now counted only four times using the coding strategy of presence or absence of bands. Similarity calculation by means of the Dice coefficient for taxa A and B is thus influenced by the physical distance between the different restriction sites on the 16S rRNA gene, as shown in Fig. 9.2.

Besides this theoretical consideration, it is possible that the two smaller fragments produced by a genetic event leading to an additional restriction site are of (almost) equal size and thus are not well separated after agarose gel electrophoresis. As a consequence, a single event might be counted only twice. Of course, a combination of these different problems may also occur. Such problems can be solved if the actual restriction sites themselves are counted instead of the restriction fragments. This can be achieved by fluorescent end-labelling during PCR, combined with partial digestion and sequencing grade electrophoresis (Tötsch *et al.,* 1996), with the latter also resulting in better separation of fragments of almost equal length.

C. *Practical considerations*

(i) Selection of restriction enzymes

An important aspect in the experimental set-up of ARDRA is the choice of restriction enzymes. Since the advantage of ARDRA compared to direct sequencing is highly dependent on the speed of the analysis and on the taxonomic resolution and phylogenetic validity of the numerical analysis of the patterns, a minimal number of well-chosen enzymes should be used. Because of the limited length of the amplified region (1500 bp), the choice is restricted to frequently cutting enzymes such as tetrameric restriction enzymes (i.e., with a four-base recognition sequence) which will produce, on average, five to six restriction fragments. It is a matter of dispute whether restriction enzymes with a longer recognition sequence which contains only four defined bases and a certain number of undefined nucleotides (N) can be used. An enzyme such as *Dde*I (recognition sequence CTNAG) will give identical restriction patterns for the sequences CTAAG, CTTAG, CTCAG and CTGAG, which seems ambiguous. However, it may be argued that a position of N within the recognition sequence has no more taxonomic implications than a position just outside the recognition sequence.

The most discriminative enzymes can be selected empirically from a larger collection of enzymes using a known set of clearly different strains (Wilson *et al.*, 1995; Vaneechoutte *et al.*, 1998a), or theoretically by a computer simulation of restriction digests on the 16S rRNA gene sequences available for the species involved (Heyndrickx *et al.*, 1996b; Moyer *et al.*, 1996). Heyndrickx *et al.* (1996b) successfully selected five restriction enzymes on the basis of a computer simulation (see D.ii, below) of the restriction sites generated by all known tetrameric enzymes (available at the daily updated "Rebase" website at *http://rebase.neb.com/rebase/rebase.html*) for the complete 16S rRNA gene sequences of *Bacillus* sensu lato (available from the Sequence Retrieval System at *http://srs.ebi.ac.uk* or the Ribosomal Database Project at *http://www.cme.msu.edu/RDP/html/index.html*). Software packages which allow this simulation include HIBIO DNASIS software (Hitachi Software Engineering America, Brisbane, CA), GeneCompar (Applied Maths, Kortrijk, Belgium), and the Genetics Computer Group (GCG) sequence analysis software (University of Wisconsin, Madison, WI).

In accordance with the findings of Heyndrickx *et al.* (1996b), a computer simulation study, testing random combinations of tetra-cutter restriction enzymes (Moyer *et al.*, 1996), showed that combinations of three or more tetra-cutter restriction enzymes detected >99% of the 'operational taxonomic units' (OTUs) – defined in the latter study as known bacterial taxa spanning the entire Bacteria domain – within a model sequence data set, and that OTUs remaining undetected had a median sequence similarity of 96.1%. Of the 10 restriction enzymes tested, the enzyme combination *Hha*I-*Rsa*I-*Bst*UI was, overall, the most efficacious at differentiating bacterial 16S rRNA sequences and at predicting correct phylogenetic affiliations, while the enzyme combination *Mbo*I-*Hin*fI-*Taq*I revealed the greatest percentage of successful affiliations to the Gram-positive phylum and the β-subdivision of the Proteobacteria.

On the other hand, the more precise use of published sequences to search for restriction enzymes that can differentiate between closely related species has been found to be problematic in several cases, since computer-guided digestion of the published sequences often results in erroneous prediction of restriction sites. The explanation that this is caused by the presence of sequencing errors in the databanks could be confirmed in a study of *Listeria*, in which re-sequencing of the 16S rRNA genes showed that the findings with ARDRA were completely reliable, whereas previously published sequences contained errors (Vaneechoutte *et al.*, 1998a). *Bacillus lautus* showed ARDRA patterns which did not correspond with the expected pattern according to the deposited sequence (Heyndrickx *et al.*, 1996a). This was also observed for the obligate insect pathogens *B. popilliae* and *B. lentimorbus* (M. Heyndrickx and P. De Vos, unpublished results). Therefore, it is sometimes necessary to screen a large number of restriction enzymes in order to establish empirically which enzymes can differentiate between the species under study. Once this is established, a more limited set of restriction enzymes can be used in the future. For example, a total of 22 enzymes had to be screened initially to enable differentiation between the six closely related *Listeria* spp., but the eventual results indicated that only five specified enzymes were sufficient to discriminate the six species (Vaneechoutte *et al.*, 1998a).

(ii) Gel electrophoresis

To construct databases of normalised patterns, the use of an agarose and gel concentration with the appropriate resolving power is important. For example, 2% (w/v) Metaphor agarose (FMC Bioproducts, Rockland, ME) or MP agarose (Life Science International, Zellik, Belgium) can be used for the high resolution separation of restriction fragments with a length between 100 and 1000 bp. Secondly, a molecular size marker should be run at regular intervals on the gel for normalisation purposes in the computer analysis. It is important that this marker spans the whole molecular size range expected. Markers can be obtained from commercial sources or a self-made marker can be used, such as an *Alu*I digest of pBR322 supplemented with a pGEM-11Zf 128-bp fragment containing two additional *Alu*I restriction sites (Heyndrickx *et al.*, 1996c). A self-made marker will not change in the future, whereas this cannot be guaranteed for commercial preparations.

Pattern analysis software then enables normalisation of the patterns by using the external molecular size markers to compensate for electrophoretic variations within and between gels. Normalisation using a combination of external and internal markers is also possible and will give an even better normalisation result. DNA fragments added to all samples can be used as internal markers. Some automated electrophoresis equipment allows the use of four-colour fluorescent labelling technology, so that an internal standard, labelled differently to the sample, can be added to each lane. Software such as GeneScan Analysis (Perkin Elmer) then normalises the runs automatically with great accuracy.

(iii) Digitisation and computer analysis of the patterns

For numerical analysis of ARDRA patterns, correct scoring and combination of the restriction fragments obtained with the different restriction enzymes is a crucial factor. Although this step can be done manually and visually, it is preferable to use appropriate software for this purpose (e.g., GelCompar or BioNumerics; Applied Maths). For computer-assisted analysis of the ARDRA patterns, digitisation of the gel photographs as TIFF files is necessary. As described in more detail in Chapter 3, this can be achieved by taking Polaroid photographs of the gels using conventional cameras and scanning these photographs with flatbed scanners or laser densitometers. Alternatively, the gel image can be captured directly by a digital or a video camera using charge coupled device (CCD) photography. Laser densitometers and video cameras are more expensive than flatbed scanners, but video cameras allow direct image acquisition at a high camera resolution (usually 800×600 pixels). The gel images captured as 8-bit (or 16-bit) TIFF files can then be imported into the pattern analysis software.

GelCompar also makes it possible to combine the individual patterns obtained with each restriction enzyme in a certain fixed order for each strain, as shown in Fig. 9.1b for the *Bacillus subtilis* type strain, which was analyzed with the enzyme combination *Hae*III-*Dpn*II-*Rsa*I-*Bfa*I-*Tru*9I. The bands in these combined patterns can be scored automatically (using a band search filter), but automatic band assignations should be checked visually on the digitised patterns and, if possible, on the original gel photographs. This is especially necessary for very dense bands, which may sometimes consist of two restriction fragments, and for the low molecular size bands, which have a lower intensity and which are therefore easily missed by the automatic band search option. It is possible to increase the contrast and to decrease the brightness of the normalised patterns, which often helps to visualise and assign low molecular size bands on the screen. Care also has to be taken that spots on the gel or in the background staining are not interpreted by the software as genuine bands. This may introduce artefactual bands in the digitised gels, and this phenomenon has been suggested by Vaneechoutte *et al.* (1999b) as a possible explanation for the aberrant results reported by Koeleman *et al.* (1998). In general, it is advisable not to score bands smaller than 50 bp in order to avoid primer and primer-dimer band interference.

If the database is calibrated for the molecular size in base pairs, the sum of all scored bands can be calculated and should match the length of the amplified DNA fragment. Smaller or higher sums may be explained by (i) micro-heterogeneity, resulting in additional weaker bands (see section 9.5.B.i.); (ii) problems in pattern generation caused by star activity of the restriction enzymes, i.e., the capability of restriction enzymes to cleave sequences which are similar but not identical to their defined recognition sequence; or (iii) inadequacies in the band assignments (e.g., missed scoring of weak bands or scoring double bands once only).

(iv) Calculation of similarity coefficients and clustering

It is recommended to use the Dice coefficient for the estimation of genetic divergence between organisms on the basis of restriction fragment patterns (Nei & Li, 1979). When the Dice coefficient is used for calculating the similarity coefficients between each pair of combined patterns, the band position tolerance is an important parameter. This parameter defines the tolerance limits within which the software will consider bands in different lanes to be identical. For example, a band position tolerance of 0.2% on a total resolution of 2500 points in a combined digitised pattern (resulting from the combination of five digitised patterns with a resolution of 500 points each) means that bands from different lanes which deviate by not more than five points are regarded as identical. The value assigned to this parameter influences the similarity matrix calculation and consequently the clustering result, which is normally expressed as a dendrogram by means of the UPGMA clustering algorithm from the similarity matrix.

It is advisable to perform a final quality control by checking that visually similar patterns actually cluster in the dendrogram at a high similarity level (minimal 90% with the Dice coefficient). If this is not the case, this usually indicates faulty band assignments and/or an inappropriate setting of the band position tolerance.

Some important settings in GelCompar, which are relevant for numerical analysis of ARDRA patterns, are summarised in Table 9.3.

Table 9.3. Some suggested settings in GelCompar for numerical analysis of ARDRA patterns run on agarose gels

	Parameter[a]	Settings
Conversion module	Track resolution[b]	400–500
Normalisation module	Normalised gel resolution[b]	400–500
	Background subtraction	rolling disk, intensity 14
Analysis module	Comparison coefficient	Dice (band-based comparison)
	Position tolerance	0.2%, with optimisation 'on' (0.5%)

[a] Other parameters in GelCompar will not influence the clustering of the ARDRA patterns.
[b] Based on a gel running distance of 10 cm in a 2% w/v agarose gel, which is sufficient for ARDRA patterns.

D. *Application of ARDRA for phylogenetic and taxonomic research*

(i) *Use as a rapid taxonomic classification tool*

The most logical use of ARDRA is as a rapid taxonomic screening method to classify a large set of strains into OTUs. This screening serves to select one or a few representatives from each ARDRA OTU for polyphasic taxonomy, which might include SDS-PAGE of cellular proteins, fatty acid analysis, the determination of DNA-DNA homology, and sequence determination of the 16S rRNA gene. Related to this application is the confirmation of observed (phenotypic, ecological) similarities between different strains before starting polyphasic studies. For example, it was shown that the ARDRA patterns of oil-degrading *Acinetobacter* strains, isolated independently from three different marine environments (Reisfeld *et al.*, 1972; Yamamoto & Harayama, 1996; Di Cello *et al.*, 1997) were all identical and different from other *Acinetobacter* spp. Once this was established, the species identity of these strains could be confirmed by a polyphasic approach (Vaneechoutte *et al.*, 1999a). Similarly, the synonymy of the hyphomycete *Scytalidium hyalinum* and the coelomycete *Nattrassia mangiferae* was first established by ARDRA and later confirmed by chromatographic techniques. This finding explained why both 'species' are regularly found in the same patient (Roeijmans *et al.*, 1997).

(ii) *Application in the clarification of the phylogeny and taxonomy of the genus Bacillus sensu lato*

ARDRA has been used extensively for phylogenetic and taxonomic analysis of the genus *Bacillus* sensu lato. The synonymy of the species *Paenibacillus gordonae* and *P. validus*, and of *P. pulvifaciens* and *P. larvae* (the former species being the later subjective synonym in both cases), was first indicated by ARDRA and confirmed in a further polyphasic approach (Heyndrickx *et al.*, 1995; 1996a). Fig. 9.3 shows a UPGMA clustering, based on numerical analysis using the Dice coefficient, of a combination of five ARDRA patterns obtained with the enzyme combination *Hae*III-*Dpn*II-*Rsa*I-*Bfa*I-*Tru*9I. Several species of the genus *Bacillus* and the related genera *Amphibacillus*, *Aneurinibacillus*, *Brevibacillus*, *Halobacillus*,

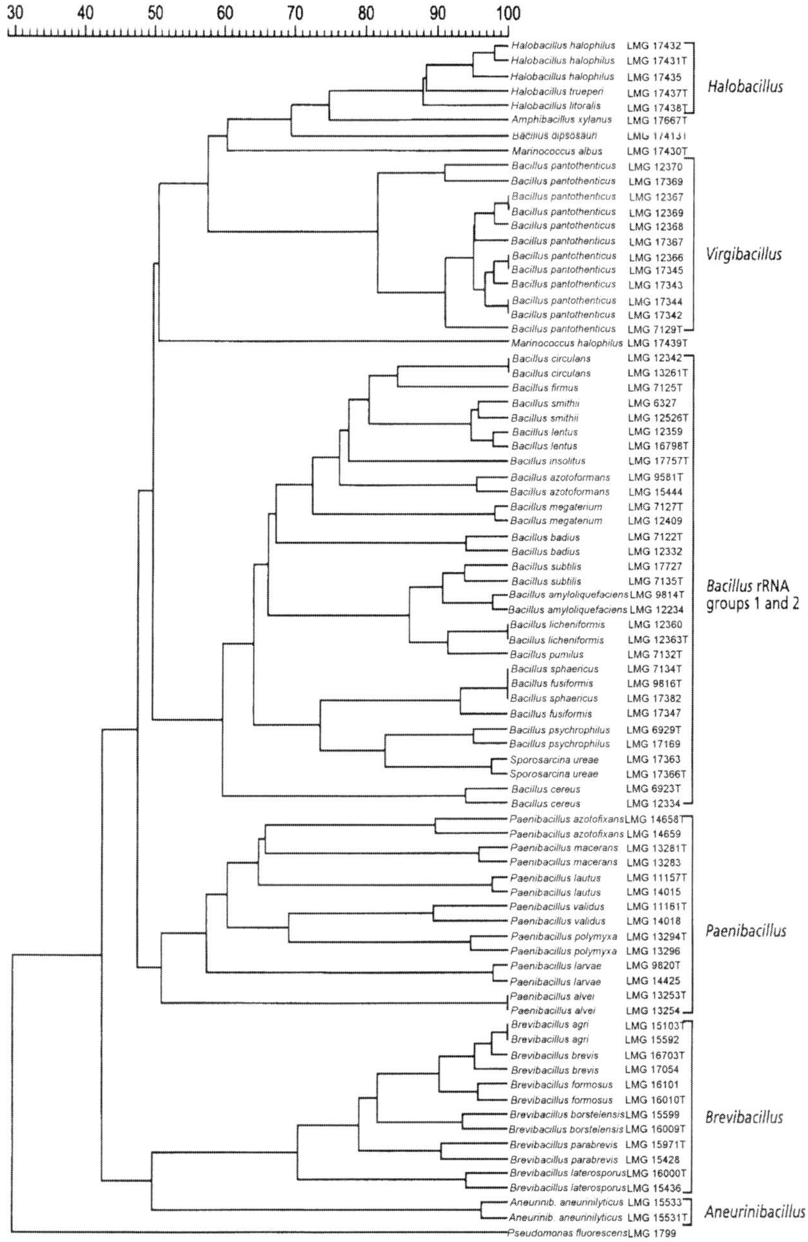

Fig. 9.3. Dendrogram based on the UPGMA clustering of the Dice similarity coefficients of normalised combined ARDRA patterns of several representatives of the allied genera *Aneurinibacillus*, *Bacillus* (rRNA groups 1 and 2), *Brevibacillus*, *Halobacillus*, *Paenibacillus*, *Marinococcus*, *Sporosarcina ureae* and *Virgibacillus*. *Pseudomonas fluorescens* was used as the outgroup. The restriction enzyme combination used was *Hae*III-*Dpn*II-*Rsa*I-*Bfa*I-*Tru*9I. T indicates the position of a type strain (reprinted from Heyndrickx *et al.*, 1998).

Sato, T., Sato, M., Matsuyama, J., Kalfas, S., Sundqvist, G. & Hoshino, E. (1998b). Restriction fragment-length polymorphism analysis of 16S rDNA from oral asaccharolytic *Eubacterium* species amplified by polymerase chain reaction. *Oral Microbiology and Immunology* **13**, 23–29.

Schmidt, T.M. (1998). Multiplicity of ribosomal RNA operons in prokaryotic genomes. In *Bacterial genomes: physical structure and analysis,* de Bruijn, F.J., Lupski, J.R. & Weinstock, G.M., eds, pp. 221–229. Chapman & Hall, New York.

Seifert, H., Dijkshoorn, L., Gerner-Smidt, P., Pelzer, N., Tjernberg, I. & Vaneechoutte, M. (1997). Distribution of *Acinetobacter* species on human skin: comparison of phenotypic and genotypic identification methods. *Journal of Clinical Microbiology* **35**, 2819–2825.

Selenska-Pobell, S., Otto, A. & Kutschke, S. (1998). Identification and discrimination of thiobacilli using ARDREA, RAPD and rep-APD. *Journal of Applied Microbiology* **84**, 1085–1091.

Sevin, E., Lamarque, D., Delchier, J.C., Soussy, C.J. & Tankovic, J. (1998). Co-detection of *Helicobacter pylori* and of its resistance to clarithromycin by PCR. *FEMS Microbiology Letters* **165**, 369–372.

Shida, O., Takagi, H., Kadowaki, K. & Komagata, K. (1996). Proposal for two new genera, *Brevibacillus* gen. nov and *Aneurinibacillus* gen. nov. *International Journal of Systematic Bacteriology* **46**, 939–946.

Shida, O., Takagi, H., Kadowaki, K., Nakamura, L.K. & Komagata, K. (1997). Transfer of *Bacillus alginolyticus, Bacillus chondroitinus, Bacillus curdlanolyticus, Bacillus glucanolyticus, Bacillus kobensis*, and *Bacillus thiaminolyticus* to the genus *Paenibacillus* and emended description of the genus *Paenibacillus*. *International Journal of Systematic Bacteriology* **47**, 289–298.

Sironi, M., Bandi, C., Novati, S. & Scaglia, M. (1997). A PCR-RFLP method for the detection and species identification of human microsporidia. *Parasitologia* **39**, 437–439.

Smit, E., Leeflang, P. & Wernars, K. (1997). Detection of shifts in microbial community structure and diversity in soil caused by copper contamination using amplified ribosomal DNA restriction analysis. *FEMS Microbiology Ecology* **23**, 249–261.

Smit, E., Leeflang, P., Glansdorf, B., van Elsas, J.D. & Wernars, K. (1999). Analysis of fungal diversity in the wheat rhizosphere by sequencing of cloned PCR-amplified genes encoding 18S rRNA and temperature gradient gel electrophoresis. *Applied and Environmental Microbiology* **65**, 2614–2621

Smith, J.K., Parry, J.D., Day, J.G. & Smith, R.J. (1998). A PCR technique based on the *Hip1* interspersed repetitive sequence distinguishes cyanobacterial species and strains. *Microbiology* **144**, 2791–2801.

Smith-Vaughan, H.C., Sriprakash, K.S., Mathews, J.D. & Kemp, D.J. (1995). Long PCR-ribotyping of nontypeable *Haemophilus influenzae*. *Journal of Clinical Microbiology* **33**, 1192–1195.

Sneath, P.H.A. (1993). Evidence from *Aeromonas* for genetic crossing-over in ribosomal sequences. *International Journal of Systematic Bacteriology* **43**, 626–629.

Sreevatsan, S., Bookout, J.B., Ringpis, F.M., Mogazeh, S.L., Kreiswirth, B.N., Pottathil, R.R. & Raj, R. (1998). Comparative evaluation of cleavase fragment length polymorphism with PCR-SSCP and PCR-RFLP to detect antimicrobial agent resistance in *Mycobacterium tuberculosis*. *Molecular Diagnosis* **3**, 81–91.

Stackebrandt, E. & Goebel, B.M. (1994). Taxonomic note: a place for DNA-DNA reassociation and 16S rRNA sequence analysis in the present species definition in bacteriology. *International Journal of Systematic Bacteriology* **44**, 846–849.

Staley, J.T. (1999). Bacterial biodiversity: a time for place. *ASM News* **10**, 681–687.

Steingrube, V.A., Brown, B.A., Gibson, J.L., Wilson, R.W., Brown, J., Blacklock, Z., Jost, K., Locke, S., Ulrich, R.F. & Wallace, R.J. (1995a) DNA amplification and restriction endonuclease analysis for differentiation of 12 species and taxa of *Nocardia*, including recognition of four new taxa within the *Nocardia asteroides* complex. *Journal of Clinical Microbiology* **33**, 3096–3101.

Steingrube, V.A., Gibson, J.L., Brown, B.A., Zhang, Y., Wilson, R.W., Rajagopalan, M. & Wallace, R.J. (1995b). PCR amplification and restriction endonuclease analysis of a 65-Kilodalton heat shock protein gene sequence for taxonomic separation of rapidly growing mycobacteria. *Jour-*

nal of Clinical Microbiology **33**, 149–153.

Stewart, G.C., Wilson, F.E. & Bott, K.F. (1982). Detailed physical mapping of the ribosomal RNA genes of *Bacillus subtilis*. *Gene* **19**, 153–162.

Stothard, J.R., Frame, I.A., Carrasco, H.J. & Miles, M.A. (1998). On the molecular taxonomy of *Trypanosoma cruzi* using riboprinting. *Parasitology* **117**, 243–247.

Telenti, A., Marchesi, F., Balz, M., Bally, F., Böttger, E.C. & Bodmer, T. (1993). Rapid identification of mycobacteria to the species level by polymerase chain reaction and restriction enzyme analysis. *Journal of Clinical Microbiology* **31**, 175–178.

Terefework, Z., Nick, G., Suomalainen, S., Paulin, L. & Lindstrom, K. (1998). Phylogeny of *Rhizobium galegae* with respect to other rhizobia and agrobacteria. *International Journal of Systematic Bacteriology* **48**, 349–356.

Torsvik, V., Daae, F.L., Sandaa, R.A. & Ovreas, L. (1998). Novel techniques for analysing microbial diversity in natural and perturbed environments. *Journal of Biotechnology* **64**, 53–62.

Tötsch, M., Brömmelkamp, E., Stücker, A., Fille, M., Gross, R., Wiesner, P., Werner Schmid, K., Böcker, W. & Dockhorn-Dworniczak, B. (1996). Identification of mycobacteria to the species level by automated restriction enzyme fragment length polymorphism analysis. *Virchows Archives* **298**, 1–5.

Urakawa, H., Kita-Tsukamoto, K. & Ohwada, K. (1997). 16S rDNA genotyping using PCR/RFLP (restriction fragment length polymorphism) analysis among the family *Vibrionaceae*. *FEMS Microbiology Letters* **152**, 125–132.

Urakawa, H., Kita-Tsukamoto, K. & Ohwada, K. (1998). A new approach to separate the genus *Photobacterium* from *Vibrio* with RFLP patterns by *Hha*I digestion of PCR-amplified 16S rDNA. *Current Microbiology* **36**, 171–174.

Vandamme, P., Heyndrickx, M., Vancanneyt, M., Hoste, B., De Vos, P., Falsen, E., Kersters, K. & Hinz, K.H. (1996). *Bordetella trematum* sp. nov., isolated from wounds and ear infections in humans, and reassessment of *Alcaligenes denitrificans* Rüger and Tan 1983. *International Journal of Systematic Bacteriology* **46**, 849–858.

Vandamme, P., Heyndrickx, M., De Roose, I., Lammens, C., De Vos, P. & Kersters, K. (1997). Characterization of *Bordetella* strains and related bacteria by amplified ribosomal DNA restriction analysis and randomly and repetitive element-primed PCR. *International Journal of Systematic Bacteriology* **47**, 802–807.

Vandamme, P., Goris, J., Coenye, T., Hoste, B., Janssens, D., Kersters, K., De Vos, P. & Falsen, E. (1999). Assignment of Centers of Disease Control group IVc-2 to the genus *Ralstonia* as *Ralstonia paucula* sp. nov. *International Journal of Systematic Bacteriology* **49**, 663–669.

Vaneechoutte, M. (1996). DNA fingerprinting techniques for microorganisms. A proposal for classification and nomenclature. *Molecular Biotechnology* **6**, 115–142.

Vaneechoutte, M., Rossau, R., De Vos, P., Gillis, M., Janssens, D., Paepe, N., De Rouck, A., Fiers, T., Claeys, G. & Kersters, K. (1992). Rapid identification of bacteria of the *Comamonadaceae* with amplified ribosomal DNA-restriction analysis (ARDRA). *FEMS Microbiology Letters* **93**, 227–234.

Vaneechoutte, M., de Beenhouwer, H., Claeys, G., Verschraegen, G., De Rouck, A., Paepe, N., Elaichouni, A. & Portaels, F. (1993). Identification of *Mycobacterium* species by using amplified rDNA-restriction analysis. *Journal of Clinical Microbiology* **31**, 2061–2065.

Vaneechoutte, M., Dijkshoorn, L., Tjernberg, I., Elaichouni, A., De Vos, P., Claeys, G. & Verschraegen, G. (1995a). Identification of *Acinetobacter* genomic species by amplified ribosomal DNA restriction analysis. *Journal of Clinical Microbiology* **33**, 11–15.

Vaneechoutte, M., Riegel, P., de Briel, D., Monteil, H., Verschraegen, G., De Rouck, A. & Claeys, G. (1995b). Evaluation of the applicability of amplified rDNA-restriction analysis to identification of species of the genus *Corynebacterium*. *Research in Microbiology* **146**, 633–641.

Vaneechoutte, M., Cartwright, C.P., Williams, E.C., Jäger, B., Tichy, H.-V., De Baere, T., De Rouck, A. & Verschraegen, G. (1996). Evaluation of 16S rRNA gene restriction analysis for the identification of cultured organisms of clinically important *Clostridium* species. *Anaerobe* **2**, 249–256.

Vaneechoutte, M., Boerlin, P., Tichy, H.-V., Bannerman, E., Jäger, B. & Bille, J. (1998a). Comparison of the value of DNA-fingerprinting techniques for the identification and taxonomical classification of *Listeria* species. *International Journal of Systematic Bacteriology* **48**, 127–139.

Vaneechoutte, M., De Bleser, D., Claeys, G., Verschraegen, G., De Baere, T., Hommez, J., Devriese, L.A. & Riegel, P. (1998b). Cardioverter-lead electrode infection due to *Corynebacterium amycolatum*. *Clinical Infectious Diseases* **21**, 1553–1554.

Vaneechoutte, M., Tjernberg, I., Baldi, F., Pepi, M., Fani, R. & Sullivan, E.R., van der Toorn, J. & Dijkshoorn, L. (1999a). The oil-degrading *Acinetobacter* strain RAG-1 and the strains described as '*Acinetobacter venetianus* sp. nov.' belong to the same genomic species. *Research in Microbiology* **150**, 69–73.

Vaneechoutte, M., Vauterin, L., van Harsselaar, B., Dijkshoorn, L. & De Vos, P. (1999b). Considerations in evaluation of the applicability of DNA fingerprinting techniques for species differentiation. *Journal of Clinical Microbiology* **37**, 3428–3429.

Vidigal, T.H.D.A., Spatz, L., Nunes, D.N., Simpson, A.J.G., Carvalho, O.S. & Neto, E.D. (1998). *Biomphalaria* spp: Identification of the intermediate snail hosts of *Schistosoma mansoni* by polymerase chain reaction amplification and restriction enzyme digestion of the ribosomal RNA gene intergenic spacer. *Journal of Experimental Parasitology* **89**, 180–187.

Vilgalys, R. & Hester, M. (1990). Rapid genetic identification and mapping of enzymatically amplified ribosomal DNA from several *Cryptococcus* species. *Journal of Bacteriology* **172**, 4238–4246.

Vodkin, M.H., Howe, D.K., Visvesvara, G.S. & McLaughlin, G.L. (1992). Identification of *Acanthamoeba* at the generic and specific levels using the polymerase chain reaction. *Journal of Protozoology* **39**, 378–385.

Wainø, M., Tindall, B.J., Schumann, P. & Ingvorsen, K. (1999). *Gracilibacillus* gen. nov., with description of *Gracilibacillus halotolerans* gen. nov., sp. nov.; transfer of *Bacillus dipsosauri* to *Gracilibacillus dipsosauri* comb. nov., and *Bacillus salexigens* to the genus *Salibacillus* gen. nov., as *Salibacillus salexigens* comb. nov. *International Journal of Systematic Bacteriology* **49**, 821–831.

Wang, G., van Dam, A.P., Le Fleche, A., Postic, D., Peter, O., Baranton, G., de Boer, R., Spanjaard, L. & Dankert, J. (1997a). Genetic and phenotypic analysis of *Borrelia valaisiana* sp. nov. (*Borrelia* genomic groups VS116 and M19). *International Journal of Systematic Bacteriology* **47**, 926–932.

Wang, Y., Zhang, Z. & Narendrakumar, R. (1997b). The actinomycete *Thermobispora bispora* contains two distinct types of transcriptionally active 16S rRNA genes. *Journal of Bacteriology* **179**, 3270–3276.

Weidner, S., Arnold, W. & Puhler, A. (1996). Diversity of uncultured microorganisms associated with the seagrass *Halophila stipulacea* estimated by restriction fragment length polymorphism analysis of PCR-amplified 16S rRNA genes. *Applied and Environmental Microbiology* **62**, 766–771.

Widjojoatmodjo, M.N., Fluit, A.C. & Verhoef, J. (1995). Molecular identification of bacteria by fluorescence-based PCR-single-strand conformation polymorphism analysis of the 16S rRNA gene. *Journal of Clinical Microbiology* **33**, 2601–2606.

Wilson, M.J., Wade, W.G. & Weightman, A.J. (1995). Restriction fragment length polymorphism analysis of PCR-amplified 16S ribosomal DNA of human *Capnocytophaga*. *Journal of Applied Bacteriology* **78**, 394–401.

Wittenbrink, M.M., Reuter, C., Baumeister, K., Schutze, H. & Krauss, H. (1998). Identification of group VS116 strains among *Borrelia burgdorferi* sensu lato grown from the hard tick, *Ixodes ricinus* (Linnaeus, 1758) by PCR-coupled restriction fragment length polymorphism analysis. *Zentralblatt fur Bakteriologie* **288**, 45–57.

Woese, C. (1987). Bacterial evolution. *Microbiological Reviews* **51**, 221–271.

Woo, T.H.S., Patel, B.K.C., Smythe, L.D., Symonds, M.L., Norris, M.A. & Dohnt, M.F. (1997). Comparison of two PCR methods for rapid identification of *Leptospira* genospecies *interro-*

gans. *FEMS Microbiology Letters* **155**, 169–177.

Wood, J., Scott, K.P., Avgustin, G., Newbold, C.J. & Flint, H.J. (1998). Estimation of the relative abundance of different *Bacteroides* and *Prevotella* ribotypes in gut samples by restriction enzyme profiling of PCR-amplified 16S rRNA gene sequences. *Applied and Environmental Microbiology* **64**, 3683–3689.

Yamamoto, S. & Harayama, S. (1996). Phylogenetic analysis of *Acinetobacter* strains based on the nucleotide sequence of *gyrB* genes and on the amino acid sequences of their products. *International Journal of Systematic Bacteriology* **46**, 506–511.

Yoon, J.-H., Lee, S.T., Kim, S.-B., Kim, W.Y., Goodfellow, M. & Park, Y.-H. (1997). Restriction fragment length polymorphism analysis of PCR-amplified 16S ribosomal DNA for rapid identification of *Saccharomonospora* strains. *Journal of Clinical Microbiology* **47**, 111–114.

Zhou, J., Davey, M.E., Figueras, J.B., Rivkina, E., Gilichinsky, D. & Tiedje, J.M. (1997). Phylogenetic diversity of a bacterial community determined from Siberian tundra soil DNA. *Microbiology* **143**, 3913–3919.

undegraded DNA are required, and the multi-step procedure is time-consuming with a significant potential for errors. In contrast, the PCR-based IS typing methods require less DNA and the quality of the DNA is less critical. Furthermore, the PCR procedure is usually simple and completed rapidly.

Several uses of PCR for IS typing have been described. The first uses one primer 'facing' outward toward one IS terminus. The second primer can be complementary to any sequence repeated in the genome of interest, including the IS itself (Otal *et al.*, 1997). In this way a PCR product is generated only when the IS is integrated close to one of the repeat sequences, which must also be in the 'correct' orientation.

An alternative method is to again use a single outward facing IS primer, but to provide the second priming site by cutting the DNA sample with a restriction endonuclease, and to then ligate oligonucleotide linkers to the digested fragments (Haas *et al.*, 1993; Palittapongarnpim *et al.*, 1993). In this 'linker-mediated' PCR, only DNA fragments carrying both types of priming site (i.e., the linker and the IS) are amplified efficiently, and a series of PCR products is produced in which each type of PCR amplicon is derived from a different IS copy.

The third approach, inverse PCR, also uses an outward facing primer within the IS. The second primer is also complementary to the IS and primes in the opposite direction. PCR amplification is only possible once the DNA has been cleaved, using an appropriate restriction endonuclease, and ligated at low concentration to form covalently closed DNA circles (Patel *et al.*, 1996; Otal *et al.*, 1997) in which the priming sites are in the correct orientation for PCR, separated by parts of the IS and some of the IS flanking sequence. The linker-mediated and inverse methods are discussed in more detail below.

C. Examples of IS typing methods

(i) IS200 typing of Salmonella enterica by Southern blotting

IS*200* was first described in *S. enterica* serovar Typhimurium strain LT2, and was originally reported to be limited to *Salmonella* (Lam & Roth, 1983). Now, IS*200*-like elements with divergent sequences have been found in *Escherichia coli* (Bisercic & Ochman, 1993), *Yersinia pestis, Y. pseudotuberculosis* and *Y. enterocolitica* (Odaert *et al.*, 1996; Simonet *et al.*, 1996). The degree of sequence divergence between the IS*200*-like elements found in the different genera suggests that they may have co-evolved *in situ* within the chromosomes of their hosts. The presence of IS*200* in different species cannot, therefore, be taken as evidence of lateral transfer of the element. It has been suggested that the lack of evidence for lateral transfer, and the infrequency of its association with plasmids, indicates that IS*200* has little affinity for extrachromosomal elements (Stanley *et al.*, 1993), and that it may therefore act as a marker to fingerprint the vertically inherited chromosome, independent of lateral transfer (Stanley & Saunders, 1996).

Analyses of IS*200* sites of insertion in *S. enterica* serovars have been performed with restriction endonucleases that either do not have sites within the element (e.g.,

*Pst*I, *Ban*I and *Pvu*II) or which cut it at a single site (e.g., *Eco*RI and *Eco*RV). An IS*200* detection probe of 692 bp can be generated conveniently by PCR (Baquar *et al.*, 1993). The probe hybridises to both fragments generated by *Eco*RI and by *Eco*RV so that the potential number of bands is doubled when these enzymes are used. This can result in a small increase in the number of different patterns obtained from a given strain collection, but also reduces the clarity of the results and somewhat complicates their analysis. The extra degree of discrimination seen after separate analysis of the sequences flanking both ends of each IS copy derives from the reduced possibility that an IS copy will be masked. When only one restriction fragment is generated for each IS copy, there is a possibility that band co-migration will result in masking of some elements, but if two fragments are analysed for each IS copy, this possibility is greatly reduced.

IS*200* typing has usually been applied in a hierarchical manner to groups of strains that have already been grouped by serovar or serovar/phagetype. In this context the technique has given useful discrimination in epidemiological studies (Pelkonen *et al.*,1994). Generally, the level of discrimination achieved by IS*200* typing is dependent upon the number of copies present in the chromosome. Serovars with five or more copies of the element are more likely to give useful epidemiological typing data (Stanley *et al.*, 1994). IS*200* typing patterns have generally been assessed and compared by eye since the number of different patterns is relatively limited.

Some attempts have been made to infer evolutionary relationships between *Salmonella* serovars from IS*200* typing data (reviewed by Stanley & Saunders, 1996), but with limited success. It has been possible to reconstruct plausible lineages for closely related clones. However, it is apparent that inferences about the relationships of strains belonging to different serovars or serogroups would require direct identification of the sites of insertion by sequencing, and would also need to take account of the possibility of *IS200* deletions as well as insertion events.

(ii) IS6110 typing of Mycobacterium tuberculosis by Southern blotting

IS*6110* is a 1355-bp element with classical terminal inverted repeats. Copies from three different strains belonging to the *M. tuberculosis* complex — one from *M. bovis* (IS*987*) and two from *M. tuberculosis* (IS*6110* and IS*986*) — have been sequenced and found to differ at only a few nucleotides (McAdam *et al.*, 1990; Thierry *et al.*, 1990; Hermans *et al.*, 1991). IS*6110* is present in the other species of the MTB-complex, which also includes *M. microti* and *M. africanum*. It is not present in more distantly related mycobacteria (Cave *et al.*, 1991). The sequence has two ORFs, the largest of which encodes a putative transposase with pronounced (51%) amino acid sequence similarity to the corresponding ORF of IS*3411*, one of the enterobacterial IS*3* family of elements (McAdam *et al.*, 1990).

The mechanism of transposition of IS*6110* was shown to be orthodox by experiments in *M. smegmatis* using artificial composite transposons comprising two copies of IS*986* flanking a kanamycin resistance cassette (Fomukong & Dale, 1993). Studies on the sites of insertion of IS*6110* have now demonstrated that the

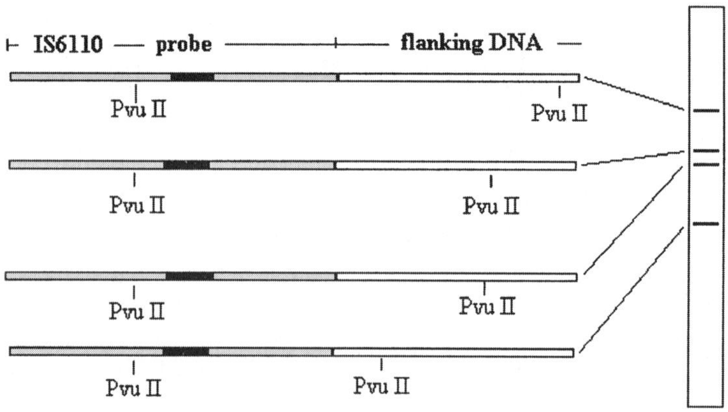

Fig. 10.2. IS*6110* typing using the method recommended by van Embden *et al.* (1993). Scheme **C** in Fig. 10.1 is used to produce a single band for each copy of the IS via digestion with *Pvu*II. The IS (medium shading) component of the fragments is constant, but the flanking sequences (unshaded) vary in length. The gel track diagram (right) shows the expected profile for a strain carrying four copies of the element in the positions shown relative to flanking restriction sites.

element integrates at preferred chromosomal loci (Fang & Forbes, 1997; Patel, 1999), although not at specific points in the nucleotide sequence. Thus, six alternative sites of integration are reported for IS*6110* at the *ipl* (IS preferential locus) sequence (Fang & Forbes, 1997). More work is required to describe the precise signals that allow IS*6110* integration at preferred loci.

To ensure that the results of IS*6110* typing in different laboratories are comparable, an international group has recommended a standardised technique (van Embden *et al.*, 1993). *Pvu*II, which has a single cleavage site in the element, is used to prepare fragments that are subjected to Southern blotting and hybridised to a 245-bp probe (Fig. 10.2). The probe hybridises only to fragments carrying the 3' end of IS*6110*. Thus, a single band on the IS*6110* profile is derived from each copy of the IS, and each band is expected to be of equal intensity since the sequence probed is identical in each case. An example is shown in Fig. 10.3.

This method has proved to be highly reproducible and the band profiles are stable when strains are passaged in guinea-pigs (Hermans *et al.*, 1990), subcultivated on laboratory media (van Soolingen *et al.*, 1991), or isolated from the same patient at intervals of up to 4.5 years (Cave *et al.*, 1994). However, minor differences in the band profiles of strains isolated from epidemiologically related cases are often encountered. This indicates that transpositions of IS*6110* occur frequently and accounts for the high diversity of the typing patterns observed.

The IS*6110* copy number of *M. tuberculosis* strains follows a bimodal distribution, with peaks of either one or 10–12 elements. IS*6110* typing of the majority of strains with five or more copies of the element is highly discriminatory, and consequently the method has been used in population-based studies. All available strains isolated within a population are analysed and the patterns compared (e.g., Chevrel-Dellagi *et al.*, 1993; Alland *et al.*, 1994; Small *et al.*, 1994). Strains with

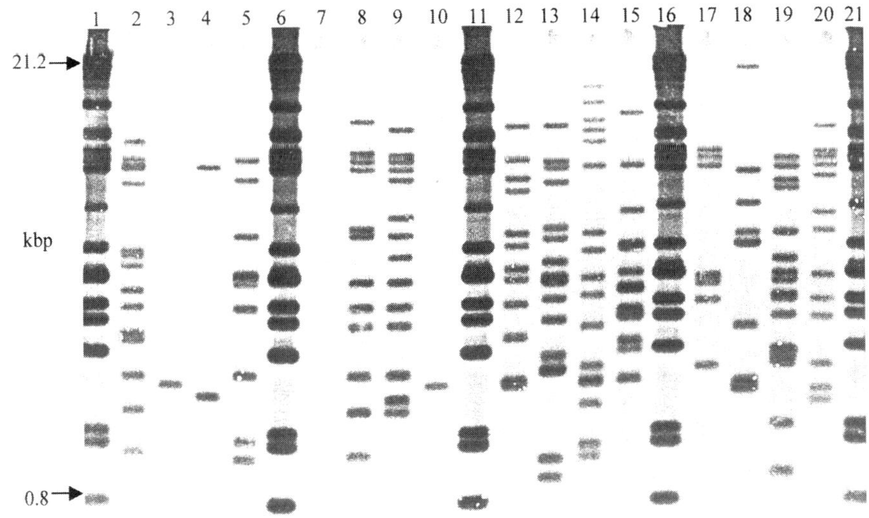

Fig. 10.3. A Southern blot of IS*6110* restriction fragments prepared using the method recommended by van Embden *et al.* (1993). The end tracks (1 and 21) and tracks 6, 11 and 16 are standards with bands from 21 kb (top) to 0.8 kb (bottom). The remaining 16 tracks show banding profiles of 16 distinct isolates. The profiles in tracks 9 and 20 which differ by a single band were from epidemiologically related cases.

indistinguishable patterns of five or more bands are generally assumed to be epidemiologically linked.

As patterns differing by a single band are often found in groups of strains with proven epidemiological associations, these are also considered as being linked. Analysis of IS*6110* typing data on the basis of these assumptions has been used to estimate the rate of active transmission of tuberculosis within different populations (Alland *et al.*, 1994; Small *et al.*, 1994). The number of strains with unique patterns is used as a measure of the rate of reactivation of old infections, while clusters are assumed to have resulted from new infections. Recently, clusters identified by IS*6110* typing have also been used to assess the transmission of *M. tuberculosis* from patients smear-negative for acid-fast bacilli (Behr *et al.*, 1999). The discriminatory power of IS*6110* typing for strains carrying four or less copies of the element is not sufficient to allow linkage of strains in the absence of other data. However, for outbreak investigation, the method can give strong confirmatory evidence that strains are from a common source even when they only have one copy of the element.

(iii) IS6110 typing of M. tuberculosis by linker-mediated PCR

Two variants of the linker-mediated PCR method have been applied to IS*6110* typing of *M. tuberculosis* (Haas *et al.*, 1993; Palittapongarnpim *et al.*, 1993). The mixed-linker PCR method (Haas *et al.*, 1993) has been the more widely used. In this technique, the adapter strand of the linker oligonucleotide, which is also the strand

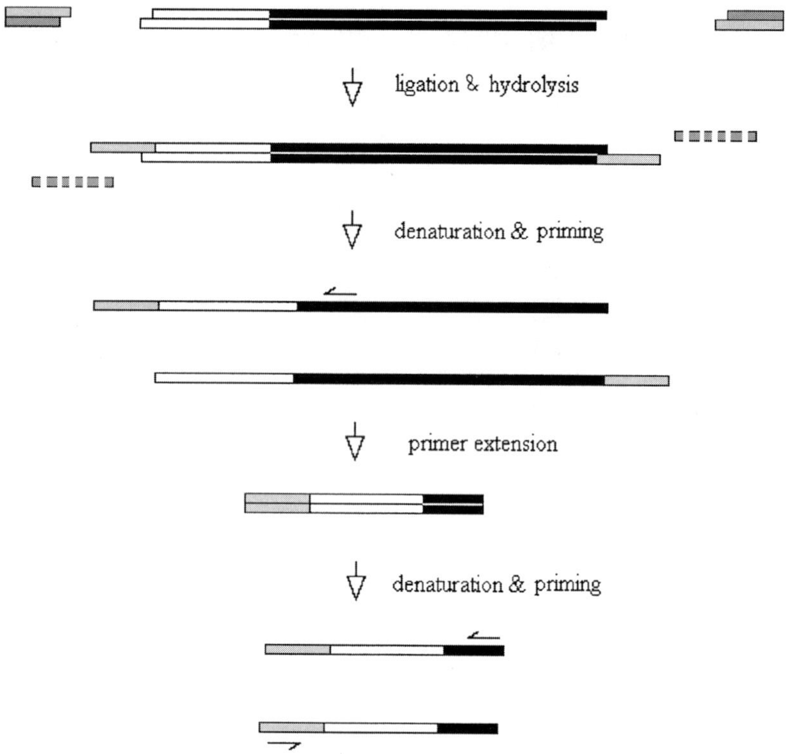

Fig. 10.4. For mixed-linker PCR, restriction fragments of the genomic DNA are ligated to the linker comprised of the uracil-modified oligonucleotide (dark shading) and the unmodified oligonucleotide (light shading). The unmodified oligonucleotide becomes covalently bound to the 5' ends of the restriction fragments, while the uracil-containing material is hydrolysed. In the first round of the subsequent PCR, the only priming sites available (indicated by arrows) are within the insertion sequence (shaded black). The product of this synthesis has both priming sites and is amplified in subsequent rounds of the PCR.

that does not become covalently bound to the target restriction fragments, has its thymidine bases replaced by uracil. Following ligation, during which the adapter is essential, the now redundant oligonucleotide is destroyed by treatment with uracil N-glycosylase (Fig. 10.4). This modification of the method was reported to give more reproducible results than when the adapter was left intact during the subsequent PCR. Even when mixed-linkers are used, it seems that some non-specific amplicons are produced (Butler *et al.*, 1996). Further reproducibility studies are required to establish fully the degree of accuracy and reproducibility that can be expected from these methods. A development of IS*6110* mixed-linker PCR, that improves the reproducibility and comparability of the method, involves analysis of the fragments on a DNA sequencer using fluorescence detection (Butler *et al.*, 1996). For this purpose, fluors are first introduced into the PCR products via end-labelled primers.

Mixed-linker PCR requires a minimal quantity of DNA, and can therefore be

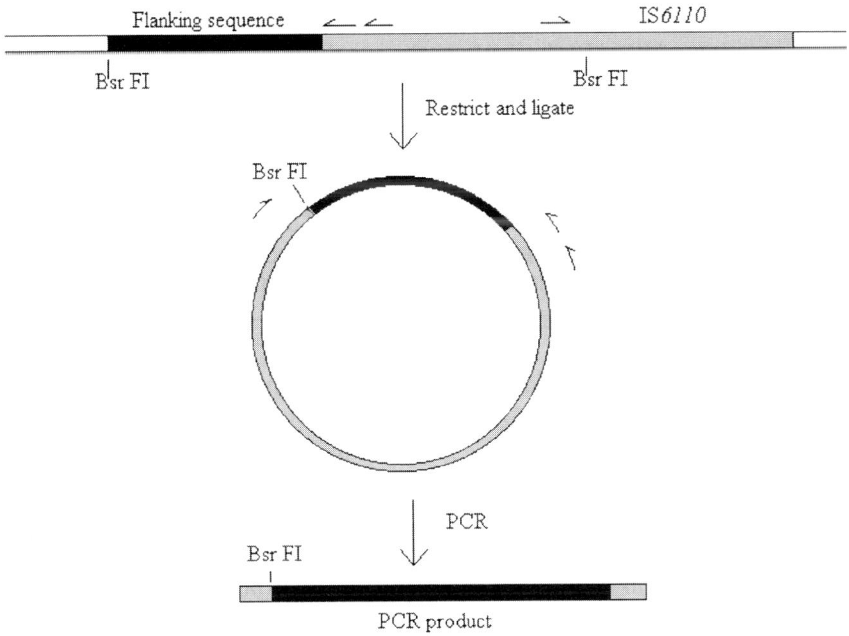

Fig. 10.5. Inverse PCR for IS*6110* is illustrated. The *M. tuberculosis* genome is cut with *Bsr*FI and then self-ligated at a low concentration of ligatable ends. Circular DNA molecules, derived from the 5'-end of *IS6110* (light shading) and its flanking sequence (dark shading), can act as a template for the PCR primers based on IS*6110* in the orientation shown. The linear PCR products, one for each copy of the IS, vary in length and can be analysed by agarose gel electrophoresis.

applied to early cultures of *M. tuberculosis*. In addition, since the length of the target sequences to be amplified is in the range 200–1500 bp, rapid methods of sample preparation resulting in significant shearing of the DNA can be employed. The method has also been used for IS*6110* fingerprinting of heat-killed cells stored on filter paper for long periods (Burger *et al*., 1998).

(iv) IS6110 typing of M. tuberculosis by inverse PCR

Inverse PCR has been applied to IS*6110* typing in two studies (Patel *et al*., 1996; Otal *et al*., 1997). The method is illustrated in Fig. 10.5. When applied to one terminus of the IS with self-ligation and amplification of short restriction fragments of up to *c*.1.5 kb (Patel *et al*., 1996), inverse PCR gave good results. Sequencing of several amplicons was used to show that they were each derived from the 5' end of IS*6110* and sequences flanking the element. For strains shown to be carrying one or two copies of the IS by Southern blotting, the same number of amplicons was produced in the inverse PCR. The discriminatory power of the inverse PCR was shown to be similar to that of the standard IS*6110* typing method (van Embden *et al*., 1993).

The other study in which inverse PCR was applied to IS*6110* used primers at either terminus of the IS (Otal *et al*., 1997). The DNA was cleaved by a restric-

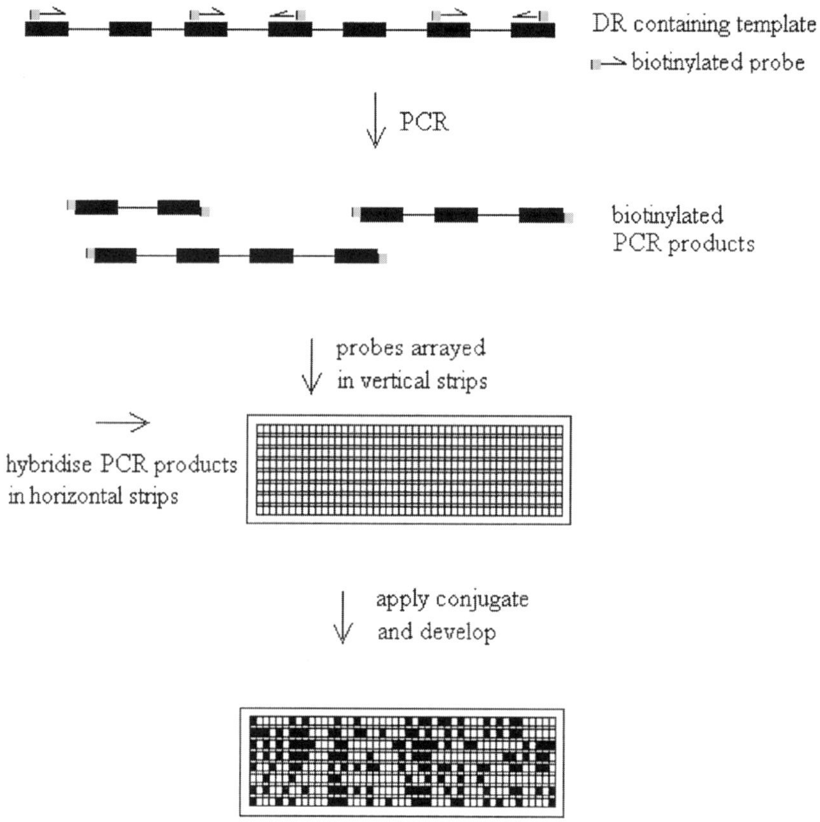

Fig. 10.6. Biotinylated PCR primers based on the direct repeat sequence (black boxes) are used to amplify the variable spacer sequences (single lines) within the DR locus. The amplicons generated carry all of the spacers for a specific strain. An array of vertical strips of different spacer sequence probes is prepared using a blotter with long slots. The PCR products for the test strains are hybridised to the filter, separately, in horizontal strips using the same apparatus but with the slots turned through 90°. The black squares represent areas of hybridisation between probe and amplicon, while light squares indicate that no hybridisation occurred due to the absence of the particular spacer sequence. A specific horizontal 'bar-code' of areas of hybridisation/non-hybridisation is generated for each strain.

C. Examples of oligotyping systems

(i) Spoligotyping for M. tuberculosis

The spoligotyping (*spacer oligotyping*) method for *M. tuberculosis* (Kamerbeek *et al.*, 1997) relies for its discriminatory ability upon variation at the direct repeat (DR) locus. This locus consists of a variable number of perfect tandem repeats of a 36-bp sequence interspersed with variable spacers of between 34 and 41 bp. There is some conservation of these sequences between strains, but comparison of pairs of strains usually reveals many differences between the inventory of spacer sequences carried.

For spoligotyping (Fig. 10.6), a series of 43 oligonucleotides, each corresponding to 25 bases of a different spacer sequence, found in either *M. tuberculosis* strain H37Rv or *M. bovis* strain P3, are covalently coupled to an activated Biodyne C membrane via a 5' aminolink. The target DNA is prepared and biotinylated by a PCR targeted at the DR region of the strains to be typed, and is then applied to the membrane. In order to allow analysis of multiple samples on a single membrane, the oligonucleotides and the target are applied to the membrane in parallel strips, at right angles, as illustrated in Fig. 10.6 (Kamerbeek *et al.*, 1997). Hybridisation of the biotinylated probes derived from the test strains is detected by chemiluminescence via a streptavidin-peroxidase conjugate. The pattern of negative and positive hybridisation signals depends upon the specific complement of spacer sequences present within the DR locus. The hybridised DNA can be stripped from the membrane following detection, and the recovered probe arrays can be re-used many times.

Spoligotyping is reproducible (Goguet de la Salmoniere *et al.*, 1997; Goyal *et al.*, 1997), convenient and rapid. It can be applied directly to extracts of clinical material without prior culture (Kamerbeek *et al.*, 1997).

(ii) Streptococcus pyogenes emm gene typing

Traditionally, typing of *S. pyogenes* has depended upon serological detection of cell wall antigens by a combined system of T- and M-antigen typing and, for opacity factor-positive cultures, the inhibition of the opacity reaction with specific antisera. M-antigen typing provides the greatest discrimination between strains, with at least 74 types recognised (Johnson & Kaplan, 1993). However, M-typing reagents are not available commercially, and maintenance of a comprehensive system is expensive and consequently restricted to a few reference centres.

The M proteins are encoded by the *emm* gene family, and specificity is conferred by the variable N-terminal domain. Many of these sequences are available in databases (GenBank, EMBL) and can be used to design type-specific oligonucleotides. PCR primers that correspond to conserved parts of the *emm* gene and amplify through the N-terminal-encoding sequence of strains of all M-types (Podbielski *et al.*, 1991) can be used to prepare and label large quantities of the type-specific sequence.

The *S. pyogenes emm* oligotyping methods rely on arrays of these M-type specific probes, bound either to filters, as described above for spoligotyping (Kaufhold *et al.*, 1994), or to 96-well microtitre trays (Saunders *et al.*, 1997). In the microtitre tray method, the biotinylated oligonucleotides are immobilised by interaction with streptavidin coating the wells. In either system, the amplified and labelled *emm* gene fragment derived from the test strain is hybridised to each of the probe sequences in the array. Hybridised probe is then detected either by chemiluminescence or colorimetrically, as appropriate. The *emm* type of test strains is revealed by positive hybridisation to one of the probes. If no positive reaction is obtained, the strain is considered non-typable by this system. Probes that hybridise to conserved sequences within the amplicon can be included in the array (Saunders *et al.*, 1997). These probes act as controls indicating that *emm* gene amplification has occurred in samples that do not hybridise to any of the type-specific probes.

Emm gene oligotyping has considerable advantages over either M-antigen serotyping, which is difficult to maintain due to the large number of sera needed, or *emm* gene sequencing (Beall *et* al., 1996), which requires more hands-on time. The microtitre tray-based arrays are easy to set up and convenient to use for assays based on up to 12 probes. However, with this number of probes, a significant proportion of strains will remain non-typable. For systems including larger numbers of probes, which are needed for a more comprehensive assay leaving few non-typable strains, the filter-bound arrays are the more promising.

D. Oligotyping: future perspectives and conclusions

The currently available oligotyping schemes have already shown the power and potential of this approach to bacterial typing. The data provided is of high quality and simple to interpret. As high density probe arrays consisting of many thousands of oligonucleotides become more widely available and accessible, it will be possible to design rapid systems for typing strains of any bacterial taxon. It is possible that arrays will be used to perform multilocus sequence typing (MLST; Maiden *et al.*, 1998). MLST relies on comparison of the sequence of parts of housekeeping genes and is therefore highly reproducible, with results being easy to compare between laboratories. The discriminatory ability of MLST depends only on the degree of divergence between the particular genes analysed in the strain collection examined. Greater discrimination can be achieved by selecting more variable genes or by adding genes to the panel. High density probe arrays are generally less accurate for sequencing than conventional methods (Kozal *et al.*, 1996). This is mostly due to the poor ability of probes to identify the sequence when two differences from the consensus occur at the same locus, especially at adjacent nucleotides. However, this limitation should not prove to be a significant drawback for MLST based on conserved bacterial housekeeping genes. The MLST approach is described in more detail in Chapter 12 of this book.

REFERENCES

Alland, D., Kalkut, G.E. & Moss, A.R. (1994). Transmission of tuberculosis in New York City. *New England Journal of Medicine* **330**, 1710–1716.

Baquar, N., Threllfall, E.J., Rowe, B. & Stanley, J. (1993). Molecular subtyping within a single *Salmonella typhimurium* phage type, DT204c, with a PCR-generated probe for IS*200*. *FEMS Microbiology Letters* **112**, 217–222.

Beall, B., Facklam, R. & Thompson, T. (1996). Sequencing *emm*-specific PCR products for routine and accurate typing of group A streptococci. *Journal of Clinical Microbiology* **34**, 953–958.

Behr, M.A., Warren, S.A., Salamon, H., Hopewell, P.C., Ponce de Leon, A., Daley, C.L. & Small, P.M. (1999). Transmission of *Mycobacterium tuberculosis* from patients smear-negative for acid-fast bacilli. *Lancet* **353**, 444–448.

Bisercic, M. & Ochman, H. (1993). Natural populations of *Escherichia coli* and *Salmonella typhimurium* harbor the same classes of insertion sequences. *Genetics* **133**, 449–454.

Borrow, R., Claus, H., Guiver, M., Smart, L., Jones, D.M., Kaczmarski, E.B., Frosch, M. & Fox, A.J. (1997). Non-culture diagnosis and serogroup determination of meningococcal B and C infection by a sialyltransferase (*sia*D) PCR ELISA. *Epidemiology and Infection* **118**, 111–117.

Burger, M., Raskin, S., Brockelt, S.R., Amthor, B., Geiss, H.K. & Haas, W.H. (1998). DNA fingerprinting of *Mycobacterium tuberculosis* complex culture isolates collected in Brazil and spotted onto filter paper. *Journal of Clinical Microbiology* **36**, 573–576.

Butler, W.R., Haas, W.H. & Crawford, J.T. (1996). Automated DNA fingerprinting analysis of *Mycobacterium tuberculosis* using fluorescent detection of PCR products. *Journal of Clinical Microbiology* **34**, 1801–1803.

Cave, M.D., Eisenach, K.D., McDermott, P.F., Bates, J.H. & Crawford, J.T. (1991). IS*6110*: conservation of sequence in the *Mycobacterium tuberculosis* complex and its utilization in DNA fingerprinting. *Molecular and Cellular Probes* **5**, 73–80.

Cave, M.D., Eisenach, K.D., Templeton, G., Salfinger, M., Mazurek, G., Bates, J.H. & Crawford, J.T. (1994). Stability of DNA fingerprint pattern produced with IS*6110* in strains of *Mycobacterium tuberculosis*. *Journal of Clinical Microbiology* **32**, 262–266.

Chevrel-Dellagi, D., Abderrahman, A., Haltiti, R., Koubaji, H., Gicquel, B. & Dellagi, K. (1993). Large-scale DNA fingerprinting of *Mycobacterium tuberculosis* strains as a tool for epidemiological studies of tuberculosis. *Journal of Clinical Microbiology* **31**, 2446–2450.

Datta, N. & Hughes, V.M. (1983). Plasmids of the same Inc groups in Enterobacteria before and after the medical use of antibiotics. *Nature* **306**, 616–617.

Egner, C. & Berg, D.E. (1981). Excision of tranposon Tn*5* is dependent on the inverted repeats but not on the transposase function of Tn*5*. *Proceedings of the National Academy of Sciences of the United States of America* **78**, 459–463.

Fang, Z. & Forbes, K.J. (1997). A *Mycobacterium tuberculosis* IS*6110* preferential locus (ipl) for insertion into the genome. *Journal of Clinical Microbiology* **35**, 479–481.

Fomukong, N.G. & Dale, J.W. (1993). Transpositional activity of IS*986* in *Mycobacterium smegmatis*. *Gene* **130**, 99–105.

Goguet de la Salmoniere, Y.O., Li, H.M., Torrea, G., Bunschoten, A., van Embden, J. & Gicquel, B. (1997). Evaluation of spoligotyping in a study of the transmission of *Mycobacterium tuberculosis*. *Journal of Clinical Microbiology* **35**, 2210–2214.

Goyal, M., Saunders, N.A., van Embden, J.D.A., Young, D.B. & Shaw, R.J. (1997). Differentiation of *Mycobacterium tuberculosis* isolates by spoligotyping and IS*6110* restriction fragment length polymorphism. *Journal of Clinical Microbiology* **35**, 647–651.

Haas, W.H., Butler, W.R., Woodley, C.L. & Crawford, J.T. (1993). Mixed-linker polymerase chain reaction: a new method for rapid fingerprinting of isolates of the *Mycobacterium tuberculosis* complex. *Journal of Clinical Microbiology* **31**, 1293–1298.

Hermans, P.W.M., van Soolingen, D., Dale, J.W., Schuitema, A.R.J., McAdam, R.A., Catty, D. & van Embden, J.D.A. (1990). Insertion element IS*986* from *Mycobacterium tuberculosis*: a useful tool for diagnosis and epidemiology of tuberculosis. *Journal of Clinical Microbiology* **28**, 2051–2058.

Hermans, P.W.M., van Soolingen, D., Bik, E.M., de Haas, P.E.W., Dale, J.W. & van Embden, J.D.A. (1991). Insertion element IS*987* from *Mycobacterium bovis* BCG is located in a hot-spot integration region for insertion elements in *Mycobacterium tuberculosis* complex strains. *Infection and Immunity* **59**, 2695–2705.

Hunter, P.R. & Gaston, M.A. (1988). Numerical index of the discriminatory ability of typing systems: an application of Simpson's index of diversity. *Journal of Clinical Microbiology* **26**, 2465–2466.

Johnson, D.R. & Kaplan, E.L. (1993). A review of the correlation of T-agglutination patterns and M-protein typing and opacity factor production in the identification of group A streptococci. *Jounal of Medical Microbiology* **38**, 311–315.

Kamerbeek, J., Schouls, L., Kolk, A., van Agterveld, M., van Soolingen, D., Kuijper, S., Bunschoten, A., Molhuizen, H., Shaw, R., Goyal, M. & van Embden, J. (1997). Simultaneous detection and strain differentiation of *Mycobacterium tuberculosis* for diagnosis and epidemiology. *Journal of Clinical Microbiology* **35**, 907–914.

Kaufhold, A., Podbielski, A., Baumgarten, G., Blakpoel, M., Top, J. & Schouls, L. (1994). Rapid typing of group A streptococci by the use of DNA amplification and non-radioactive allele-specific oligonucleotide probes. *FEMS Microbiology Letters* **119**, 19–26.

Kozal, M.J., Shah, N., Shen, N., Yang, R., Fucini, R., Merigan, T.C., Richman, D.D., Hubbell, E., Chee, M. & Gingeras, T.R. (1996). Extensive polymorphisms observed in HIV-1 clade B protease gene using high-density oligonucletide arrays. *Nature Medicine* **2**, 753–758.

Lam, S. & Roth, J.R. (1983). IS*200*: a *Salmonella*-specific insertion sequence. *Cell* **34**, 951–960.

Maiden, M.C., Bygraves, J.A., Feil, E., Morelli, G., Russell, J.E., Urwin, R., Zhang, Q., Zhou, J., Zurth, K., Caugant, D.A., Feavers, I.M. Achtman, M. & Spratt, B.G. (1998). Multilocus sequence typing: a portable approach to the identification of clones within populations of pathogenic microorganisms. *Proceedings of the National Academy of Sciences of the United States of America* **95**, 3140–3145.

McAdam, R.A., Hermans, P.W.M., van Soolingen, D., Zainuddin, Z.F., Catty, D., van Embden, J.D.A. & Dale, J.W. (1990). Characterization of a *Mycobacterium tuberculosis* insertion sequence belonging to the IS*3* family. *Molecular Microbiology* **4**, 1607–1613.

Odaert, M., Berche, P. & Simonet, M. (1996). Molecular typing of *Yersinia pseudotuberculosis* by using an IS*200*-like element. *Journal of Clinical Microbiology* **34**, 2231–2235.

Otal, I., Samper, S., Asensio, M.P., Vitoria, M.A., Rubio, M.C. Gómez-Lus, R. & Martin, C. (1997). Use of a PCR method based on IS*6110* polymorphism for typing *Mycobacterium tuberculosis* strains from BACTEC cultures. *Journal of Clinical Microbiology* **35**, 273–277.

Palittapongarnpim, P., Chomyc, S., Fanning, A. & Kunimoto, D. (1993). DNA fingerprinting of *Mycobacterium tuberculosis* isolates by ligation-mediated polymerase chain reaction. *Nucleic Acids Research* **21**, 761–762.

Patel, S. (1999). *Molecular typing and identification of Mycobacteria*. PhD thesis, University of London.

Patel, S., Wall, S. & Saunders, N.A. (1996). A hemi-nested inverse PCR for typing of sequences flanking the 5'-end of IS*6110* from *Mycobacterium tuberculosis* strains. *Journal of Clinical Microbiology* **34**, 1686–1690.

Pease, A.C., Solas, D., Sullivan, E.J., Cronin, M.T., Holmes, C.P. & Fodor, S.A. (1994). Light-generated oligonucleotide arrays for rapid DNA sequence analysis. *Proceedings of the National Academy of Sciences of the United States of America* **91**, 5022–5026.

Pelkonen, S., Romppanen, E-L., Siitoen, A. & Pelkonen, J. (1994). Differentiation of *Salmonella* serovar *infantis* isolates from human and animal sources by fingerprinting IS*200* and 16S *rrn* loci. *Journal of Clinical Microbiology* **32**, 2128–2133.

Podbielski, A., Melzer, B. & Lütticken, R. (1991). Application of the polymerase chain reaction to study of the M protein(-like) gene family in beta-hemolytic streptococci. *Medical Microbiology and Immunology* **180**, 213–227.

Saunders, N.A., Hallas, G., Gaworzewska, E.T., Metherell, L., Efstratiou, A., Hookey, J.V. & George, R.C. (1997). PCR-enzyme-linked immunosorbent assay and sequencing: an alternative to serology for M typing of *Streptococcus pyogenes*. *Journal of Clinical Microbiology* **35**, 2689–2691.

Simonet, M., Riot, N., Fortineau, N. & Berche, P. (1996). Invasin production by *Yersinia pestis* is abolished by insertion of an IS*200*-like element within the *inv* gene. *Infection and Immunity* **64**, 375–379.

Small, P.M., Hopewell, P.C., Singh, S.P., Paz, A., Parsonnet, J., Ruston, D.C., Schecter, G.F., Daley, C.L. & Schoolnik, G.K. (1994). The epidemiology of tuberculosis in San Francisco: a population-based study using conventional and molecular methods. *New England Journal of Medicine* **330**, 1703–1709.

Stanley, J. & Saunders, N.A. (1996). DNA insertion sequences and the molecular epidemiology of *Salmonella* and *Mycobacterium*. *Journal of Medical Microbiology* **45**, 236–251.

Stanley, J., Baquar, N. & Threllfall, E.J. (1993). Genotypes and phylogenetic relationships of *Salmo-*

nella typhimurium are defined by molecular fingerprinting of IS*200* and 16S *rrn* loci. *Journal of General Microbiology* **139**, 1133–1140.

Stanley, J., Powell, N., Jones, C. & Burnens, A.P. (1994). A framework for IS*200*, 16S rRNA gene and plasmid-profile analysis in *Salmonella* serogroup D1. *Journal of Medical Microbiology* **41**, 112–119.

Thierry, D., Brisson-Noel, A., Vincent-Levy-Frebault, V., Nguyen, S., Guesdon, J. & Gicquel B. (1990). Characterization of a *Mycobacterium tuberculosis* insertion sequence, IS*6110*, and its application in diagnosis. *Journal of Clinical Microbiology* **28**, 2668–2673.

van Embden, J.D.A., Cave, M.D., Crawford, J.T., Dale, J.W., Eisenach, K.D., Gicquel, B., Hermans, P., Martin, C., McAdam, R.A., Shinnick, T.M. & Small, P.M. (1993). Strain identification of *Mycobacterium tuberculosis* by DNA fingerprinting: recommendations for a standardized methodology. *Journal of Clinical Microbiology* **31**, 406–409.

van Soolingen, D., Hermans, P.W.M., de Haas, P.E.W., Soll, D.R. & van Embden, J.D.A. (1991). Occurrence and stability of insertion sequences in *Mycobacterium tuberculosis* complex strains: evaluation of an insertion sequence-dependent DNA polymorphism as a tool in the epidemiology of tuberculosis. *Journal of Clinical Microbiology* **29**, 2578–2586.

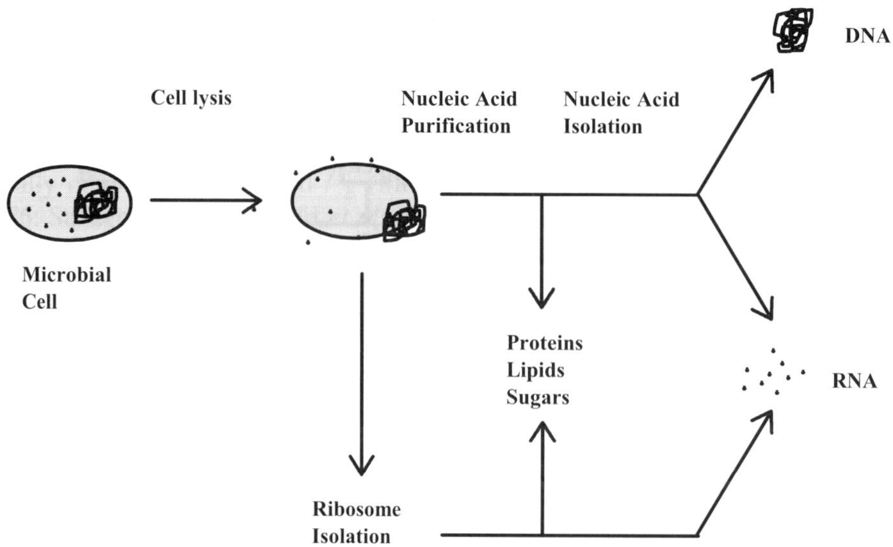

Fig. 11.2. Schematic representation of the procedures used to isolate DNA and RNA from a mixture of bacterial cells.

eral, procedures for the isolation of nucleic acids from microorganisms or environmental samples consist of three steps that will be discussed below: cell lysis, purification of nucleic acids, and isolation of nucleic acids (Fig. 11.2).

(i) Cell lysis and purification of nucleic acids

One of the important steps in the extraction of nucleic acids from an environmental sample is the lysis of microbial cells. Equally efficient lysis of all cells in an ecosystem is necessary to obtain a reliable picture of the microbial community. Efficient cell lysis may be hampered by the different cell envelope composition of various microorganisms. Hence, a protocol which is suitable for one species may not necessarily be suitable for another species. Microbial cells can be lysed chemically, enzymatically or mechanically. Various Gram-negative bacteria can be lysed chemically by treatment with detergents, such as sodium dodecyl sulphate (SDS). Disruption of the cell envelope of Gram-positive bacteria by detergents needs prior treatment with enzymes such as lyzozyme, N-acetylmuramidase or other muramidases. Most of these lytic enzymes are restricted to a certain range of microorganisms, because the cell-envelope composition differs for each species (Johnson, 1991). As a consequence, it is very difficult to develop chemical or enzymatic based lysis methods for complex communities. Therefore, procedures which include mechanical cell lysis by French press disruption, freeze/thaw incubations, sonication, or bead beating are preferred (Maniatis *et al.,* 1989; Johnson, 1991; Akkermans, *et al.,* 1998).

Bead beating is a widely used method to lyse bacterial communities, sometimes in combination with different chemical, enzymatic, or other mechanical lysis procedures. Bead beating has been shown to be successful in all kinds of samples, varying from soil systems to the mammalian GI tract (Stahl *et al.*, 1988; Harmsen *et al.*, 1995; Felske *et al.*, 1996; Ramírez-Saad *et al.*, 1996; Wilson & Blitchington, 1996; Doré *et al.*, 1998; Zoetendal *et al.*, 1998). In this procedure, glass or zirconium beads are added to an environmental sample in a buffered solution and shaken vigorously (3000 to 5000 rpm). The beads collide during this treatment, thereby facilitating the disruption of cells between the beads. Phenol can be added to the sample to prevent enzymatic degradation of the nucleic acids during the bead beating procedure.

A disadvantage of this mechanical cell lysis is that nucleic acids are partly sheared, especially when fragile bacteria such as some Gram-negative species are involved. Sheared nucleic acids cannot be used for genetic fingerprinting methods based on intact genomic DNA, such as RAPD and RFLP. Shearing of nucleic acids may also result in a reduced recovery of bacterial DNA or RNA, or increase the formation of chimeric structures during amplification of certain genes (Kopczynski *et al.*, 1994). Determination of the optimal conditions for efficient cell disruption is therefore very important. The easiest way is to examine the sample under the microscope before and after the lysis procedure (Apajalathi *et al.*, 1998). However, it may be difficult to distinguish between pieces of lysed cells and intact cells. Plate counting analysis of easily cultivable bacteria may give some indication of lysis efficiency when non-detrimental treatments are used during the lysis procedure (Felske *et al.*, 1996). However, this may be difficult or cumbersome, notably when anoxic environments such as the GI tract are analysed.

Another way to check for lysis efficiency is to determine the nucleic acid concentration before and after treatment. A disadvantage of this calculation method is that the genome size and ribosome number may vary according to cell and species, making the calculation less reliable. TGGE of 16S rDNA amplicons has been used to check the lysis efficiency of GI tract samples following different periods of bead beating (Zoetendal *et al.*, 1998; see section 11.2.Dii). In this way, shearing of nucleic acids could be minimised by determining the minimal time required for maximal disruption of the cells. It was shown that at least 3 min of bead beating was necessary to lyse a *Ruminococcus*-like species in a faecal sample.

Following cell lysis, nucleic acids have to be purified, because most analytical procedures involve enzymes and require relatively pure DNA or RNA. Most proteins, carbohydrates and lipids can be removed using phenol and chloroform extractions (Maniatis *et al.*, 1989). This purification procedure can be enhanced by addition of cetyltrimethylammonium bromide (CTAB), which forms complexes with the nucleic acids. It has been shown that CTAB extraction following bead beating of actinorrhisal nodules facilitated the recovery of DNA (Ramírez-Saad *et al.*, 1996). High molecular size DNA can also be separated from contaminants by CsCl centrifugation. Additional steps for purification can be added to the protocol, although it should be kept in mind that each additional step results in a decreased yield of extracted nucleic acids. An alternative step is the addition of specific proteins, such as bovine serum albumine or protein gp32 to the DNA sample. These proteins have been shown to enhance the amplification efficiency of template DNA containing PCR-inhibiting compounds (Kreader, 1996). Recently, improved DNA recovery from ancient faecal samples has been reported in which cross-links between reducing sugars and amino groups could be cleaved by adding N-phenacylthiazoliumbromide, thereby allowing for the amplification of DNA sequences (Poinar *et al.*, 1998).

Following purification, nucleic acids can be concentrated by polyethyleneglycol, isoamylalcohol or ethanol. Addition of sodium acetate facilitates the precipitation of nucleic acid fragments in ethanol. Instead of precipitation, nucleic acids can also be concentrated by binding to glass fibres or silica materials. Most commercial nucleic acid isolation kits rely on this procedure.

When RNA is isolated, all procedures should be performed with special care. RNA is more sensitive to degradation than DNA because its ribose contains a 2' hydroxyl group which makes RNA chemically less stable, especially under alkaline conditions. In addition, the double helix B structure found in DNA cannot be formed by RNA. Besides the chemical differences, RNase is much more stable than DNase, making removal of RNase more difficult and contamination with RNase easier. If possible, every step should be performed at 4°C or on ice, and equipment should be RNase free. Most procedures to isolate DNA and RNA are based on the principle of lysing the cells followed by direct purification of the nucleic acids. Another approach to obtain rRNA is based on the isolation of ribosomes (Felske *et al.*, 1996, 1998c; Fig. 11.2). In this procedure, cells from soil samples are mechanically lysed, followed by the isolation of intact ribosomes by ultracentrifugation. RNA is then subsequently isolated from the ribosome collection and purified. This method, which may also be applicable to other ecosystems such as the GI tract, resulted in a high yield of purified 16S rRNA and 23S rRNA which could be used directly for RT-PCR.

A number of studies have reported the extraction of nucleic acids from faecal and rumen samples (Stahl *et al.*, 1988; Wilson & Blitchington, 1996; Apajalathi *et al.*, 1998, Doré *et al.*, 1998; Zoetendal *et al.*, 1998). Most protocols are based on bead beating cell lysis in phenol, followed by phenol/chloroform extraction. A different approach has also been described in which bacterial cells from faeces are lysed by boiling in a phosphate-buffered saline (PBS) solution containing 1% Triton X100 (Wang *et al.*, 1996a). Following cell lysis, the solution was used directly as a template for PCR amplification. Amplicons derived from a variety of Gram-positive and Gram-negative species known to be present in faecal samples could be detected following amplification with specific primers. Although the number of strains tested is limited, this fast PCR approach seems to be accurate and may be useful for analysing samples containing a low number of microorganisms.

(ii) Quantification of nucleic acids

Several methods to visualise and quantify nucleic acids have been developed. DNA fragments separated on agarose or polyacrylamide gels are usually visualised with fluorescent dyes, such as ethidium bromide (Maniatis *et al.,* 1989), or by silver staining (Bassam *et al.,* 1991; Sanguinetti *et al.,* 1994). Staining is used to quantify the yield of nucleic acids following addition of a concentration standard. Quantification works fine when RNA and DNA are not sheared. The correlation between the concentration of nucleic acids and the signal of the stain is normally not linear and extrapolation is not possible. Instead of quantification by gel electrophoresis, DNA and RNA can also be quantified by Southern, northern or dot blot hybridisation using a universal probe for a gene of interest. A third way to measure the nucleic acid concentration is to determine the UV light absorption at 260 nm and 280 nm. Additionally, the ratio A_{260}/A_{280} gives an indication of purity. DNA is relatively pure when this ratio is between 1.80 and 2.00.

Another less well-known procedure for measuring the DNA and RNA concentration is use of High-Performance Liquid Chromatography (HPLC). HPLC has been used to determine the copy number of plasmids in recombinant yeast or *Escherichia coli* cells. Chromosomal DNA, plasmid DNA, rRNA and tRNA could be separated using HPLC (Coppella *et al.,* 1987a, b). When the various methods were compared for nucleic acids isolated from marine sediments (Dell'Anno *et al.,* 1998), it was found that the yield of DNA appeared to be similar with spectrometric and HPLC measurements, but was significantly lower when the yield was determined by the fluorescent method. This might be due to the fact that the fluorescent stain is dependent on the structure, the size and the composition of the nucleic acids. Another finding was that RNA and DNA could be separated by HPLC, so that RNA measurement was not biased by DNA and *vice versa,* which might not be the case for the other two techniques.

B. RT-PCR/PCR of 16S rRNA/rDNA

In order to gain an insight into the microbial structure in different ecosystems, various methods have been developed based on the nucleic acid sequences of small subunit (ssu) rRNA or rDNA because these molecules are ideal phylogenetic and taxonomic markers (Woese, 1987; Amann *et al.,* 1995). There are various reasons to use rRNA and rDNA genes as markers, including (1) their presence in all cells; (2) their high degree of sequence conservation which facilitates their detection; (3) the presence of highly variable regions in their sequences which makes them useful to discriminate at (sub)species to higher phylogenetic levels; and (4) the presence of databases containing up to 20,000 ssu rRNA sequences (M. Wagner, personal communication) from different taxa that facilitates the phylogenetic characterisation of cultured and uncultured microbes. Moreover, rDNA can be amplified by PCR *in vitro* (Saiki *et al.,* 1988).

The principle of PCR is that cycles of DNA melting, primer annealing and elongation using a thermostable polymerase are repeated, resulting in an exponential increase of amplified genes. In addition, rRNA can also be amplified, but it has first to be converted into DNA by reverse transcription. This can be done by reverse transcriptase using an oligonucleotide primer which targets the RNA template (a

Table 11.1. Overview of some general artifacts concerned with (RT-)PCR and some solutions to minimise these artifacts

Factors causing PCR artifacts	Some bias-preventing solutions	Reference
Nucleic acid purity	Additional purification steps	—
	Use of BSA or protein gp32	Kreader (1996)
High G+C content of template	Increase denaturing time	—
Secondary structure of template	Increase denaturing time	—
	Use DMSO	Baskaran *et al.* (1996)
Preferential amplification	Decrease PCR cycles	Polz & Cavanough (1998)
	Mix replicate reactions	Polz & Cavanough (1998)
	Exclude degenerated primers	Polz & Cavanough (1998)
	Use high template concentration	Polz & Cavanough (1998)
Formation of chimeric constructs	Longer elongation time	Wang & Wang (1996)

procedure termed RT-PCR). Although DNA and RNA can be amplified with other techniques (Carrino & Lee, 1995), this section will only focus on RT- and regular PCR. Some important factors which may influence the amplification procedure, notably when mixtures of DNA or RNA from different organisms are amplified, include the purity of DNA, the G+C content of the target, the secondary structure of the target, preferential amplification, and formation of chimeric structures. Several methods minimising these factors have been reported (Table 11.1; Baskaran *et al.,* 1996; Kreader, 1996; Wang & Wang, 1996; Polz & Cavanough, 1998).

(i) Preferential amplification

Some sequences may be preferentially amplified in a mixture of different sequences from comparable genes. For 16S rDNA it has regularly been reported that variations in primer pairs result in biased amplification when using mixtures of DNA as a template (Reysenbach *et al.,* 1992; Suzuki & Giovannoni, 1996; Wilson & Blitchington, 1996; Polz & Cavanough, 1998). Equal amplification efficiency of 16S rDNA is necessary to get an insight into the microbial composition of an ecosystem. It was suggested that the bias in amplification observed with the canonical universal primers 27F and 1492 (Lane, 1991) can be decreased by (1) decreasing the number of amplification cycles; (2) mixing several replicate PCR amplifications; (3) using high template concentrations; and (4) excluding degenerate primers (Polz & Cavanough, 1998). A disadvantage of high template concentrations might be a high risk of the formation of chimera, consisting of PCR fragments originating from more than one target gene. Chimera formation during the amplification of 16S rDNA from an environmental sample results in an overestimation of the biodiversity. Since the homology between different 16S rRNA genes is relatively high, chimeras are thought to arise from reannealing of different 16S rDNA genes during PCR (Liesack *et al.,* 1991).

Multiple competitive PCR and quantitative RT-PCR have been used to test the universal bacterial primers U968-GC and L1401 when used to amplify 16S rRNA from soil (Nübel *et al.*, 1996; Felske *et al.*, 1998b). It was found that 16S rDNA clones and bacterial 16S rRNA sequences from different phylogenetic groups were not preferentially amplified, although some target sequences have some minor sequence differences at the annealing sites of the primers (Felske *et al.*, 1998b). It has been shown that there are no differences in the TGGE patterns of DNA amplified for different numbers of cycles with the same primers (Zoetendal *et al.*, 1998). However, it was observed that the primers which were used to amplify complete 16S rDNA were preferentially amplifying *Prevotella*-like sequences. This was specifically noted when amplified 16S rDNA was reamplified using the primer pair which amplified the V6 to V8 regions. Reamplification of amplicons using another primer pair is called nested PCR. Although primer pairs may show limited preferential amplification, this undesired bias can never be excluded. For example, if target DNA or rRNA from an unknown, uncultured microbe is not amplified during the first PCR cycles then it will stay undetected forever.

(ii) Quantitative (RT-)PCR

Several studies have described the quantification of microbial 16S rDNA or rRNA amplicons by PCR and RT-PCR, respectively. It should be remembered that quantification of amplicons only reflects the relative number of ribosomes or corresponding genes in a community and not the relative frequency of a species. The number of ribosomes per cell depends on the type and activity of a species. A positive correlation between the activity of a cell and the amount of rRNA has been described (Wagner, 1994). However, it has also been shown for two food-associated pathogens that this correlation was only found under extreme heat conditions (McKillip *et al.*, 1998). Moreover, the number of 16S rRNA genes per genome varies between species. For example, seven 16S rRNA genes were found in *E. coli* (Bachmann, 1990), five to six genes in *Streptococcus* spp. (Bentley & Leigh, 1995), and four genes in *S. pneumoniae* (Bacot & Reeves, 1991).

Besides differing copy numbers of 16S rRNA genes per genome, the genome sizes of bacteria are also different. It has been shown that differences in genome size and 16S rDNA copy number influence the ratio of amplicons when mixtures of target DNA from *E. coli*, *Pseudomonas aeruginosa*, *Bacillus subtilis* and *Thermus thermophilus* are mixed in equal molarities (Farrelly *et al.*, 1995). From this study it was concluded that the number of bacterial cells could not be calculated exactly when both parameters are unknown.

Despite the fact that relative cell numbers cannot be extrapolated from (RT-)PCR data, changes in the structure and activity of a microbial community can be analysed when 16S rRNA or rDNA amplicons are quantified. *E. coli* and *P. aeruginosa* 16S rRNA genes from mixtures of these bacteria with ratios of 1:100 could be quantified using the Perkin Elmer QPCR™ system 5000 (Blok *et al.*, 1997). Biotinylated PCR products were captured on streptavidin-coated paramagnetic beads after different PCR cycles, and specific PCR products were quantified by measuring the electrochemoluminescent signals from the specific reporter probes directed against the different amplicons.

Another way of quantifying 16S rRNA genes is a so-called most probable number (MPN)-PCR (Sykes *et al.,* 1992). In this method the target DNA for PCR is diluted to extinction, followed by analysis of PCR products by agarose gel electrophoresis. This approach was used to quantify the relative amounts of 16S rDNA derived from different groups of bacteria in faecal samples using different primer combinations (Wang *et al.,* 1996a). Although this form of multiplex PCR has the potential to be useful, the data have to be analysed carefully, since the PCR conditions may not be quantitative as the primer pairs and product sizes are different.

Another approach involves the use of competitive PCR for the quantification of mRNA (Wang *et al.,* 1989). In this method a specific standard of known concentration is added in different amounts to the target. The different sizes of the standard and target allows for differentiation and subsequent quantification following agarose or polyacrylamide gel electrophoresis.

The combination of TGGE and competitive RT-PCR resulted in the development of a new quantification method called multiple competitive RT-PCR (Felske *et al.,* 1998b). In this approach, changes in amplification conditions were minimised because the products were amplified with the same primer pairs and had the same size as the added standard. TGGE was used to separate and quantify the different products. It was found that the 20 most abundant sequences, which derived mainly from Gram-positive species of low G+C content, represented about 50% of the total microbial community in the Drentse A grassland soils.

C. Cloning, sequencing and phylogenetic analysis

To get an overview of the complexity of ecosystems, it is essential to classify the individuals from a population. Classification is used to clarify the relationships between different organisms. It has to be emphasised that there is no single unifying classification of organisms. This section will focus on classification based on phylogenetic relationships between organisms.

Phylogeny is determined by scoring for the presence or absence of homologous morphological or physiological characteristics across operational taxonomic units (OTUs), which can be populations, species or strains. Both physiological and genetic characteristics can be used for phylogenetic analysis. The principle of phylogenetic analysis is the assumption that all life forms have evolved from a common origin. The common ancestor of two closely related organisms disappeared more recently than that of two more distantly related organisms. It is believed that evolution follows a pattern of successive branchings into populations in which further evolutionary changes subsequently proceed independently. Phylogeny involves the determination and analysis of these branching patterns.

Cloning, sequencing and phylogenetic analysis of 16S rDNA sequences have become powerful tools in microbial ecology, particularly since it was discovered that the majority of microorganisms in environmental samples are unknown

(Amann *et al.*, 1995). The highly conserved, but discriminative 16S rDNA molecule makes it possible to identify a species in an ecosystem without the use of unreliable culturing methods. Cloning and sequencing of 16S rDNA amplicons has become a standard procedure in molecular ecology and provides information about the genetic diversity and phylogenetic relationships between microorganisms in an ecosystem. Since up to 20,000 different ssu rRNA sequences are available in different databases, the comparison of new sequences is reliable. However, a clone library (a collection of clones from a DNA sample) has to be very large to give a reliable picture of the genetic diversity in complex ecosystems. This makes the approach expensive and time consuming. Therefore, a rapid approach for screening 16S rDNA clone libraries has recently been developed (Mau & Timmis, 1998). Habitat-based probes were designed using subtractive hybridisation. These habitat-based probes were used to screen a 16S rDNA library generated from the same habitat. It was shown that this screening method prevents sequencing many similar or identical clones of the dominant members in sediments.

(i) Cloning

To construct a clone library, a mixture of 16S rDNA amplicons is first generated by PCR using bacterial or universal primers to amplify 16S rDNA from an environmental sample. A cloning strategy is necessary to sequence individual amplicons derived from the DNA of a complex microbial community. There are different strategies to create a clone library. DNA fragments can be cloned into a plasmid vector or a bacteriophage. Detailed principles, possibilities and procedures for cloning DNA fragments have been described elsewhere (Maniatis *et al.*, 1989; Old & Primrose, 1989). Amplicons are usually cloned into a sequencing vector, which is then transformed into an *E. coli* strain. Although a great variety of different cloning vectors are available, they all show some common characteristics. One common feature is that amplicons are inserted into a gene with many restriction sites (e.g., the polylinker in that part of the *lac*Z gene coding for the α-peptide). The amplicons and the vectors have to be restricted with the same restriction enzyme(s) for cloning. Some vectors contain a 3'T-overhang at the insertion sites, and these are particularly useful for amplicons produced by certain polymerases (e.g., *Taq* polymerase) which sometimes make 3'A-overhangs (Clark, 1988). Amplicons of 16S rDNA are cloned into the vector with a ligase. After ligation, the vectors containing a single 16S rDNA insert are transformed into a competent *E. coli* strain. The cells are grown on selective plates and single transformants are screened for the presence of vectors containing a 16S rDNA insert by means of PCR or colony hybridisation with 16S rDNA-specific probes. The vectors containing a 16S rDNA insert can be isolated after regrowing the positive transformants and can then be subjected to sequence analysis or fingerprinting.

(ii) Sequence analysis

Sequence analysis is used to provide information about the nucleotide sequence of a cloned amplicon. There are different methods to determine the sequence of

Table 11.2. Overview of possibilities for calculating phylogenetic relationships and making a phylogenetic tree

Tree construction	Basis for calculation	Refs.
Distance matrix	Each nucleotide change equal	Jukes & Cantor (1969)
	Differing substitution rates	Tajima & Nei (1984)
	Differing substitution rates and transition/transversion correction	Kimura (1980)
Discrete character data	Maximum parsimony	Edwards & Cavilla-Sforza (1963)
	Maximum likelihood	Felsenstein (1981)

a DNA fragment, but these will not be discussed here. Several programs to determine the closest relative of the DNA sequence are available on internet sites. Mostly, these programs use homology searches provided by BLAST (Altschul *et al.,* 1990; *http://www.ncbi.nlm. nih. gov/BLAST/*) or FASTA (Pearson & Lipman, 1988; *http://biogate.mlg.co.jp/tssfree/Fasta.html*). The benefit of these programs is that the search for homology is fast and reliable, and several DNA databases can be used for comparison. When DNA sequences are compared, the alignment of sequences for highest homology is a crucial step. The alignment of sequences is performed by giving homologous or conserved parts the same nucleotide position. The variable regions in between are compared in such a way that the highest homology is found. Gaps in a nucleotide sequence are also included in the alignment, but the number of gaps should be minimised. Each position in a sequence can be one of the four nucleotide bases or a gap. This alignment of nucleotide sequences is necessary in order to construct phylogenetic trees and to develop oligonucleotide probes (Lane, 1991; Stahl & Amann, 1991; see below).

(iii) Tree construction

It is difficult to visualise phylogenetic relationships between species from numerical values based on multiple pairwise comparisons, particularly when many different sequences are compared. An alternative way to visualise phylogenetic relationships is by the construction of a phylogenetic tree based on the identity values. The calculations for construction of phylogenetic trees can be handled in two ways: by distance matrix or from discrete character data (Table 11.2). In the first calculation, data based on evolutionary distances are set in a distance matrix. Most calculation methods do not weight each nucleotide mutation equally. The DNA structure plays an important role in the calculation procedures. It has been postulated that transversions are more easily recognised by the DNA repair system than are transitions because of the spherical DNA helix distortions (Kimura, 1980). These changes are therefore considered to be less frequent and result in a lower substitution rate, which can be taken into account when calculating distance values.

Another example of differences in substitution rates is postulated for protein-coding genes. Substitution rates of the third nucleotide position in a triplet coding for an amino acid are usually higher than in the other two nucleotide positions (Shoemaker & Fitch, 1989).

A more complex feature is the formation of gaps. The cost of introducing a gap in an alignment is generally higher than the introduction of a base substitution. Although the introduction of gaps is necessary to align sequences, it is often omitted from distance calculations because it is difficult to verify how the gap has originated. The most frequently used distance calculation models are those developed by Jukes & Cantor (1969), Kimura (1980), and Tajima & Nei (1984). The Jukes and Cantor model does not discriminate between different nucleotide substitutions, in contrast to that of Tajima and Nei which, however, does not correct for nucleotide transitions or transversions, as does the model of Kimura. The model of Jukes and Cantor has probably been applied most frequently in evolutionary studies because it performed well in most studies simulating the evolution of nucleic acid sequences (van de Peer & de Wachter, 1995).

Phylogenetic trees can be plotted from distance matrices. Commonly used models which calculate distance trees are the unweighted pair group method using arithmetic averages (UPGMA; Sokal & Michener, 1958) and neighbour joining method (Saitou & Nei, 1987). UPGMA is a clustering method which pairs the least distant sequences into a node, and subsequently pairs two nodes into a new node. The neighbour joining method uses a simplified algorithm to calculate branch lengths and tree topologies.

Discrete character data calculations are not based on evolutionary distances, but consider each character state of the nucleotide position in the sequence separately. Trees can be constructed from each nucleotide position. The data can be handled in two ways. The first way is based on the maximum parsimony principle in which the true tree is the one which requires the fewest number of mutational changes to explain the differences observed between the gene sequences (Edwards & Cavilla-Sforza, 1963). Only so called 'informative nucleotides' (a common nucleotide position in a set of sequence positions which favour only some of all possible trees) are used. In general, this means that a constant base in all sequences and a variable base which does not favour one tree over all others are not informative. The second way to handle discrete character data is called the maximum likelihood phylogeny. This calculation uses statistical models to calculate the probability that one sequence is converted into another sequence by mutation over time (Felsenstein, 1981). More detailed explanations and comparison of the methods have been described extensively elsewhere (Sneath, 1989; Priest & Austin, 1993; Wolters, 1998).

It has to be realised that evolutionary events cannot be checked for and that phylogenetic trees therefore only represent a systematic ordering of genes. Furthermore, calculations based on different DNA sequences or different genes may result in completely different trees. As a consequence, it is difficult to choose which tree-constructing approach is most optimal. For 16S rDNA sequences from cultured and uncultured *Frankia* strains, it was found that trees constructed by methods based on discrete character data or distance matrices were roughly the same (Wolters, 1998). The choice of the program might therefore depend on the speed, ease and possibilities of the different programs and on the applications of the user. A comparison of different methods can be used to demonstrate the robustness of the phylogenetic tree generated.

(iv) Phylogeny of 16S rRNA genes

Phylogeny based on 16S and 23S rRNA analysis has led to the construction of phylogenetic trees which illustrate the evolutionary relationship between different organisms. This has resulted in a division of all life into three main domains: Archaea, Bacteria and Eucarya (Woese, 1987; Woese *et al.,* 1990). The increasing number of 16S rDNA sequences of bacterial isolates has allowed phylogenetic analysis of 16S rDNA to be applied to bacterial taxonomy. The threshold for species determination is set at 70% DNA-DNA hybridisation between the genomes of different strains (Wayne *et al.,* 1987). Strains showing values above this threshold are considered to be the same species and this threshold has been translated into a 16S rRNA value (Stackebrandt & Goebel, 1994). It was estimated that strains with less than 97% 16S rDNA homology have less than 70% DNA-DNA hybridisation values. This threshold can be used to determine whether two strains do not belong to the same species, but cannot be used as the only characteristic for species determination. Indeed, some *Bacillus* spp. have less than 70% DNA-DNA hybridisation, but more than 99.5% 16S rRNA homology (Fox *et al.,* 1992).

Sometimes traditional taxonomic methods can be compared to the 16S rRNA phylogeny. Some genera in the GI tract that have been characterised physiologically (e.g., bifidobacteria) form a monophyletic group in the 16S rRNA tree. However, many bacterial genera in the GI tract do not form monophyletic clusters in the 16S rRNA phylogenetic tree. In particular, the genera *Clostridium* and *Eubacterium* are mixed and divide into different distinguishable clusters (Collins *et al.,* 1994). Other genera like *Ruminococcus* and *Butyrivibrio* are mixed in these clusters, making identification quite difficult. Additionally, it was found that the 16S rRNA sequences of two strains, identified as *Fusobacterium prausnitzii* by physiological characteristics, were not phylogenetically related to other *Fusobacterium* strains, but grouped in one of the *Clostridium* clusters (Wang *et al.,* 1996b). In such cases, physiological characteristics cannot be translated from phylogenetic characters when the species are not closely related.

The use of 16S rRNA sequence analysis in taxonomy has also resulted in proposals for renaming bacterial species. Based on their 16S rRNA sequences and their physiological characteristics, it was proposed to redesignate *Peptostreptococcus productus* and *Streptococcus hansenii* as *Ruminococcus* species (Ezaki *et al.,* 1994). In conclusion, it seems that phylogenetic and physiological analysis are both necessary for a reliable identification.

Following the increase in the number of DNA sequences, databases such as EMBL and Genbank have become available on internet sites. These databases contain up to 20,000 ssu rRNA sequences and are suitable for comparing, downloading and the deposition of sequences. The most commonly used database for rRNA sequence analysis is that of the Ribosomal Database Project (RDP; Maidak *et al.*, 1997; *http://www.cme.msu.edu/RDP/analysis.html*). The RDP contains an aligned database of 16S rRNA sequences, which are present in a phylogenetic tree. Another software package containing a rRNA database is the ARB software package (Strunk & Ludwig, 1995). This program is comparable to the RDP, but the package needs a powerful computer for calculations. The secondary structure of the rRNA molecule is used for the similarity calculation and is visualised in the alignment program of the ARB package. This facilitates sequence analysis and is ideal when checking for sequencing errors. The phylogenetic trees in both programs are comparable.

16S rRNA databases are not only used for strain identification, but are also used to study the bacterial diversity in an ecosystem. The development of molecular methods in microbial ecology has resulted in an increasing number of sequences from cloned amplicons derived from different types of environments. Several papers on intestinal samples from different types of animals have shown that many of the cloned 16S rDNA sequences show identity below the 97% threshold to their closest culturable relative in the DNA databases (Ohkuma & Kudo, 1996; Wilson & Blitchington, 1996; Whitford *et al.*, 1998; Zoetendal *et al.*, 1998, Suau *et al.*, 1999). This means that the species from which the sequences have been derived have not yet been cultured or, alternatively, are present in a culture collection but their 16S rDNA has not yet been sequenced. Fig. 11.3 shows a phylogenetic tree built from 16S rDNA sequences of bacterial clones from human faeces and their closest cultured relatives found in the 16S rDNA database of the ARB software package. These results reinforce the concept that most bacteria in existence have not yet been cultured (Amann *et al.*, 1995). For the GI tract, this means that our knowledge about the role of the microbial community in the intestine is limited. Therefore, cloning and sequencing of faecal or intestinal clones is needed to determine the microbial diversity and to study the structure of the community in the GI tract.

D. Fingerprinting

Sequence analysis of 16S rDNA/rRNA clone libraries gives reliable information about the genetic diversity of an ecosystem. However, this approach is expensive, time-consuming and not suitable for monitoring complex ecosystems. It will be necessary to study complex ecosystems using alternative methods that are better suited for studying the composition and temporal variation in ecosystems, probably based on sequence differences of the nucleic acids. Fingerprinting techniques are suitable to describe the genetic diversity at different levels of a microbial community. There are many types of fingerprinting techniques which can be useful at the community, species and even strain level. The next section describes SSCP, DGGE and TGGE fingerprinting techniques which are commonly used to study the microbial diversity of different ecosystems. These techniques are based on differences in the nucleotide sequence of amplicons of similar size and are suitable

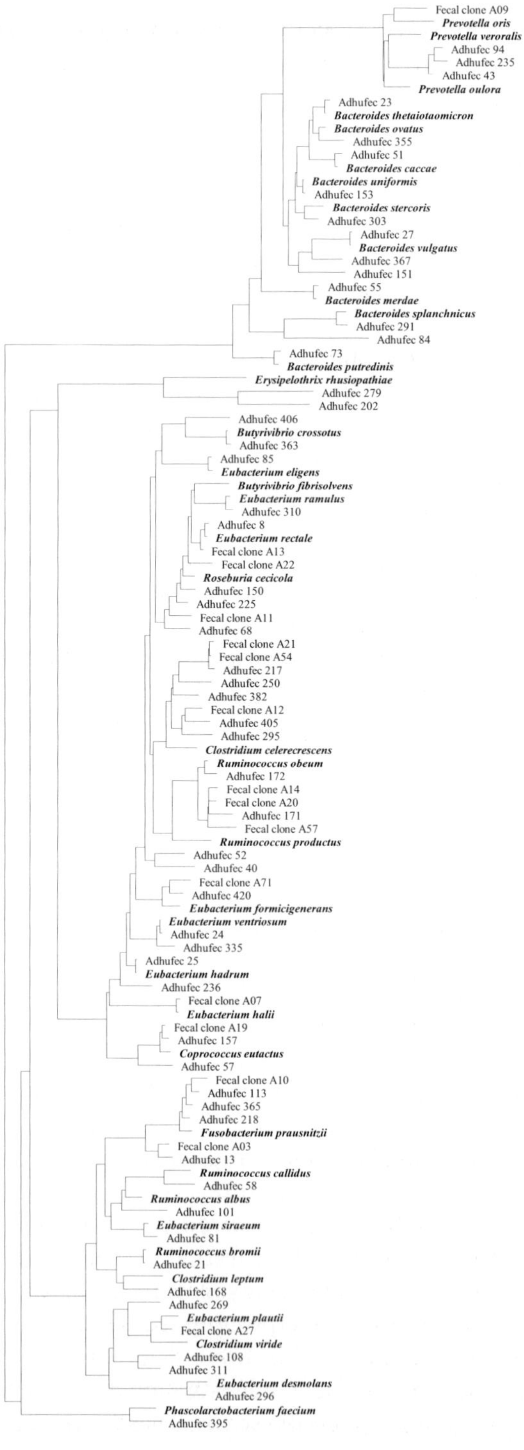
Fecal clone A09
Prevotella oris
Prevotella veroralis
Adhufec 94
Adhufec 235
Adhufec 43
Prevotella oulora
Adhufec 23
Bacteroides thetaiotaomicron
Bacteroides ovatus
Adhufec 355
Adhufec 51
Bacteroides caccae
Bacteroides uniformis
Adhufec 153
Bacteroides stercoris
Adhufec 303
Adhufec 27
Bacteroides vulgatus
Adhufec 367
Adhufec 151
Adhufec 55
Bacteroides merdae
Bacteroides splanchnicus
Adhufec 291
Adhufec 84
Adhufec 73
Bacteroides putredinis
Erysipelothrix rhusiopathiae
Adhufec 279
Adhufec 202
Adhufec 406
Butyrivibrio crossotus
Adhufec 363
Adhufec 85
Eubacterium eligens
Butyrivibrio fibrisolvens
Eubacterium ramulus
Adhufec 310
Adhufec 8
Eubacterium rectale
Fecal clone A13
Fecal clone A22
Roseburia cecicola
Adhufec 150
Adhufec 225
Fecal clone A11
Adhufec 68
Fecal clone A21
Fecal clone A54
Adhufec 217
Adhufec 250
Adhufec 382
Fecal clone A12
Adhufec 405
Adhufec 295
Clostridium celerecrescens
Ruminococcus obeum
Adhufec 172
Fecal clone A14
Fecal clone A20
Adhufec 171
Fecal clone A57
Ruminococcus productus
Adhufec 52
Adhufec 40
Fecal clone A71
Adhufec 420
Eubacterium formicigenerans
Eubacterium ventriosum
Adhufec 24
Adhufec 335
Adhufec 25
Eubacterium hadrum
Adhufec 236
Fecal clone A07
Eubacterium halii
Fecal clone A19
Adhufec 157
Coprococcus eutactus
Adhufec 57
Fecal clone A10
Adhufec 113
Adhufec 365
Adhufec 218
Fusobacterium prausnitzii
Fecal clone A03
Adhufec 13
Ruminococcus callidus
Adhufec 58
Ruminococcus albus
Adhufec 101
Eubacterium siraeum
Adhufec 81
Ruminococcus bromii
Adhufec 21
Clostridium leptum
Adhufec 168
Adhufec 269
Eubacterium plautii
Fecal clone A27
Clostridium viride
Adhufec 108
Adhufec 311
Eubacterium desmolans
Adhufec 296
Phascolarctobacterium faecium
Adhufec 395
0.10

for describing ecosystems at the species level. The sequence-specific separation of PCR amplicons is an essential element, but differs between the techniques. SSCP relies on the secondary structures of the single strands, while the other techniques rely on the melting behaviour of the double stranded amplicon.

(i) SSCP

SSCP is an electrophoretic technique which has been developed for the detection of mutations in genes and has been used widely in the field of human genetics (Orita *et al.,* 1989; Hayashi, 1991). The principle of SSCP is that the mobility of a single-stranded DNA fragment is dependent on the secondary structure of the fragment. The secondary structure is determined by the nucleotide sequence and the physiological environment (e.g., temperature, pH and ionic strength). SSCP has been shown to be able to detect single base differences in 99% of amplicons which are up to 300 bases in size. This detection limit drops using longer fragments (Hayashi, 1991; Hayashi & Yandell, 1993). A typical SSCP profile consists of two single-stranded DNA fragments and one double-stranded fragment, although different conformations from one strand are also possible. This technique has only been used occasionally in ecological studies. SSCP of the 16S-23S rRNA spacer has been used to analyse mixtures of bacteria (Scheinert *et al.,* 1996), and SSCP of different regions in the rRNA operon has been used to differentiate between root-associated fungi (Clapp, 1999).

SSCP has been used to analyse microbial communities in a few studies. For example, the V3 region of the 16S rRNA gene was used for SSCP fingerprinting of bacterial strains and environmental samples (Lee *et al.,* 1996). The problem of bands caused by heteroduplex formation in mixed DNA samples could be solved by removing glycerol from the gel, but this removal resulted in a lower separation efficiency of the single strands. It has been reported that the bands in the profiles of the environmental samples did not correspond to bands in the profiles of those bacteria that could be cultivated. The 16S rDNA sequence of a bacterium making up about 1.5% of a community could be visualised with this technique.

Recently, a new approach for SSCP analysis was reported (Schwieger & Tebbe, 1998). In this study, amplicons containing the V4 to V5 region were used. One of the primers was phosphorylated at the 5' end. After amplification the phosphorylated strand could be digested selectively with lambda exonuclease. Using this technique, the number of bands per species could be minimised and heteroduplex formation in mixed DNA samples could be prevented. Clear banding patterns

---

Fig. 11.3. Phylogenetic tree showing the phylogenetic relationships between cloned 16S rDNA from faeces and the closest culturable relatives found in 16S rDNA databases. Complete and partial 16S rDNA sequences from faecal samples were added to the phylogenetic tree of the ARB software. Sequences called faecal clone A03–A71 and adhufec 8–420 were retrieved from Zoetendal *et al.* (1998) and Suau *et al.* (1999) respectively. These sequences and the sequences from the closest cultivable relatives were marked, and the remaining sequences were removed from the tree. Bold-marked sequences represent the closest cultivable relative found in the database; the bar indicates the calculated genetic distance between the sequences.

could be obtained from environmental samples.

(ii) DGGE and TGGE

The separation of 16S rDNA amplicons is based either on a linear gradient of denaturants, at constant temperature in the case of DGGE (Fischer & Lerman, 1979), or on a linear temperature gradient parallel to the running direction in the case of TGGE (Rosenbaum & Riesner, 1987), or on increasing temperature with time in the case of TTGE (also called TSGE; Yoshino *et al.,* 1991). These fingerprinting techniques are frequently used to study the microbial diversity of different ecosystems. At the time of writing, TTGE has been used occasionally in ecological studies, but there are no published articles describing its use. In DGGE, TGGE and TTGE, amplicons of the same length with different nucleotide sequences can be separated on polyacrylamide gels containing denaturants (urea and formamide). During electrophoresis the amplicons start to melt in so-called melting domains with identical melting behaviour. The size of these domains may vary between 50 and 300 bp (Myers *et al.,* 1987). In this way, the electric mobility of amplicons which contain the double helix structure and disordered single-stranded regions drops. Sequence variations within such domains causes the different melting behavior of the amplicons.

Attachment of a GC-clamp makes it possible that sequence variations in the most stable domains can also be separated (Myers *et al.,* 1985; Sheffield *et al.,* 1985). This GC-clamp is a G + C rich domain which is attached to the amplicons by adding it at the 5' end of one of the primers, and which prevents complete melting of the amplicons. In principle, all single base differences at each position of the amplicons can be separated for amplicons of up to 500 bp (Sheffield *et al.,* 1985). The final position of the amplicons in the gel depends on the melting behaviour of the amplicons (and, therefore, the nucleotide sequence) and the running time. A simplified representation of DGGE and TGGE analysis of amplicons is shown in Fig. 11.4.

The introduction of DGGE into ecological studies was originally designed to separate amplified V3 regions of 16S rDNA from marine ecosystems. Amplicons derived from sulphate-reducing bacteria could be detected after blotting the DGGE profiles with a specific probe (Muyzer *et al.,* 1993). Following this study, the application of these techniques to ecological studies increased enormously. Different ecosystems have been analysed by separation of different amplified regions from 16S rDNA and 16S rRNA using these fingerprinting techniques. These techniques have not only been used for analysing the composition and stability of different ecosystems, but have also been used for comparing DNA extraction protocols, screening of clone libraries, determining 16S rRNA sequence heterogeneities, monitoring enrichment and isolation procedures, and determining biases introduced by PCR and cloning. Recent overviews of the use of these and other methods for studying different ecosystems are available (Muyzer & Smalla, 1998; Muyzer, 1999).

To increase the separation efficiency, an optimal gradient has to be chosen.

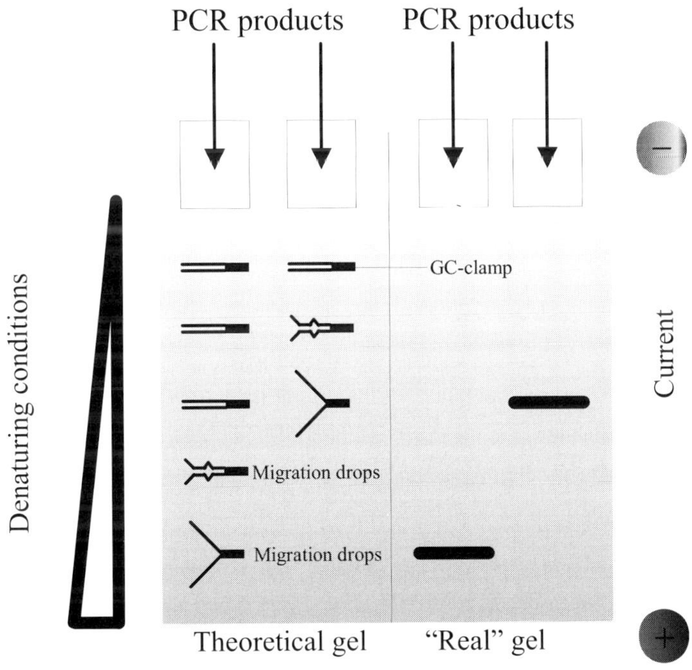

Fig. 11.4. Schematic representation of a polyacrylamide gel which explains the principles of DGGE and TGGE.

This can be done by applying the gradient perpendicular to the running direction (Muyzer *et al.,* 1993). For TTGE, the optimal temperature gradient has to be calculated from known sequences. Amplicons with only one nucleotide difference can be separated when an optimal gradient is applied (Myers *et al.,* 1985; Nübel *et al.,* 1996). Additionally, it has been shown that a wobble base (either C or T) in the reverse primer may result in two distinct bands (Kowalchuk *et al.,* 1997). The opposite, however, also takes place. Sometimes, 16S rDNA amplicons cannot be separated although they differ in nucleotide sequence (Buchholz-Cleven *et al.,* 1997; Vallaeys *et al.,* 1997).

TGGE and DGGE of 16S rDNA and rRNA amplicons have been used to describe the microbial composition of several ecosystems. In these studies, different universal primer pairs have been used to describe dominant populations. An MPN (RT-)PCR can be used to check if the dominant community is visualised on TGGE (Fig. 11.5). In general, 16S rDNA, but also 16S rRNA, is used as a target for analysing microbial diversity (Fig. 11.5). Profiles derived from 16S rRNA represent the relative number of different ribosomes in an environmental sample, which reflects the active fraction of a community. Comparing rRNA- and rDNA-derived amplicons may give information about the activity in the microbial community of a certain group (Felske *et al.,* 1996; Teske *et al.,* 1996; Zoetendal *et al.,* 1998), but it has to be realised that the number of ribosomes and rRNA genes may differ per species. Several studies have shown that sequences derived from a bacterium

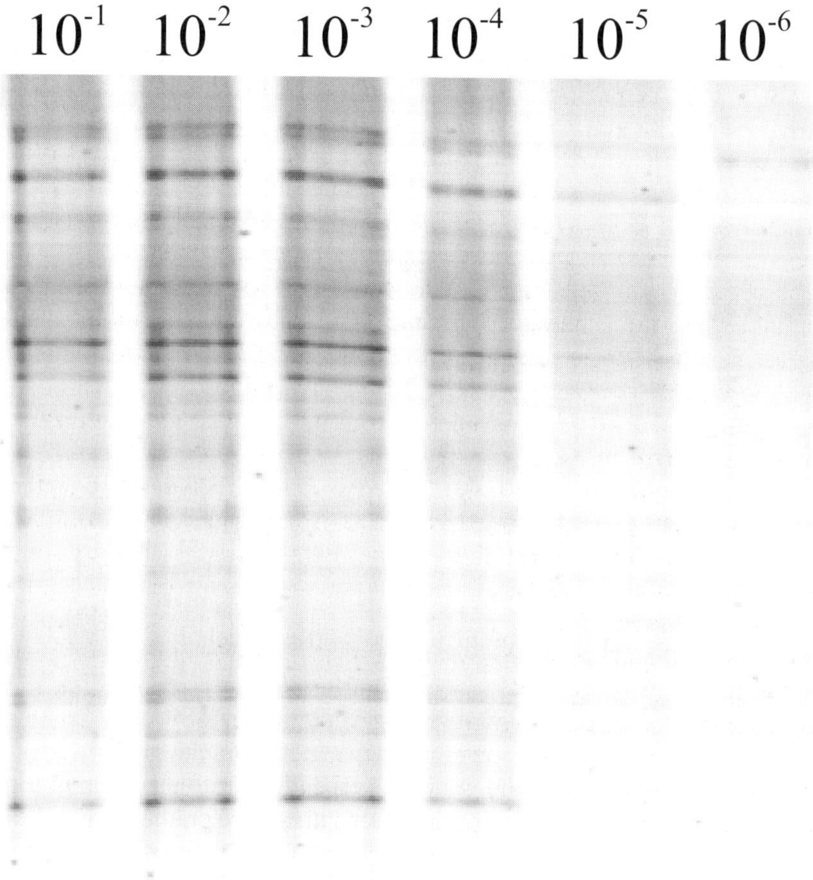

Fig. 11.5. TGGE profiles of amplified V6 to V8 regions of 16S rRNA from a faecal sample using the MPN RT-PCR approach. 10^{-x} represents the dilution of template RNA used for amplification.

which makes up about 1% of a microbial community can still be visualised using TGGE and DGGE (Muyzer *et al.*, 1993; Murray *et al.*, 1996; Zoetendal *et al.*, 1998), which is similar to the sensitivity of SSCP analysis. Instead of using universal primers, group- or species-specific primers can be used to focus on particular groups. The genetic diversity of uncultured ammonia-oxidising bacteria (Kowalchuk *et al.*, 1997) and cyanobacteria (Nübel *et al.*, 1997) has been studied using specific 16S rDNA primers for both groups. DGGE and TGGE have also been used to describe the expression of functional genes such as the [NiFe] hydrogenases from *Desulfovibrio* populations (Wawer *et al.*, 1997). The combination of D/TGGE analysis of 16S rDNA and a functional gene may be used to study relationships between the structures and functions of ecosystems.

TGGE and DGGE analysis of amplicons is semi-quantitive, i.e., an intensive band is more abundant than a weak band in a profile. When an appropriate standard template of known concentration is added to the nucleic acids extracted from a microbial community, different genes or ribosome fractions can be quantified. Bands for which the intensity is identical to the intensity of the standard can be quantified. This approach is called multiple competitive (RT-)PCR (Felske *et al.*, 1998b). Different ribosome fractions from Drentse A soils could be quantified this way. Equal amplification of different ribosomes is necessary for quantification. In case of the Drentse A soils, the primers used did not preferentially amplify specific cloned 16S rDNA amplicons or ribosomes from strains of different phylogenetic clusters, although the primers did not match 100% to the target of any of the strains and clones tested. However, preferential amplification cannot be ruled out completely because species can always be missed during amplification and will therefore not be analysed.

TGGE and DGGE analysis of amplicons is a quick and reliable method for studying the dynamics of ecosystems, but the identification of single bands in the profiles is very time-consuming. Identification can be done by cutting out the bands in a profile followed by reamplification and sequence analysis. This approach has been applied successfully to ethidium bromide-stained gels (Muyzer & de Waal, 1994) and silver-stained gels (Ramírez-Saad, 1999; F. Schut, personal communication), but the disadvantage of this method is that a maximum of 500 bp can be used for sequence analysis. Identification can also be done by screening a clone library for dominant band positions, followed by sequence analysis. In this way, complete sequences could be retrieved, thereby making the phylogenetic analysis more reliable. This approach has been introduced by Felske *et al.* (1997; 1998a). The identity of the bands can be checked by sequencing more clones with identical motility or by blotting the profiles and using specific probes (Muyzer *et al.*, 1993; Felske *et al.*, 1997; Kowalchuk *et al.*, 1997).

Despite the increasing number of applications in molecular ecology, only a few studies have been performed with GI tract ecosystems. DGGE analysis of the V3 regions of 16S rDNA was used to study the role of uncultured bacteria in pre-term infants with and without necrotising enterocolitis (NEC) (Millar *et al.*, 1996). It was found that the number of uncultured bacteria in faecal samples from children with NEC was not more frequent than in faecal samples from children without NEC. TGGE based on the V6 to V8 regions of amplified 16S rDNA and 16S rRNA has been used to study the bacterial composition of different faecal samples (Zoetendal *et al.*, 1998). This study showed that each adult individual has his own faecal microbiota, which is relatively stable over time. Only a few amplicons were shared by all faecal samples. It was found that the faecal community in one person remained stable for almost 2 years (Fig. 11.6). A band corresponding to a cloned *Fusobacterium prausnitzii*-like ribotype increased slightly over this period. Furthermore, it was found that most dominant amplicons in an individual's profile derived from species that have not been cultured. Recently, the microbial community in the porcine GI tract has been studied using DGGE analysis of the V3 region of 16S rDNA. It was observed that unique bands were found in the fingerprints of faecal samples from pigs differing in age, and that the profiles were most similar

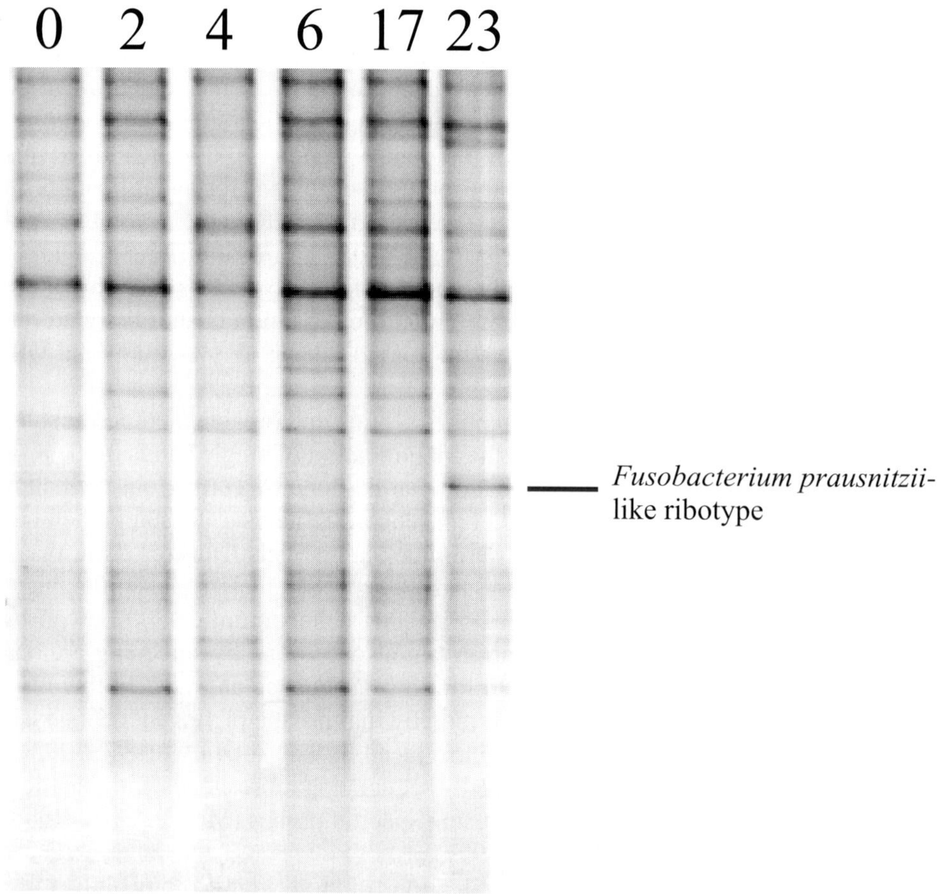

Fig. 11.6. DGGE profiles of the V6 to V8 regions of 16S rDNA from different faecal samples of one individual taken over a period of 23 months. A band originating from a *Fusobacterium prausnitzii*-like ribotype which increased over time is indicated.

within a single GI tract compartment and between adjacent ones (Simpson *et al.*, 1999).

(iii) Quantitative fingerprint analysis

The use of TGGE and DGGE to study complex ecosystems can be enhanced by quantifying profile similarities. Computer analysis of scanned fingerprints can be used to calculate similarity indices between fingerprints. These indices can be used to determine the stability of microbial communities or to monitor the effect of certain conditions which may change the composition of a microbial community. The calculation of similarity indices of DGGE profiles has been used to monitor the spatial and seasonal diversity of antarctic picoplankton assemblages (Murray *et al.*, 1998). A similar approach was used for samples from the porcine GI tract (Simpson *et al.*, 1999). The highest similarity indices were found within a single

A

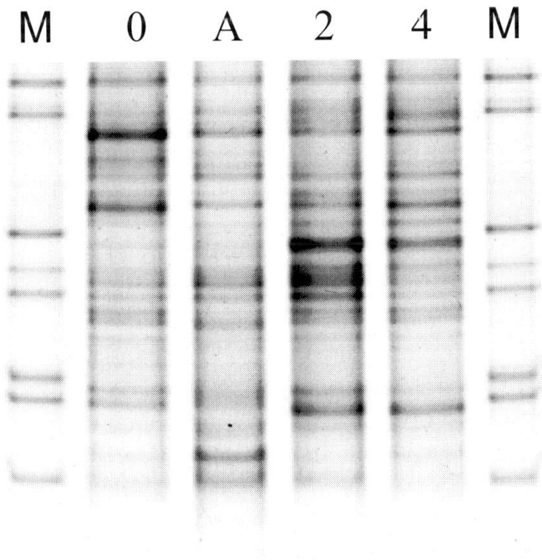

B

	0	A	2	4
0	100.0			
A	50.4	100.0		
2	33.5	42.7	100.0	
4	62.1	37.5	73.3	100.0

Fig. 11.7. DGGE profiles (A) of the V6–V8 regions of 16S rDNA, and (B) a similarity matrix expressed in Pearson correlations (x 100) of the DGGE profiles from faecal samples of one individual taken before (0), during (A), and after (2 and 4 months) treatment with doxycycline for 1 week. Lane M contains a marker consisting of cloned V6–V8 amplicons.

compartment and between adjacent compartments, indicating that the microbial communities were quite similar in these compartments.

Fig. 11.7 illustrates how densitometric curves of DGGE profiles can be used with the Molecular Analyst software (BioRad) to quantify the effect of treatment for 1 week with the antibiotic doxycycline on the dominant microbial community in faeces. A relatively stable microbial community was recovered 2 months after the treatment. The matrix illustrates these changes in a quantitative way and shows the high similarity between the faecal samples taken after the antibiotic treatment, thereby indicating the recovery of a stable microbial community. However, the community structures before and after treatment were not identical. These examples illustrate that the use of quantitative DGGE analysis is a reliable method to monitor changes in microbial communities and should be preferred above the subjective comparison made by eye.

Another method for quantifying DGGE profiles has recently been published (Nübel *et al.,* 1999). The Shannon-Weaver indices (which are the most common diversity indices and richness estimates of DGGE profiles) and two other cultivation-independent methods were used to quantify the microbial diversity and richness in different hypersaline microbial mats. A similar approach was used to study the effect of chlorobenzoates on the microbial community in soil (Ramírez-Saad, 1999). It was clearly demonstrated that the genetic diversity in the contaminated soils decreased significantly. In the case of the GI tract, the role of the host, food and antibiotics on the bacterial composition can be quantified by the methods described above. This fingerprinting analysis, in combination with multiple competitive PCR, quantitative profile comparison, or the calculation of diversity indices, can be used to determine and quantify changes in microbial composition caused by exposure to antibiotics in intestinal samples.

11.3 OLIGONUCLEOTIDE CHIP TECHNOLOGY

One of the new approaches that can be used to analyse environmental samples is the application of oligonucleotide microchips or microarrays (reveiwed by Lemieux *et al.,* 1998; Schena *et al.,* 1998). Microchips consist of oligonucleotides which are immobilised in a polyacrylamide gel matrix bound to a glass slide. Labelled target DNA or RNA can be added to the microchip, and the subsequent hybridisation signal can be detected and quantified using a computer-regulated camera connected to the microscope. Microarrays consist of numerous cloned or amplified DNA fragments rather than synthesised oligonucleotide probes, but the principle of microarrays is the same as that of microchips. Oligonucleotide microchips and microarrays have already been used several times for nucleotide sequencing. This sequencing by hybridisation (SBH) has been proposed as an alternative technique for genome sequencing (Drmanac *et al.,* 1993; Broude *et al.,* 1994; Yershov *et al.,* 1996). Microarrays have also been used to study the expression of certain genes, and also nucleotide variability between genes (Schena *et al.,* 1995; 1996; Chee *et al.,* 1996).

Recently, microarrays have been used to identify unique DNA regions present in a pathogenic strain of *Pseudomonas aeruginosa* which appeared to be missing in a control strain (Bangera *et al.,* 1999). The use of microchips could eventually be relatively cheap, because the concentration of probes per chip is very low, the information per analysis is high, and the chips can be used 20 to 30 times (Guschin *et al.,* 1997). The use of microarrays could be even cheaper because cloned amplicons are produced more economically than oligonucleotide probes. The application of these microchips or microarrays to answer microbial ecological questions looks promising. Environmental samples can be screened on microchips containing hundreds or thousands of 16S rRNA targeted probes, or on microarrays containing many cloned 16S rDNA fragments. However, one of the difficulties in using these approaches is the optimisation of the hybridisation conditions for the different immobilised DNA fragments (Kelly *et al.,* 1999). Recently, it has been reported that nitrifying bacteria and *Bacillus* spp. could be detected and identified at the rRNA level using oligonucleotide microchips (Guschin *et al.,* 1997; Kelly *et al.,* 1999). These approaches look promising for wider applications in microbial ecology. In the near future it might be possible to monitor expression of ribosomal and functional genes of an ecosystem with a single microchip or microarray.

11.4 CONCLUSIONS AND PERSPECTIVES

The application of TGGE, DGGE, and SSCP to studies in microbial ecology is growing and the future perspectives are promising. The combination of (RT-) PCR, cloning and fingerprinting of environmental samples may give an accurate description of a community. Despite some pitfalls concerned with biases in nucleic acid extraction and amplification methods, these techniques have been shown to be useful for describing the microbial composition of different ecosystems. The power of these techniques is their reliability, speed and ease of use. TGGE, DGGE and SSCP are ideal for studying the temporal and spatial variation in microbial communities, both qualitatively and quantitatively. The only time-consuming aspect is the identification of specific amplicons in the profiles. The approach of combining 16S rDNA profiles and profiles of functional genes may enable the structure to be related to the function of an ecosystem. Another variant of this approach is the use of oligonucleotide microchips. Extensive data can be obtained from a single analysis and may be quantitative, although the different T_m values might be a problem in quantification.

In studying the role of the microbial community in the intestine of man and other animals, the approaches described in this chapter should be applied instead of unreliable plate counting analysis to describe the microbial composition. The fingerprinting approach has already demonstrated that the dominant microbial community in adults is quite stable with time and differs for each individual. This approach can also be used to monitor the fate of certain bacteria, such as probiotic strains in the intestine (de Vos *et al.,* 1999). The impact of these strains on the microbial community in the GI tract can be analysed by quantifying similarities between the profiles, or calculating microbial diversity and richness indices from profiles. Changes in band positions can be identified using cloning and sequencing analysis. The introduction of internal standards to the profiles may help the changes to be monitored quantitatively. These approaches will definitely help in gaining an understanding of some aspects of the microbial community in the intestine.

It is evident that the use of PCR-based fingerprinting techniques is useful in answering ecological questions. Although the use of these techniques is still in development, their application has already been shown to be a powerful tool for determining the structure of microbial communities in different environments and monitoring changes in microbial communities without unreliable cultivation procedures.

ACKNOWLEDGEMENTS

Part of this work was supported by The Wageningen Centre for Food Sciences, the EU FAIR CT-97-3035, the PROBDEMO-Project Fair CT96-1028, and the VTT Biotechnology and Food Research.

REFERENCES

Akkermans, A.D.L., van Elsas, J.D. & de Bruijn F.J., eds. (1999). *Molecular microbial ecology manual.* Kluwer, Dordrecht.

Altschul, S.F., Gish, W., Miller, W., Meyers, E.W. & Lipman, D.J. (1990). Basic local alignment search tool. *Journal of Molecular Biology* **215**, 403–410.

Amann, R.I., Ludwig, W. & Schleifer, K-H. (1995). Phylogenetic identification and in situ detection of individual cells without cultivation. *Microbiological Reviews* **59**, 143–169.

Apajalathi, J.H., Särkilahti, L.K., Mäki, B.R.E., Heikkinen, J.P., Nurminen, P.H. & Holben, W.E. (1998). Effective recovery of bacterial DNA and percent-guanine-plus-cytosine based analysis of community structure in the gastrointestinal tract of broiler chickens. *Applied and Environmental Microbiology* **64**, 4084–4088.

Bachmann, B.J. (1990). Linkage map of *Escherichia coli* K-12, edition 8. *Microbiological Reviews* **54**, 130–197.

Bacot, C.M. & Reeves, R.H. (1991). Novel tRNA gene organization in the 16S-23S intergenic spacer of the *Streptococcus pneumoniae* rRNA gene cluster. *Journal of Bacteriology* **173**, 4234–4236.

Bangera, M.G., Norris, A., Lorry, S. & Olsen, M. (1999). Pathogenicity determinants in *Pseudomonas aeruginosa* associated with cystic fibrosis. In *Abstracts of the 99th general meeting of the American Society for Microbiology,* p.32. American Society for Microbiology, Washington DC.

Bassam, B.J., Caetano-Anollés, G. & Gresshoff, P.M. (1991). Fast and rapid silver staining of DNA in polyacrylamide gels. *Analytical Biochemistry* **196**, 80–83.

Baskaran, N., Kandpal, R.P., Bhargava, A.K., Glynn, M.W., Bale, A. & Weissman, S.M. (1996). Uniform amplification of a mixture of deoxyribonucleic acids with varying GC content. *Genome Research* **6**, 663–668.

Bentley, R.W. & Leigh, J.A. (1995). Determination of 16S ribosomal RNA gene copy number in *Streptococcus uberis, S. agalactiae, S. dysgalactiae* and *S. pauberis*. *FEMS Immunology and Medical Microbiology* **12**, 1–8.

Blok, H.J., Gohlke, A.M. & Akkermans, A.D.L. (1997). Quantitative analysis of 16S rDNA using competitive PCR and the QPCR™ system 5000. *BioTechniques* **22**, 700–704.

Broude, N.E., Sano, T., Smith, C.L. & Cantor, C.R. (1994). Enhanced sequencing by hybridization. *Proceedings of the National Academy of Sciences of the United States of America* **91**,

3072–3076.
Buchholz-Cleven, B.E.E., Rattunde, B. & Straub, K.L. (1997). Screening for genetic diversity of isolates of anaerobic Fe(II)-oxidizing bacteria using DGGE and whole cell hybridization. *Systematic and Applied Microbiology* **20**, 301–309.
Carrino, J.J. & Lee, H.H. (1995). Nucleic acid amplification methods. *Journal of Microbiological Methods* **23**, 3–20.
Chee, M., Yang, R., Hubbell, E., Berno, A., Huang, X.C., Stern, D., Winkler, J., Lockhart, D.J., Morris, M.S. & Fodor, S.P.A. (1996). Accessing genetic information with high-density DNA arrays. *Science* **274**, 610–614.
Clark, J.M. (1988). Novel non-templated nucleotide addition reactions catalyzed by procariotic and eukariotic DNA polymerases. *Nucleic Acids Research* **16**, 9677–9686.
Clapp, J.P. (1999). The identification of root-associated fungi by polymerase chain reaction-single strand conformational polymorphism (PCR-SSCP). In *Molecular microbial ecology manual,* Akkermans, A.D.L., van Elsas, J.D. & de Bruijn, F.J., eds, 3.4.7, pp.1–18. Kluwer, Dordrecht.
Collins, M.D., Willems, A., Cordoba, J.J., Fernandez-Garayzabal, J., Garcia, P., Cal, J., Hippe, H. & Farrow, J.A.E. (1994). The phylogeny of the genus *Clostridium*: proposal of five new genera and eleven new species combinations. *International Journal of Systematic Bacteriology* **44**, 812–826.
Coppella, S.J., Acheson, C.M. & Dhurjati, P. (1987a). Measurement of copy number using HPLC. *Biotechechnology and Bioengineering.* **29**, 646–647.
Coppella, S.J., Acheson, C.M. & Dhurjati, P. (1987b). Isolation of high-molecular weight nucleic acids for copy number analysis using high-performance liquid chromatography. *Journal of Chromatography* **402**, 189–199.
de Vos, W.M., Zoetendal, E.G., Poelwijk, E., Heilig, H. & Akkermans, A.D.L. (1999). Molecular tools for analyzing the functionality of probiotic properties of microorganisms. In *Proceedings of the 25th international dairy congress,* pp. 323–328. Danish National Committee of the International Dairy Federation, Aarhus.
Dell'Anno, A., Fabiano, M., Duineveld, G.C.A., Kok, A. & Danovaro, R. (1998). Nucleic acid (DNA, RNA) quantification and RNA/DNA ratio determination in marine sediments: comparison of spectrophotometric, fluorometric, and high-performance liquid chromatography methods and estimation of detrital DNA. *Applied and Environmental Microbiology* **64**, 3238–3245.
Doré, J., Sghir, A., Hennequart-Gramet, G., Corthier, G. & Pochart, P. (1998). Design and evaluation of a 16S rRNA-targeted oligonucleotide probe for specific detection and quantitation of human faecal *Bacteroides* populations. *Systematic and Applied Microbiology* **21**, 65–71.
Drmanac, R., Drmanac, S., Strezosca, Z., Paunesku, T., Labat, I., Zeremski, M., Snoddy, J., Funkhouser, W.K., Koop, B., Hood, L. & Crkvanjakov, R. (1993). DNA sequence determination by hybridization: a strategy for efficient large-scale sequencing. *Science* **260**, 1649–1652.
Edwards, A.W.F. & Cavilla-Sforza, L.L. (1963). The reconstruction of evolution. *Heredity* **18**, 553.
Ezaki, T., Li, N., Hashimoto, Y., Miura, H. & Yamamoto, H. (1994). 16S ribosomal DNA sequences of anaerobic cocci and proposal of *Ruminococcus hansenii* comb. nov. and *Ruminococcus productus* comb. nov. *International Journal of Systematic Bacteriology* **44**, 130–136.
Farrelly, V., Rainey, F.A. & Stackebrandt, E. (1995). Effect of genome sizes and *rrn* gene copy number on PCR amplification of 16S rRNA genes from a mixture of bacterial species. *Applied and Environmental Microbiology* **61**, 2798–2801.
Felsenstein, J. (1981). Evolutionary trees from DNA sequences: a maximum likelihood approach. *Journal of Molecular Evolution* **17**, 368–376.
Felske, A., Engelen, B., Nübel, U. & Backhaus, H. (1996). Direct ribosome isolation from soil to extract bacterial rRNA for community analysis. *Applied and Environmental Microbiology* **62**, 4162–4167.
Felske, A., Rheims, H., Wolterink, A., Stackebrandt, E. & Akkermans, A.D.L. (1997). Ribosome analysis reveals prominent activity of an uncultured member of the class Actinobacteria in grassland soils. *Microbiology* **143**, 2983–2989.

Felske, A., Wolterink, A., van Lis, R. & Akkermans, A.D.L. (1998a). Phylogeny of the main bacterial 16S rRNA sequences in Drentse A Grassland soils. *Applied and Environmental Microbiology* **64**, 871–879.

Felske, A., Akkermans, A.D.L. & de Vos, W.M. (1998b). Quantification of 16S rRNAs in complex bacterial communities by multiple competitive reverse transcription-PCR in temperature gradient gel electrophoresis. *Applied and Environmental Microbiology* **64**, 4581–4587.

Felske, A., Backhaus, H. & Akkermans, A.D.L. (1998c). Direct ribosome isolation from soil. In *Molecular microbial ecology manual,* Akkermans, A.D.L., van Elsas, J.D. & de Bruijn, F.J., eds, 1.2.4, pp. 1–10, Kluwer, Dordrecht.

Finegold, S.M., Sutter, V.L. & Mathisen, G.E. (1983). Normal indigenous intestinal flora. In *Human intestinal microflora in health and disease,* Hentges D.J., ed., pp. 3–31. Academic Press, New York.

Fischer, S.G. & Lerman, L.S. (1979). Length-independent separation of DNA restriction fragments in two-dimensional gel electrophoresis. *Cell* **16**, 191–200.

Fox, G.E., Wisotzkey, J.D. & Jurtshuk, P. (1992). How close is close: 16S rRNA sequence identity may not be sufficient to guarantee species identity. *International Journal of Systematic Bacteriology* **42**, 166–170.

Franks, A.H., Harmsen, H.J.M., Raangs, G.C., Jansen, G.J., Schut, F. & Welling, G.J. (1998). Variations of bacterial populations in human feces measured by fluorescent in situ hybridization with group-specific 16S rRNA-targeted oligonucleotide probes. *Applied and Environmental Microbiology* **64**, 3336–3345.

Guschin, D.Y., Mobarry, B.K., Proudnikov, D., Stahl, D.A., Rittmann, B.E. & Mirzabekov, A.D. (1997). Oligonucleotide microchips as genosensors for determinative and environmental studies in microbiology. *Applied and Environmental Microbiology* **63**, 2397–2402.

Harmsen H.J.M., Stams, A.J.M., Akkermans, A.D.L. & de Vos, W.M. (1995). Phylogenetic analysis of two syntrophic propionate-oxidizing bacteria in enrichment cultures. *Systematic and Applied Microbiology* **18**, 67–73.

Hayashi, K. (1991). PCR-SSCP: a simple and sensitive method for detection of mutations in the genomic DNA. *PCR Methods and Applications* **1**, 34–38.

Hayashi, K. & Yandell, D.W. (1993). How sensitive is PCR-SSCP? *Human Mutation* **2**, 38–346.

Head, I.M., Saunders, J.R. & Pickup, R.W. (1998). Microbial evolution, diversity, and ecology: a decade of ribosomal RNA analysis of uncultivated microorganisms. *Microbial Ecology* **35**, 1–21.

Holdeman, L.V., Good, I.J. & Moore, W.E.C. (1976). Human fecal flora: Variation in bacterial composition within individuals and a possible effect of emotional stress. *Applied and Environmental Microbiology* **31**, 359–375.

Johnson, J.L. (1991). Isolation and purification of nucleic acids. In *Nucleic acid techniques in bacterial systematics,* Stackebrandt, E. & Goodfellow, M., eds, pp. 1–19. Wiley, Chichester.

Jukes, T.H. & Cantor, C.R. (1969). Evolution of protein molecules. In *Mammalian protein metabolism,* Munro, H.N., ed., pp. 21–132. Academic Press, New York.

Kelly, J.J., Sappelsa, L., Bavykin, S.G., Mirzabekov, A.D. & Stahl, D.A. (1999). Optimization of DNA microarrays for the rapid characterization of microbial community population structure. In *Abstracts of the 99th general meeting of the American Society for Microbiology*, pp. 470–471. American Society for Microbiology, Washington D.C.

Kimura, M. (1980). A simple method for estimating evolutionary rates of base substitution through comparative studies of nucleotide sequences. *Journal of Molecular Evolution* **16**, 111–120.

Kopczynski, E.D., Bateson, M.M. & Ward, D.M. (1994). Recognition of chimeric small-subunit ribosomal DNAs composed of genes from uncultivated micro-organisms. *Applied and Environmental Microbiology* **60**, 746–748.

Kowalchuk, G.A., Stephen, J.R., de Boer, W., Prosser, J.I., Embley, T.M. & Woldendorp, J.W. (1997). Analysis of ammonia-oxidizing bacteria of the β subdivision of the class *Proteobacteria* in coastal sand dunes by denaturing gel electrophoresis and sequencing of PCR-amplified 16S

ribosomal DNA fragments. *Applied and Environmental Microbiology* **63**, 1489–1497.
Kreader, C.A. (1996). Relief of amplification in PCR with bovine serum albumin or T4 gene 32 protein. *Applied and Environmental Microbiology* **62**, 1102–1106.
Kuritza, A.P. & Salyers, A.A. (1985). Use of a species-specific DNA hybridization probe for enumeration of *Bacteroides vulgatus* in human feces. *Applied and Environmental Microbiology* **50**, 958–964.
Lane, D.J. (1991). 16S/23S rRNA sequencing. In *Nucleic acid techniques in bacterial systematics*, Stackebrandt, E. & Goodfellow, M., eds, pp. 115–175. Wiley, Chichester.
Langendijk, P.S., Schut, F., Jansen, G.J., Raangs, G.C., Camphuis, G.R., Wilkinson, M.F. & Welling, G.J. (1995). Quantitative fluorescence in situ hybridization of *Bifidobacterium* spp. with genus-specific 16S rRNA-targeted probes and its application in fecal samples. *Applied and Environmental Microbiology* **61**, 3069–3075.
Lee, D-H., Zo, Y-G. & Kim, S-J. (1996). Nonradioactive method to study genetic profiles of natural bacterial communities by PCR-single-strand-conformation polymorphism. *Applied and Environmental Microbiology* **62**, 3112–3120.
Lemieux, B., Aharoni, A. & Schena, M. (1998). Overview of DNA chip technology. *Molecular Breeding* **4**, 277–289.
Liesack, W., Weyland, H. & Stackebrandt, E. (1991). Potential risks of gene amplification by PCR as determined by 16S rDNA analysis of a mixed-culture of strict barophilic bacteria. *Microbial Ecology* **21**, 191–198.
Maidak, B.L., Olsen, G.J., Larsen, N., Overbeek, R., McCaughey, M.J. & Woese, C. (1997). The RDP (Ribosomal Database Project). *Nucleic Acids Research* **25**, 109–111.
Maniatis, T., Fritsch, E.F. & Sambrook, J. (1989). *Molecular cloning: a laboratory manual*, 2nd edn. Cold Spring Harbor Laboratory, New York.
Mau, M. & Timmis, K.N. (1998). Use of subtractive hybridization to design habitat-based oligonucleotide probes for investigation of natural bacterial communities. *Applied and Environmental Microbiology* **64**, 185–191.
McCartney, A.L., Wenzhi, W. & Tannock, G.W. (1996). Molecular analysis of the composition of the Bifidobacterial and *Lactobacillus* microflora of humans. *Applied and Environmental Microbiology* **62**, 4608–4613.
McFarlene, G.T. & Gibson, G.R. (1994). Metabolic activities of the normal colonic microflora. In *Human health: contribution of microorganisms*, Gibson, S.A.W., ed., pp. 17–38. Springer, Frankfurt.
McKillip, J.L., Jaykus, L-A. & Drake, M. (1998). rRNA stability in heat-killed and UV-irradiated enterotoxigenic *Staphylococcus aureus* and *Escherichia coli* O157:H7. *Applied and Environmental Microbiology* **64**, 4264–4268.
Millar, M.R., Linton, C.J., Cade, A., Glancy, D., Hall, M. & Jalal, H. (1996). Application of 16S rRNA gene PCR to study bowel flora of preterm infants with and without necrotizing enterocolitis. *Journal of Clinical Microbiology* **34**, 2506–2510.
Moore, W.E.C. & Holdeman, L.V. (1974). Human fecal flora: The normal flora of 20 Japanese-Hawaiians. *Applied and Environmental Microbiology* **27**, 961–979.
Murray, A.E., Hollibaugh, J.T. & Orrego, C. (1996). Phylogenetic compositions of bacterioplankton from two California estuaries compared by denaturing gradient gel electrophoresis of 16S rDNA fragments. *Applied and Environmental Microbiology* **62**, 2676–2680.
Murray, A.E., Preston, C.M., Massana, R., Taylor, L.T., Blakis, A., Wu, K. & de Long, E.F. (1998). Seasonal and spatial variability of bacterial and archaeal assemblages in the coastal waters near Avers island, Antarctica. *Applied and Environmental Microbiology* **64**, 2585–2595.
Muyzer, G. (1999). DGGE/TGGE a method for identifying genes from natural ecosystems. *Current Opinion in Microbiology* **2**, 317–322.
Muyzer, G. & de Waal, E.C. (1994). Determination of the genetic diversity of microbial communities using DGGE analysis of PCR-amplified 16S rRNA. *NATO ASI Series* **G35**, 207–214.
Muyzer, G. & Smalla, K. (1998). Application of denaturing gradient gel electrophoresis (DGGE) and

teria in a stratified fjord (Mariager Fjord, Denmark) as evaluated by most-probable-number counts and denaturating gradient gel electrophoresis of PCR-amplified ribosomal DNA fragments. *Applied and Environmental Microbiology* **62**, 1405–1415.

Vallaeys, T., Topp, E., Muyzer, G., Macheret, V., Laguerre, G. & Soulas, G. (1997). Evaluation of denaturing gradient gel electrophoresis in the detection of 16S rDNA sequence variation in Rhizobia and Methanotrophs. *FEMS Microbiology Ecology* **24**, 279–285.

van de Peer, Y. & de Wachter, R.(1995). Investigation of fungal phylogeny on the basis of small ribosomal subunit RNA sequences. In *Molecular microbial ecology manual,* Akkermans, A.D.L., van Elsas, J.D. & de Bruijn, F.J., eds, 3.3.4, pp.1–12. Kluwer, Dordrecht,

Wagner, R. (1994). The regulation of ribosomal RNA synthesis and bacterial cell growth. *Archives of Microbiology* **161**, 100–106.

Wang, G.C.Y. & Wang Y. (1996). The frequency of chimeric molecules as a consequence of PCR co-amplification of 16S rRNA genes from different bacterial species. *Microbiology* **142**, 1107–1114.

Wang, A.M., Doyle, M.V. & Mark, D.F. (1989). Quantification of mRNA by the polymerase chain reaction. *Proceedings of the National Academy of Sciences of the United States of America* **86**, 9717–9721.

Wang, R-F., Cao, W-W. & Cerniglia, C.E. (1996a). PCR detection of predominant anaerobic bacteria in human and animal fecal samples. *Applied and Environmental Microbiology* **62**, 1242–1247.

Wang, R-F., Cao, W-W. & Cerniglia, C. E. (1996b). Phylogenetic analysis of *Fusobacterium prausnitzii* based upon the 16S rRNA gene sequence and PCR confirmation. *International Journal of Systematic Bacteriology* **6**, 341–343.

Wawer, C., Jetten, M.S.M. & Muyzer, G. (1997). Genetic diversity and expression of the [NiFe] hydrogenase large subunit gene of *Desulfovibrio* spp. in environmental samples. *Applied and Environmental Microbiology* **63**, 4360–4369.

Wayne, L.G., Brenner, D.J., Colwell, R.R., Grimont, P.A.D., Kandler, O., Krichevski, M.I., Moore, L.H., Moore W.E.C., Murray, R.G.E., Stackebrandt, E., Starr, M.P. & Trüper, H.G. (1987). Report of the ad hoc committee on reconciliation of approaches to bacterial systematics. *International Journal of Systematic Bacteriology* **37**, 463–464.

Whitford, M.F., Forster, R.J., Beard, C.E., Gong, J. & Teather, R.M. (1998). Phylogenetic analysis of rumen bacteria by comparative sequence analysis of cloned 16S rRNA genes. *Anaerobe* **4**, 153–163.

Wilson, K.H. & Blitchington, R.H. (1996) Human colonic biota studied by ribosomal DNA sequence analysis. *Applied and Environmental Microbiology* **62**, 2273–2278.

Wintzingrode, F. v., Göbel, U. B. & Stackebrandt, E. (1997). Determination of microbial diversity in environmental samples: pitfalls of PCR-based rRNA analysis. *FEMS Microbiology Reviews* **21**, 213–229.

Woese, C.R. (1987). Bacterial evolution. *Microbiological Reviews* **51**, 221–271.

Woese, C.R., Kandler, O. & Wheelis, M.L. (1990). A definition of the domains *Archaea, Bacteria,* and *Eucarya* in terms of small subunit ribosomal characteristics. *Systematic and Applied Microbiology* **14**, 305–310.

Wolters D.J. (1998). *Ineffective Frankia in wet alder soils.* PhD Thesis, University of Groningen.

Yershov, G., Barsky, V., Belgovskiy, A., Kirillov, E., Kreidlin, E., Ivanov, I., Parinov, S., Guschin, D., Drobishev, A., Dubiley, S. & Mirzabekov, A. (1996). DNA analysis and diagnostics on oligonucleotide microchips. *Proceedings of the National Academy of Sciences of the United States of America* **93**, 4913–4918.

Yoshino, K., Nishigaki, K. & Husimi,Y. (1991). Temperature sweep gel electrophoresis: a simple method to detect point mutations. *Nucleic Acids Research* **19**, 3153.

Zoetendal, E.G., Akkermans, A.D.L. & de Vos, W.M. (1998). Temperature gradient gel electrophoresis analysis from human fecal samples reveals stable and host-specific communities of active bacteria. *Applied and Environmental Microbiology* **64**, 3854–3859.

12 From Multilocus Enzyme Electrophoresis to Multilocus Sequence Typing

Dominique A. Caugant[1,2]
[1]*Department of Bacteriology, WHO Collaborating Centre for Reference and Research on Meningococci, National Institute of Public Health, PO Box 4404 Nydalen, N-0403 Oslo, Norway;* [2]*Institute for Oral Biology, PO Box 1052 Blindern, N-0316 Oslo, Norway*

CONTENTS

12.1 INTRODUCTION

Differentiation and classification of bacterial strains at the sub-species level requires methods that are highly reproducible, little affected by the experimental conditions, and that give effective discrimination of epidemiologically unrelated strains. Bacteria derived from a single precursor through clonal expansion are genetically identical or closely related. To provide significant biological information, a typing method should reflect the amount of genetic variation that has accumulated over time since the divergence of the microbial organisms from a common ancestor.

The two main factors that determine the extent to which different microbial organisms have diverged from a common cell line are the time elapsed since their divergence and their potential for evolutionary changes. The potential for evolution of bacteria depends upon the genetic mechanisms that they have available (mutation, recombination, inversion, DNA repair, etc.), and their respective rates, as well as upon the genetic structure and the ecology of their population. Bacteria that are able to incorporate foreign DNA into their genome via (e.g.) transformation are less stable than species relying mainly on mutation. Thus, differentiation from the parental type in transformable organisms will generally occur much more rapidly than in non-transformable ones.

Depending on the type of questions to be addressed, different methods may be more or less suitable for molecular typing. The level of discrimination provided by a given technique varies with the degree of genetic diversity of the species, and the appropriateness of the variation depends on the epidemiological questions that are posed (local versus global epidemiology). It is essential, however, that any typing technique, regardless of the narrowness of the problem to be addressed, should be validated in comparison to a genotyping method using a multilocus approach.

Multilocus enzyme electrophoresis (abbreviated MEE or MLEE) has been used extensively in studies of population biology and phylogenetics of both eukaryotes and prokaryotes, and has proved to be the gold standard for studying the population genetics of bacteria (Boerlin, 1997). The impact of the method on our knowledge of bacterial genetics, systematics and epidemiology has been considerable since the beginning of the 1980s, essentially as a result of the pioneering work of Selander's group (Selander & Levin, 1980; Caugant *et al.*, 1981; Ochman *et al.*, 1983). In the following years, MEE was successfully applied to numerous bacterial species, and some of the most significant studies for medical and veterinary bacteriology are listed in Table 12.1. The application of the method has contributed to the knowledge of the natural history of infectious diseases and has practical consequences for disease control strategies (Caugant, 1998).

The technique is based on the analysis of electrophoretic variation in a set of chromosomally encoded cytoplasmic enzymes. Although the allelic variation is measured indirectly, it is a fully validated genotyping method (Struelens *et al.*, 1996). For several species of bacteria, the genetic relationships obtained by MEE have been shown to be strongly correlated with those obtained from DNA-DNA

Table 12.1. Bacterial species analysed using MEE, together with selected references. References marked in bold are the source of the listed data.

Species	No. isolates	No. loci	Mean no. alleles/locus	No. ETs[a]	H_{ETs}[b]	References
Acinetobacter baumanii	65	13		14		**Thurm & Ritter, 1993**
Actinobacillus acinomycetemcomitans	97	14	3.1	50	0.34	Caugant *et al.*, 1990; **Poulsen *et al.*, 1994**; Haubek *et al.*, 1995; 1997
Actinobacillus pleuropneumoniae	250	18	2.9	37	0.247	Musser *et al.*, 1987; **Møller *et al.*, 1992**
Aeromonas spp.	153	11	6.6	122	0.645	**Altwegg *et al.*, 1991**; Kokka *et al.*, 1991; Boyd *et al.*, 1994
Bacillus cereus/thuringiensis	36	15	5.3	27	0.553	**Carlson *et al.*, 1994**; Helgason *et al.*, 1998; Zahner *et al.*, 1989
Bacillus subtilis	60	13	5.9	55	0.93	**Istock *et al.*, 1992**
Bordetella spp.	188	15	2.7	38	0.178	Musser *et al.*, 1996; 1987; **Van der Zee *et al.*, 1997**
Borrelia spp.	50	11	5.9	35	0.673	**Boerlin *et al.*, 1992**; Balmelli & Piffaretti, 1995; 1996;
Burkholderia cepacia	213	8	4.4	164	0.574	**Wise *et al.*, 1995**
Burkholderia pseudomallei	18	10		9		**Norton *et al.*, 1998**
Campylobacter spp.	125	9	6.9	64	0.634	**Aeschbacher & Piffaretti, 1989**
Citrobacter diversus	42	20	2.2	16	0.244	**Li *et al.*, 1990**
Corynebacterium diphteriae	156	23		76		**Popovic *et al.*, 1996**
Enterobacter cloacae	62	12	3.7			**Gaston & Warner, 1989**
Enterococcus faecalis	65	15	2.7	26	0.264	**Tomayko & Murray, 1995**
Escherichia coli	1608	12	9.3	301	0.517	Selander & Levin, 1980; Caugant *et al.*, 1981;1985; **Ochman *et al.*, 1983**; Ochman & Selander, 1983; Achtman *et al.*, 1989; Selander *et al.*, 1986; 1987; Maslow, 1985; Beutin *et al.*, 1990; Tzabar & Pennington, 1991; Whittam *et al.*, 1983a, b; 1993. Ngeleka *et al.*, 1996; Rodrigues *et al.*, 1996; Gordon, 1997; Pupo *et al.*, 1997; Feng *et al.*, 1998; Martinez *et al.*, 1999

(Continued.)

Table 12.1. Continued.

Species	No. isolates	No. loci	Mean no. alleles/locus	No. ETs[a]	H_{ETs}[b]	References
Haemophilus influenzae	2209	17	6.4	280	0.467	**Musser *et al.*,** 1985; 1986; 1988a; 1988b; **1990**; Porras *et al.*, 1986a, b; Weinberg *et al.*, 1989; Musser & Selander, 1990; Quentin *et al.*, 1990; 1993; Lagos *et al.*, 1991; Fusté *et al.*, 1996; van Alphen *et al.*, 1997
Haemophilus parasuis	40	17	3.1	34	0.405	**Blackall *et al.*, 1997**
Helicobacter pylori	74	6	11.2	73	0.735	**Go *et al.*, 1996**
Klebsiella pneumoniae	19	9	2.3	11	0.378	**Nouvellon *et al.*, 1994**
Legionella spp.	292	22	4.3	62	0.380	**Selander *et al.*, 1985**; Woods *et al.*, 1988; Lanser *et al.*, 1990; Marques *et al.*, 1995
Listeria monocytogenes	175	16	3.6	45	0.424	**Pifaretti *et al.*, 1989**; Bibb *et al.*, 1989; 1990; Boerlin & Pifaretti, 1991; Boerlin *et al.*, 1991; 1996; 1997; Farber *et al.*, 1991; Kolstad *et al.*, 1992; Baxter *et al.*, 1993; Nørrung & Gerner-Schmidt, 1993; Nørrung & Skovgaard, 1993; Trott *et al.*, 1993; Graves *et al.*, 1994; Harvey & Gilmour, 1994; Lawrence & Gilmour, 1995; Rørvik *et al.*, 1995; Caugant *et al.*, 1996; Flint *et al.*, 1996; Nesbakken *et al.*, 1996
Mycobacterium tuberculosis complex	76	26	1.1	5	0.1	**Feizabadi *et al.*, 1996**
Mycobacterium avium	115	17	3.8	58	0.29	Wasem *et al.*, 1991; Arbeit *et al.*, 1993; **Feizabadi *et al.*, 1997**
Mycobacterium spp.	97	17	4.2	74	0.38	Wallace *et al.*, 1989; **Feizabadi *et al.*, 1996**
Neisseria meningitidis	688	15	7.2	331	0.547	**Caugant *et al.*,** 1986; **1987**; 1988; 1994; 1998; Crowe *et al.*, 1987; Olyhoek *et al.*, 1987; Moore *et al.*, 1989; Achtman 1990; 1994; 1995; 1997; Ashton *et al.*, 1991; Sacchi *et al.*, 1992a, b; Wang *et al.*, 1992; 1993; Woods *et al.*, 1992
Neisseria gonorrhoeae	227	9		89	0.41	Gutjar *et al.*, 1997; **O'Rourke & Stevens, 1993**; 1994; Vasquez *et al.*, 1993

(Continued.)

Table 12.1. Continued.

Species	No. isolates	No. loci	Mean no. alleles/locus	No. ETs[a]	H_{ETs}[b]	References
Ornithobacterium rhinotracheale	55	18		6		**Amonsin *et al.*, 1997**
Pasteurella haemolytica	178	18	2.5	22	0.297	Angen *et al.*, 1997; **Davies *et al.*, 1997a**
Pasteurella multocida	100	18	3.4	71	0.302	**Blackall *et al.*, 1998**
Pasteurella trehalosi	60	19	2.6	20	0.289	**Davies *et al.*, 1997b**
Porphyromonas gingivalis	100	17	4.5	78	0.384	**Loos *et al.*, 1993**
Prevotella spp.	51	14	3.9	46	0.52	**Frandsen *et al.*, 1995**
Pseudomonas aeroginosa	314	18		17	0.138	Levin *et al.*, 1984; **Martin *et al.*, 1999**
Pseudomonas syringae	23	26	3.5	10	0.683	**Denny *et al.*, 1988**
Rhizobium spp.	147	15	7.7	75	0.669	Young 1985; Young & Wexler, 1988; Pinero *et al.*, 1988; Harrison *et al.*, 1989; Segovia *et al.*, 1991; **Eardly *et al.***, 1990; **1995**
Salmonella enteritidis	96	24		80	0.627	Beltran *et al.*, 1988; 1991; Reeves *et al.*, 1989; Selander *et al.*, 1990a, b; 1992; **Boyd *et al.***, 1993; **1996**; Cox *et al.*, 1996
Serpulina hyodysenteriae	231	15	2.7	50	0.29	Lymbery *et al.*, 1990; **Trott *et al.*, 1997**
Serpulina pilosicoli	164	15	1.5	33	0.18	**Trott *et al.*, 1997**; 1998; Oxberry *et al.*, 1998
Serratia marcescens	99	9	3.0	33	0.376	**Gargallo-Viola, 1989**
Serpulina spp.	56	15		40	0.587	Stanton *et al.*, 1996; Trott *et al.*, 1996; **McLaren *et al.*, 1997**
Staphylococcus aureus	315	20	3.2	49	0.267	**Musser *et al.*, 1990**; Musser & Kapur, 1992; Kapur *et al.*, 1995b; Fitzgerald *et al.*, 1997
Streptococcus agalactiae	128	11	2.5	19	0.304	**Musser *et al.*, 1989**; Helmig *et al.*, 1993; Quentin *et al.*, 1995; Hauge *et al.*, 1996

(Continued.)

Table 12.1. Continued.

Species	No. isolates	No. loci	Mean no. alleles/locus	No. ETs[a]	H_{ETs}[b]	References
Streptococcus pyogenes	108	12	4.4	33	0.420	**Musser *et al.*, 1991**; 1992; 1993a, b; 1995; Martin & Single, 1993; Haase *et al.*, 1994; Reda *et al.*, 1994; Whatmore *et al.*, 1994
Streptococcus pneumoniae	66	17	2.2	28	0.247	Munoz *et al.*, 1991; 1992; Martin *et al.*, 1992; **Sibold *et al.*, 1992**; Soares *et al.*, 1993; Versalovic *et al.*, 1993; Klugman *et al.*, 1994; Lomholt, 1995; Coffey *et al.*, 1995; McDougal *et al.*, 1995; Hall *et al.*, 1996; Pons *et al.*, 1996; Takala *et al.*, 1996
Streptococcus suis	124			17		Hampson *et al.*, 1993; **Mwaniki *et al.*, 1994**
Streptococcus spp.	50	16	12.2	40	0.857	**Gilmour *et al.*, 1987**; Bert *et al.*, 1995
Treponema spp.	34	10	7.8	34	0.751	**Dahle *et al.*, 1995**
Vibrio cholerae	397	17	9.5	279	0.441	Wachsmuth *et al.*, 1993; Evins *et al.*, 1995; Popovic *et al.*, 1995; **Beltran *et al.*, 1999**
Yersinia spp.	244	18	6.7	137	0.531	Schill *et al.*, 1984; Caugant *et al.*, 1989; **Dolina & Peduzzi, 1993**

[a] ETs, electrophoretic types.
[b] H_{ETs}, mean genetic diversity among ETs at the loci studied.

hybridisation analysis (Ochman *et al.*, 1983; Selander *et al.*, 1985; Gilmour *et al.*, 1987; Boerlin *et al.*, 1991). This demonstrates that the analysis of 10–20 enzymes can adequately index the variation in the whole bacterial genome and give a representative measure of genetic relatedness among isolates. This also provides evidence for the relevance of the method to taxonomic applications.

The cytoplasmic enzymes that are analysed in MEE are essential for bacterial metabolism and are coded by so-called housekeeping genes. Allelic variation in such genes is usually neutral or very nearly so, which means that strains possessing either one or another allele will have basically the same fitness. Convergence to the same allele through adaptative evolution is thus unlikely to happen. Differences between strains in a set of housekeeping genes reflect the genetic events that have occurred overall in their genome since their divergence from a common ancestor, making possible phylogenetic interpretations of the data (Musser, 1996). Bacteria with identical multilocus enzyme genotypes are assumed to descend from a single ancestral cell line and to be members of the same clone. As a result, MEE analyses have often been designated as clonal analyses.

12.2 MEE METHODOLOGY

The technique identifies naturally occurring allelic variation in chromosomal housekeeping genes by indexing differential electrophoretic migration of the gene products (the enzymes) in a support, usually a starch gel. The migration of a protein in such gels is a function of its molecular mass, its electrophoretic charge and the conformation of the molecule. Mutations, especially those affecting the net electrostatic charges of the proteins, are reflected by differential migration of the individual enzymes in the electrophoretic field. Evidently, only a fraction of all the base substitutions occurring in the nucleotide sequence of genes can be detected in this manner; silent substitutions at the nucleotide level, for example, will not be revealed, as they are not affecting the structure of the protein. Up to 50% of the mutations resulting in an amino acid change can be detected by ordinary electrophoresis, but by using sequential electrophoresis in which the pH, buffer system and pore size of the gel are varied, nearly all amino acid sequence variation can be revealed (Selander *et al.*, 1986a).

The level of expression of certain enzymes may vary with the growth temperature or the composition of the culture medium (e.g., detection of β-galactosidase activity in *Escherichia coli* requires induction of the enzyme). The electrophoretic mobilities (electromorphs) of individual enzymes are, however, unaffected by the culture conditions, methods of storage, number of passages in the laboratory and other environmental factors (Selander *et al.*, 1986a; Musser *et al.*, 1987a, b). Consequently, all bacterial isolates can be unequivocally characterised by their combination of electromorphs over the loci studied, and all strains are typable.

To provide a reliable index of the overall genome and enable appropriate interpretation of the data for epidemiological studies, an analysis of at least 10 enzyme loci is necessary. For taxonomic and population genetics analyses, somewhat larger numbers of chromosomal loci are required. In some studies, to establish the strength of the technique, more than 30 loci have been assayed for limited numbers of isolates (Selander *et al.,* 1987b). Because of its relative rapidity, the method employed by most investigators has been starch gel electrophoresis (Selander *et al.,* 1986a; Boerlin & Pifaretti, 1995).

A. Preparation of the bacterial extracts

A pure culture of each isolate is grown under appropriate conditions to obtain a sufficient number of cells (about 10^{11}) for the preparation of the enzyme lysate. The cells can be scraped directly from agar plates or slants (Feizabadi *et al.*, 1997), but an overnight culture in broth is usually preferred, as large quantity of cells may be obtained at lower cost in that way. For most organisms that are easy to grow, a 100-ml broth culture will give a suitable cell quantity. For difficult organisms, such as *Borrelia burgdorferi,* 2-L cultures have been used to prepare the protein extracts (Boerlin *et al.,* 1992). If a broth culture is used, the cells are collected by centrifugation.

After harvest, the cells are suspended in a small volume (1–2 ml) of buffer solution (e.g., 10 mM Tris-HCl, 1mM EDTA, pH 6.8) and lysed by a suitable method. Sonication, vortexing with glass beads, repeated freezing and thawing, or a combination of these methods can be used. The essential element is that the proteins must not be denatured during the lysis of the cells. Thus, the bacterial suspension must be kept cool at all time during the lysis process and thereafter. When beginning the study of a new organism, it is recommended to test several methods of protein extraction, with a limited set of isolates, for enzymes commonly found in most species, in order to identify the most convenient method providing sufficient enzyme activity. After lysis, the cellular debris are pelleted by centrifugation at 20,000 *g* for 20 min at 4°C. For pathogenic bacteria, filtration of the supernatant through a 0.45 μm membrane is recommended, followed by storage of the lysate at –70°C until required for electrophoresis.

B. Gel preparation

Starch gels are convenient because they can be sliced in their thickness and different enzymes may be stained on each slice of a gel. However, other types of gels such as acrylamide (Gaston & Warner, 1989; Fusté *et al.*, 1996), cellulose acetate (Wise *et al.*, 1995; Souza *et al.*, 1999) or agarose may be employed. The resolution obtained using different supports has not been evaluated in detail, but improved resolution and reproducibility has been reported using polyacrylamide instead of starch (John & Hussain, 1994; Flint *et al.*, 1996). However, a significant advantage of starch over polyacrylamide is its absence of toxicity.

The quality of the starch is critical to permit a regular slicing of the gel in the thickness and assure a good resolution of the electromorphs. Hydrolysed starch from Connaught Laboratories (Ontario, Canada) has proved reliable through many

Table 12.2. Six buffer systems commonly used for MEE

System	Electrode buffer	Gel buffer	Voltage (V)
A	Tris-citrate, pH 8.0 83.20 g Tris, 33.09 g citric acid monohydrate, 1 L water	Tris-citrate, pH 8.0 Electrode buffer diluted 1:29	130
B	Tris-citrate, pH 6.3 27.00 g Tris, 18.07 g citric acid monohydrate, 1 L water; pH adjusted with NaOH	Tris-citrate, pH 6.7 0.97 g Tris, 0.63 g citric acid monohydrate, 1 L water; pH adjusted with NaOH	150
C	Borate, pH 8.2 18.50 g boric acid, 2.40 g NaOH, 1 L water	Tris-citrate, pH 8.7 9.21 g Tris, 1.05 g citric acid monohydrate, 1 L water	250
D	Lithium hydroxide, pH 8.1 1.20 g LiOH monohydrate, 11.89 g boric acid, 1 L water	Lithium hydroxide, pH 8.3 Electrode buffer diluted 1:9 in 6.20 g Tris, 1.60 g citric acid monohydrate, 1 L water	325
F	Tris-maleate, pH 8.2 12.10 g Tris, 11.60 g maleic acid, 3.72 g Na_2 EDTA, 2.03 g $MgCl_2 \cdot 6H_2O$, 1 L water, pH adjusted with 5.15 g NaOH	Tris-maleate, pH 8.2 Electrode buffer diluted 1:9	100
G	Potassium phosphate, pH 6.7 18.14 g KH_2PO_4, 2.39 g NaOH, 1 L water	Potassium phosphate, pH 7.0 1.06 g KH_2PO_4, 0.25 g citric acid monohydrate, 1 L water	100

years of use, with little quality variation from batch to batch. Starch at a concentration of 11–12% is suspended in an appropriate volume of gel buffer (e.g., 48 g starch for 420 ml buffer) in a 1-L Erlenmeyer flask with thick walls, tolerating heat and vacuum. Some commonly used gel buffers and the corresponding electrode buffers are listed in Table 12.2.

The mixture is heated over a Bunsen burner with continuous and vigorous hand swirling until the suspension starts boiling and develops large air bubbles. The boiling time may need to be adjusted depending on the gel buffer and the batch of the starch. Constant swirling is necessary to avoid burning of the starch at the bottom of the flask. After boiling, the gel is degassed for 1 min and immediately poured into a gel mould, positioned on a perfectly level surface. A gel mould with a size of 18 × 20 × 1 cm is suitable for a volume of 420 ml of gel buffer and can be used for the electrophoresis of 20 samples. If air bubbles are visible in the gel after pouring, they should be quickly removed by aspiration with a Pasteur pipette before the starch begins to solidify. The surface of the gel should be level and even. A gel that sticks to the flask during pouring and has an uneven surface is undercooked. The gel is left to solidify either at room temperature for 2 h or at room temperature for 30 min, followed by 30 min at 4°C, before wrapping in plastic film to prevent desiccation. To keep the surface of the gel smooth, no air bubbles should be trapped under the plastic film. For optimal results, a gel must be used within 24 h of its preparation. Normally, gels are stored overnight at room temperature before electrophoresis; the gels may be used on the day they are made, but they will be more difficult to slice.

C. Electrophoresis

The protein extracts of the bacteria to be analysed are thawed and immediately put on ice. The gel is unwrapped and a slit is cut through it with a scalpel at a distance of 5 cm from the shorter side of the gel mould. Pieces of Whatman filter paper no. 3 (6×9 mm) are used to load the lysate into the gel. Using forceps, a piece of filter paper is dipped in the cell lysate, blotted on a filter paper to eliminate excess liquid, then inserted into the slit in the gel, leaving a 1 cm space from the left end of the gel. The paper must be placed perfectly straight, with its short side in contact with the bottom of the gel form. The same procedure is repeated with the next lysate, leaving a 2–3 mm space between the filter papers. Up to 20 samples can be loaded on a 18-cm wide gel. Pieces of filter paper, dipped in amaranth dye (100 mg amaranth dissolved in 1 ml ethanol, plus 19 ml water), are placed in the spaces remaining at each end of the slit to mark the migration of the buffer front during electrophoresis. The two pieces of the gel are then carefully pressed together to eliminate the air between the filter papers and avoid denaturation of the proteins by drying. The gel slit containing the filter papers is then covered again with the plastic film, which is then folded back in a straight line at 2.5 cm from the end of the gel.

The electrophoresis tanks are filled with about 250 ml of appropriate buffer (see Table 12.2). Two thin sponge wicks, 10 cm apart on each side of the gel, provide contact between the gel and the buffer. One wick is lined up with the row of samples, separated by the plastic film, and the other wick is placed directly on the gel. Both are covered again with the plastic film. During electrophoresis, the gel is cooled by a pan of ice supported by a glass plate. A constant voltage is maintained during electrophoresis (Table 12.2). The duration of electrophoresis in these conditions varies from 4–8 h, depending on the buffer. An alternative is to perform the electrophoresis at a lower voltage, running the gels overnight. In this case, the electrophoresis should be carried out in a cool room. Standardisation of the migration between gels is assured by measuring the migration of the amaranth dye.

After electrophoresis, three to four slices (1–2 mm thick) may be cut from each gel using a thin wire drawn through the gel placed on a slicing tray. Each slice is carefully placed in a plastic box, labelled with the gel number and the name of the enzyme to be stained. The quality of the slices decreases from the bottom to the top of the gel. It is thus recommended to always stain the same enzymes on the same slice level to assure reproducibility of the reading.

D. *Staining procedures*

The gels are immersed in a freshly prepared solution containing the substrate for the enzyme, coenzyme, intermediary catalyst and dye, and possibly other ingredients such as intermediary enzymes. Staining methods commonly used with bacteria for 31 enzymes are given in Table 12.3. The enzymatic reactions involved are illustrated in the handbook of Harris & Hopkinson (1976), where staining methods for additional enzymes can also be found. When intermediary enzymes are needed to reveal a specific enzyme activity (e.g., for an isomerase, such as phosphoglucose isomerase, or a transferase, such as hexokinase), it is usually necessary to mix the staining ingredients in an agar overlay to ensure the sharpness of the bands of enzymatic activity.

The gels are incubated at 37°C, usually in the dark, until appearance of the enzyme reaction (dark bands). Indophenol oxidase and catalase are stained in the light, at room temperature. Enzyme activity for these two enzymes shows as white bands on a dark background. Depending on the enzyme and the bacterial species being analysed, the staining reaction may take from a few minutes to several hours. In some cases, the gel may be incubated overnight. After staining, the solution is poured off and the gel slice rinsed in water if no agar overlay was used. The gels are then fixed in a 1:5:5 solution of acetic acid, methanol and water. Gels stained for indophenol oxidase and catalase should not be fixed, but can be kept in water.

For each bacterial species, the optimal electrophoretic conditions for each individual enzyme need to be determined. An enzyme that may appear monomorphic in one buffer system may present numerous electromorphs in another buffer. Therefore, electrophoresis of a small number of strains, using the various buffer systems given in Table 12.2, possibly with different times of migration or different gel concentrations, needs to be performed for each enzyme. The bands of enzyme activity must be narrow and clearly visible to assure good resolution. Fig. 12.1 shows gel slices in which the same 20 isolates of *Neisseria meningitidis* have been electrophoresed and stained for glucose 6-phosphate dehydrogenase, isocitrate dehydrogenase and alkaline phosphatase, respectively.

E. *Interpretation of the gels*

Relative mobilities of each enzyme from different isolates must be compared visually against one another on the same gel. Distinctive electromorphs are numbered in order of decreasing anodal mobility, i.e., the electromorph that has migrated the farthest is assigned the number 1. In Fig. 12.1, the gel slice stained for glucose 6-phosphate dehydrogenase revealed two electromorphs, while the gel slice stained for isocitrate dehydrogenase from the same 20 bacteria revealed four electromorphs. In some bacterial species, more than 20 electromorphs of an individual enzyme may be detected by ordinary starch gel electrophoresis, illustrating the discriminatory power of the technique.

In bacteria, genes coding for an enzyme activity usually exist as a single copy.

rium may result from the occurrence of an additional copy of the gene carried on a plasmid (Caugant *et al.*, 1981).

For some enzymes, several bands of activity may appear in all isolates of a given species. This may be due to the existence of several conformational forms of certain enzymes (i.e., different multimeric structures of the same protein showing enzymatic activity) or to a low specificity of the staining method. The later eventuality is exemplified by the esterases, where several genetic loci code for esterases differing in their affinity for various substrates. To ensure that the recorded polymorphism originates from a single genetic locus, substrates that can be utilised only by more specific enzymes should be chosen (e.g., naphthyl-propionate instead of naphthyl-acetate). The use of esterase inhibitors may also eliminate some bands of activity and permit the identification of electromorphs coded by the same locus. Enzyme activities that are not being specifically stained for may sometimes be detected. In *N. meningitidis*, an additional band of activity appears at a similar location on most gels stained for a dehydrogenase. This unknown dehydrogenase is clearly polymorphic and has been incorporated as one of the enzymes routinely scored for epidemiological analysis of the meningococcus (Caugant *et al.*, 1990a). However, it is important that allelic variation in such enzymes is recorded only once, although they may be revealed on gels stained for different enzymes. An analogous situation sometimes occurs with the peptidases. The detection of peptidases encoded by different genes can be achieved by using different peptides (see Table 12.3). However, in some species, a peptidase may have broad specificity and hydrolyse several peptides, e.g., phenylalanyl-leucine and leucyl-glycyl-glycine, although, in most bacteria, this is achieved by enzymes encoded by two distinct loci.

One or a few strains of a bacterial species may occasionally lack activity for an enzyme. Although housekeeping genes code for the stained metabolic enzymes, they are not always needed for survival, especially when the strain has been subcultured in the laboratory for a long time. Null alleles are recorded as such. It is essential, however, to assess that they do not result from a poor preparation or storage of the protein lysate.

Because the electromorphs of an enzyme detected within a species are often very close to each other, measuring the distance of migration from the origin is not sufficient. All electromorphs must be identified by side by side comparison on the same gel slice. In practice, this means that the protein extract from a single bacterium will be electrophoresed and stained for the same enzyme many times before its electromorph can be unambiguously identified. When analysing a new set of strains, it is recommended to include on the gel at least the extracts from two reference strains of the same species, if possible presenting distinct electromorphs for most enzymes assayed. For a gel allowing electrophoresis of 20 samples, the reference extracts should be put in positions 7 and 14 in the gel. The first migration permits a rough evaluation of the electromorphs of the unknown strains in comparison to the two standards. Accurate electromorph assignment for each individual enzyme should be obtained by repeatedly running the strains side by side with different reference electromorphs, until identity is ascertained. For species with only a few alleles at individual loci, identification of the electromorphs is an easy task. For highly polymorphic species, numerous successive runs are necessary to identify the electromorphs. Consequently, the amount of work involved in the analysis of 100 strains by MEE increases significantly in relation to the degree of polymorphism of the species.

F. Analysis of the data

For each isolate, an electromorph is assigned to each of the 10–20 enzyme loci. This combination of electromorphs represents the allelic profile of the bacterium or its multilocus genotype (Table 12.4). Distinct allelic profiles are also designated as electrophoretic types (ETs). Thus, each isolate can be characterised by its ET, and bacteria can readily be classified as having identical or different ETs.

Further analyses of the data are performed using computers. Programs for analysing MEE data for bacterial strains have been developed and made available for general use by Dr T.S. Whittam (Selander *et al.*, 1986a). These programs have been further improved in the past decade and are now available through the Internet at the home page of Dr Whittam's laboratory (*http://www.bio.psu.edu/People/Faculty/Whittam/Lab/programs*).

The statistical package includes five programs: ETDIV, ETCLUS, ETMEGA, ETLINK and ETBOOT.

To analyse the electrophoretic data for a set of strains, each electromorph must be given as an integer. Null alleles are coded as "0" and are treated as missing information. Consequently, strains with null alleles are assigned to the same ET as strains with similar enzyme profiles, but which express enzyme activity at the locus for which a null allele is found. If it is wished to distinguish the strains with null alleles, the electromorph should be assigned an integer different from the electromorphs represented in the collection of strains. The input data files need to be saved as text files.

ETDIV identifies ETs within the group of bacterial isolates analysed. It provides a list of the distinct allelic profiles, indicates the number of isolates with each ET, and lists the isolates belonging to each ET represented by more than one strain. For each enzyme locus, the alleles identified are indicated together with their frequency, and from the allele frequencies the genetic diversity is calculated as $h = (1 - \Sigma x_i^2)\,[n/(n\text{-}1)]$, where x_i is the frequency of the ith allele and n is the number of ETs (Nei, 1978). Mean diversity per locus (H) is the arithmetic average of h values over the loci studied. The strains to be analysed can be grouped beforehand in defined populations, according to specific parameters, such as their geographical origin, clinical source, serotype, biotype, etc. ETDIV will analyse the allele frequencies and genetic diversity concomitantly for isolates in each population and for the whole sample. A table will then be generated indicating the ET diversity within the population and in the whole sample, together with the coefficient of genetic differentiation for each locus and averaged over all loci (Nei, 1977).

C. *Molecular epidemiology*

Distinct clones within a species are frequently characterised by unique combinations of virulence genes or alleles of virulence genes. Thus, the identification of pathogenic clones has important implications for our understanding and control of infectious diseases.

MEE measures variation that is accumulating very slowly in the population. Thus, for short-term or local molecular epidemiology of the highly clonal species, the method presents relatively moderate discriminatory power. For some salmonellae, for example, MEE may not be sufficient to ascertain whether or not strains belong to the same outbreak, or whether a special food source may be linked to an outbreak. In contrast, for species with significant rates of recombination in nature, such as the transformable species *N. meningitidis* and *S. pneumoniae*, MEE has proved very effective for short-term or local epidemiology. For long-term or global epidemiology of nearly all bacterial species, MEE has been, until recently, the most appropriate method. As will be seen from the examples below, although MEE has been applied to the study of bacteria for less than two decades, it has already provided considerable insight into the molecular mechanisms of temporal and geographical variation in disease frequency, the adaptation of clonal lineages to the environment, and the relationship between disease severity and specific naturally occurring bacterial clones. However, because it is somewhat cumbersome, the method has only been used by a limited number of specialised laboratories

(i) Neisseria meningitidis

Extensive work has been devoted in the last 15 years to analysing the molecular epidemiology of *N. meningitidis*, predominantly using MEE. Thousands of meningococcal strains from patients and healthy carriers in all continents have been analysed and many hundreds of clones have been identified (Caugant *et al.*, 1986; 1990a; Olyhoek *et al.*, 1987; Achtman *et al.*, 1990; 1995; Ashton *et al.*, 1991; Woods *et al.*, 1992; Caugant, 1998; Maiden *et al.*, 1998). In spite of tremendous clonal diversity, most meningococcal disease in the world is caused by a handful of lineages of very closely related clones, reflecting the epidemic population structure of the organism.

One such lineage (the ET-5 complex), which has no close genetic relationship to other lineages of the species, was responsible for an epidemic of serogroup B meningococcal disease that began in Norway in the early 1970s, and subsequently spread through much of Europe. Clones of the ET-5 complex have been traced to all continents, where they have caused outbreaks and epidemics of invasive disease (Caugant, 1998).

MEE analysis has provided evidence of the dynamics of meningococcal clones causing disease. In a study of serogroup B isolates that caused invasive disease in The Netherlands between 1958 and 1986, significant temporal variation in the clonal composition of meningococcal populations was found (Caugant *et al.*, 1990a). Starting in 1980, a new clone complex (called lineage III) was identified and, in the subsequent years, its prevalence increased to reach 20% of the disease isolates, becoming the most prevalent clone causing disease in the population (Scholten *et al.*, 1994). Thereafter, lineage III spread to many European countries, then reached New Zealand in the mid-1990s where a severe epidemic is ongoing, affecting especially the Maoris and Pacific Islanders (Martin *et al.*, 1998a, b).

An increase of invasive disease due to serogroup C *N. meningitidis* strains has been recently reported in several countries, including Canada (Ashton *et al.*, 1991), various regions of the United States (Jackson *et al.*, 1995), the Czech Republic (Kriz *et al.*, 1999), the United Kingdom (Kaczmarski, 1997) and Australia (Jelfs *et al.*, 1998). The organism responsible for these outbreaks is a new variant of an "old" clone-complex of *N. meningitidis* that can be differentiated from the ancestral clone at a single enzyme locus, the fumarase gene, and was designated ET-15 (Ashton *et al.,* 1991). ET-15 organisms have a significantly higher case-fatality ratio than other invasive meningococcal disease isolates, which may be due to lower herd immunity to the newly emerged clone (Whalen *et al.*, 1995).

Clonal analysis of *N. meningitidis* strains expressing the serogroup A capsular polysaccharide has shown that they are a restricted phylogenetic subpopulation of the species (Caugant *et al.*, 1987). Serogroup A meningococcal strains are unusual in that they may cause large epidemics, with an incidence of >500/100,000, sometimes encompassing several countries or continents. Since the Second World War, such epidemics have been restricted to China and the Sahel region of sub-Saharan Africa. Achtman and co-workers have assembled and characterised serogroup A meningococcal isolates representing the organisms responsible for most epidemics or outbreaks since the 1960s (Olyhoek *et al.*, 1987; Wang *et al.*, 1992). Most epidemics were due to a single clone, and the same clone was often responsible for epidemics in contiguous countries. One such clone, designated clone III-1, has been responsible for two pandemics that both started in China, 15 years apart (Fig. 12.4). The first pandemic moved from China in the 1960s to Romania, Russia and Scandinavia in 1969, then to Brazil at the beginning of the 1970s. The second clone III-1 pandemic started in the early 1980s, again in China. In August 1987, 7000 cases of meningococcal disease occurred during the annual Haj pilgrimage to Mecca, Saudi Arabia, probably carried by pilgrims from South Asia (Moore *et al.*, 1989). The clone was then spread worldwide by returning pilgrims. Cases were reported in the United States, England and France among pilgrims and their close-contacts, but the strain did not spread further and no epidemics developed in these countries (Achtman, 1995). Clone III-1 strains were then introduced for the first time to the African continent. Major epidemics followed in Chad, Ethiopia, Sudan and Kenya in 1988 and 1989 (Moore *et al.*, 1989; Tekle Haimanot *et al.*, 1990; Salih *et al.*, 1990; Pinner *et al.*, 1992), followed by Niger, the Central African Republic, Burundi, Guinea, Mali, Zambia, Cameroon, Uganda and Rwanda (Caugant, 1998). In 1996, the sub-Saharan region of Africa was again affected by a clone III-1 epidemic of unprecedented scale with over 150,000 reported cases and 16,000 deaths, of which nearly 80% occurred in two countries, Burkina Faso and Nigeria. Meningitis outbreaks caused by clone III-1 are still ongoing in sub-Saha-

1989).

In the food industry, it was generally believed that contamination of products with *L. monocytogenes* resulted from the spread of bacteria that originated from animal or fish sources. However, the studies of Boerlin & Pifaretti (1991), Harvey & Gilmour (1994), Rørvik *et al.* (1995), Nesbakken *et al.* (1996) and Boerlin *et al.* (1997) have shown that contamination of meat and fish products with *L. monocytogenes* originates mainly from the processing environment rather than from animal or fish sources.

12.4 MLST METHODOLOGY

Recent developments in automation and nucleic acid sequencing chemistries have made possible the use of large-scale DNA sequencing techniques. These can be used to rapidly and unambiguously identify a causative infectious agent, and confirm or refute the identity of isolates recovered from temporally-linked patients thought to be involved in a disease outbreak. Accurate strain identification using a newly developed multilocus typing scheme based on DNA sequencing, designated multilocus sequence typing (MLST), has improved epidemiological surveillance of major bacterial pathogens, such as *N. meningitidis* and *S. pneumoniae*.

MLST and MEE are based on the same principles, i.e., the analysis of allelic variation in multiple, randomly selected, housekeeping genes that diversify through random accumulation of neutral variation. While the technique of MEE identifies naturally occurring allelic variation in chromosomal housekeeping genes indirectly by indexing the variation at the protein level, MLST uses variations in fragments of the nucleic acid sequence (about 500 bases) that make-up these housekeeping genes.

The principles and an evaluation of MLST were recently reviewed by Spratt (1999). MLST clearly presents a number of advantages over MEE. The technique is highly automated and the sequence data are unambiguous. Thus, data can be readily compared between laboratories. No reference strains are needed to standardise the results within or between laboratories. This is a significant advantage because, for a very polymorphic organism such as *N. meningitidis*, more than 50 reference strains are necessary to include all the alleles at the 14 loci studied. Potentially, the MLST method can be utilised even without prior cultivation of the organism. Due to its electronic portability via the Internet and the possibilities for creation of a global database, the technique has considerable importance for international epidemiological surveillance.

Currently, MLST schemes have been developed for three organisms: *N. meningitidis* (Maiden *et al.*, 1998), *S. pneumoniae* (Enright & Spratt, 1998) and *Helicobacter pylori* (Achtman *et al.*, 1999). In each case, seven housekeeping gene loci have been used, which appears to be the minimal set required to obtain a good representation of the overall genome. The genes sequenced should be widely separated on the chromosome and should not be adjacent to genes that might be under selective pressure. While working with a more limited number of genes than with

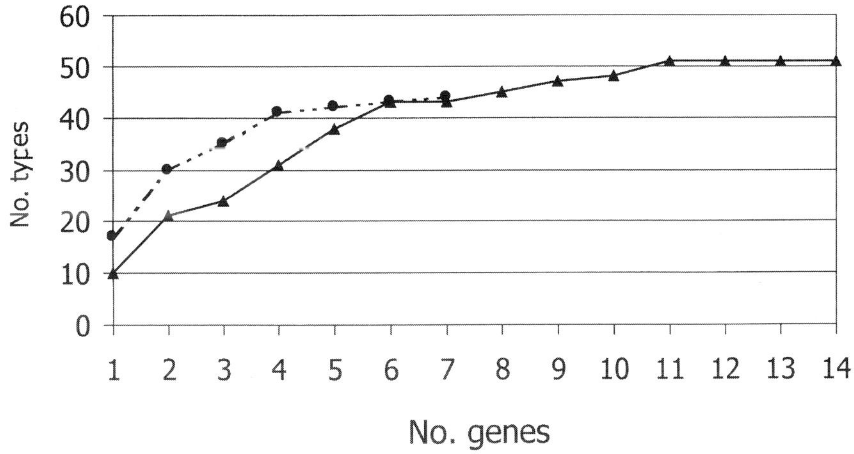

Fig. 12.5. Comparison of the discriminatory power of MEE and MLST in a collection of *N. meningitidis* isolates from patients. The cumulative number of types distinguished by addition of genes are indicated by circles for the MLST scheme and triangles for the MEE scheme.

MEE, the degree of resolution at individual loci is much higher; thus, the discriminatory potential of MLST is extremely high. For meningococci, it has been estimated that the typing scheme could resolve over 24 million sequence types (STs) (Maiden *et al.*, 1998).

Fig. 12.5 compares the cumulative number of types obtained in a collection of 70 *N. meningitidis* strains from patients by MEE and MLST, respectively. The individual genes are ordered according to the number of discriminated alleles, starting with the more polymorphic. While MLST provided a higher degree of discrimination with the first loci, the number of types discriminated by each method was very similar after reaching seven loci, and was increased only slightly by the use of seven additional loci in MEE. Consequently, MLST provides very comparable information to that obtained with MEE, and it can accordingly be used as a general tool for studies of population genetics and the molecular epidemiology of bacteria.

The one inconvenient aspect of MLST compared with MEE is that some *a priori* knowledge of the nucleotide sequence of the organism to be studied is necessary to determine which genes will be suitable for the typing scheme and to enable the design of appropriate primers.

A. *Preparation of PCR products*

A pure culture of each isolate is grown on solid medium under appropriate conditions. One loopful of cells is scraped from the agar plate or slant and suspended in 100 μl of Tris-EDTA buffer and boiled for 5 min. After centrifugation, the supernatant is used as the DNA source. For *H. pylori* (Achtman *et al.*, 1999), DNA was extracted using the CTAB method of Ausubel *et al.* (1994).

Table 12.5. PCR and sequencing primers for meningococcal genes used in MLST

PCR primer	PCR primer sequence	PCR primer	Sequencing primer sequence
abcZ-P1	5'-AATCGTTTATGTACCGCAGG-3'	abcZ-S1	same as abcZ-P1
abcZ-P2	5'-GTTGATTTCTGCCTGTTCGG-3'	abcZ-S2	5'-GAGAACGAGCCGGGATAGGA-3'
adk-P1	5'-ATGGCAGTTTGTGCAGTTGG-3'	adk-S1	5'-AGGCTGGCACGCCCTTGG-3'
adk-P2	5'-GATTTAAACAGCGATTGCCC-3'	adk-S2	5'-CAATACTTCGGCTTTCACGG-3'
aroE-P1	5'-ACGCATTTGCGCCGACATC-3'	aroE-S1	5'-GCGGTCAAC/TACGCTGATT-3'
aroE-P2	5'-ATCAGGGCTTTTTTCAGGTT-3'	aroE-S2	5'-ATGATGTTGCCGTACACATA-3'
fumC-A1	5'-CACCGAACACGACACGAT GG-3'	fumC-S1	5'-TCGGCACGGGTTTGAACAGC-3'
fumC-A2	5'-ACGACCAGTTCG TCAAAC TC-3'	fumC-S2	5'-CAACGGCGGTTTCGCGCAAC-3'
gdh-P1	5'-ATCAATACCGATGTGGCGCGT-3'	gdh-S1	5'-CCTTGGCAAAGAAAGCCTGC-3'
gdh-P2	5'-GGTTTTCATCTGCGTATAGAG-3'	gdh-S2	5'-GCGCACGGATTCATATGG-3'
pdhC-P1	5'-GGTTTCCAACGTATCGGCGAC-3'	pdhC-S1	5'-TCTACTACATCACCCTGATG-3'
pdhC-P2	5'-ATCGGCTTTGATGCCGTATTT-3'	pdhC-S2	same as pdhC-P2
pgm-P1	5'-CTTCAAAGCCTACGACATCCG-3'	pgm-S1	5'-CGGCGATGCCGACCGCTTGG-3'
pgm-P2	5'-CGGATTGCTTTCGATGACGGC-3'	pgm-S2	5'-GGTGATGATTTCGGTTGCGCC-3'

Table 12.6. PCR and sequencing primers for pneumococcal genes used in MLST

Primer	PCR/sequencing primer sequence
AroE-up	5'-GCCTTTGAGGCGACAGC-3'
AroE-dn	5'-TGCAGTTCA(G/A)AAACAT(A/T)TTCTAA-3'
Gdh-up	5'-ATGGACAAACCAGC(G/A/T/C)AG(C/T)TT-3'
Gdh-dn	5'-GCTTGAGGTCCCAT(G/A)CT(G/A/T/C)CC-3'
Gki-up	5'-GGCATTGGAATGGGATCACC-3'
Gki-dn	5'-TCTCCCGCAGCTGACAC-3'
RecP-up	5'-GCCAACTCAGGTCATCCAGG-3'
RecP-dn	5'-TGCAACCGTAGCATTGTAAC-3'
Spi-up	5'-TTATTCCTCCTGATTCTGTC-3'
Spi-dn	5'-GTGATTGGCCAGAAGCGGAA-3'
Xpt-up	5'-TTATTAGAAGAGCGCATCCT-3'
Xpt-dn	5'-AGATCTGCCTCCTTAAATAC-3'
Ddl-up	5'-TGC(C/T)CAAGTTCCTTATGTGG-3'
Ddl-dn	5'-CACTGGGT(G/A)AAACC(A/T)GGCAT-3'

The meningococcal MLST scheme uses internal fragments of the following seven housekeeping genes: putative ABC transporter *(abcZ)*, adenylate kinase (*adk*), shikimate dehydrogenase *(aroE)*, fumarate dehydrogenase (*fumC*), glucose-6-phosphate dehydrogenase (*gdh*), pyruvate dehydrogenase subunit (*pdhC*) and phosphoglucomutase (*pgm*). The pneumococcal MLST scheme uses internal fragments of shikimate dehydrogenase (*aroE*), glucose-6-phosphate dehydrogenase (*gdh*), glucose kinase (*gki*), transketolase *(recP)*, signal peptidase I (*spi*), xanthine phosphoribosyltransferase *(xpt)* and D-alanine-D-alanine ligase *(ddl)*. Fragments of the following seven genes are used for *H. pylori*: urease accessory protein (*ureI*), A/G-specific adenine glycosylase (*mutY*), elongation factor EF-P (*efp*), inorganic pyrophosphatase (*ppa*), GTPase (*yphC*), ATP synthase-F1α (*atpA*) and anthranilate isomerase (*trpC*). The primer pairs used for PCR amplification of internal fragments of the genes and sequencing of the PCR products are listed in Tables 12.5–12.7 for *N. meningitidis, S. pneumoniae* and *H. pylori*, respectively.

PCR amplification is carried out with 1 µl of the chromosomal DNA preparation, 5 µl of 10 × PCR buffer, 0.2 µM of each PCR primer, 200 µM of each dNTP mix, 0.5 U *Taq* polymerase and H_2O to a total volume of 50 µl. The annealing temperature should be determined for each different set of primers and may depend on the PCR equipment; primers are generally designed to have optimal annealing temperatures of 52–60°C. After completion of the PCR, 5 µl portions of the end-products are electrophoresed on an agarose gel, together with size standards, to check for a successful reaction.

Table 12.7. PCR and sequencing primers for *H. pylori* genes used in MLST

PCR primer	PCR primer sequence	PCR primer	Sequencing primer sequence
atp-A1	5'-GCTTAAATGGTGTGATGTCG-3'	atp-A7	5'-CGCTTTGGGTGAGCCTATTG-3'
atp-A6	5'-AATGGGCAAGGGCGAATAAG-3'	atp-A4	5'-TGCCCGTCTGTAATAGAAATG-3'
efp-F01	5'-GGCAATTGGGATGAGCGAGCTC-3'	efp-F02	5'-GGGCTTGAAAATTGAATTGGGCGG -3'
efp-R02	5'-CTTCACCTTTTCAAGATACTC-3'	efp-R01	5'-GTATTGACTTTAATGATCTCACCC-3'
mutY-101	5'-AGCGAAGTGATGAGCCAACAAAC-3'	mutY-101	5'-AGCGAAGTGATGAGCCAACAAAC-3'
mutY-102	5'-AAAGGGCAAATCGCACATTTGGG-3'	mutY-102	5'-AAAGGGCAAATCGCACATTTGGG-3'
ppa1 (+)	5'-GTGAGCCATGACGCTGATTCTTTGT-3'	ppa1 (+)	5'-GTGAGCCATGACGCTGATTCTTTGT-3'
ppa2 (–)	5'-GCCTTGATAGGCTTTTATCGCTTTCT-3'	ppa2 (–)	5'-GCCTTGATAGGCTTTTATCGCTTTCT-3'
ppa3 (–)	5'-GCCATTTCACACCAACACCCAAT-3'	ppa3 (–)	5'-GCCATTTCACACCAACACCCAAT-3'
HptrpC-1 (+)	5'-CAAGCTCCTAGAAGTCTCTG-3'	HptrpC-9 (+)	5'-CGCTTGCTCAA(AG)CTCCAATACGAC-3'
HptrpC-5 (–)	5'-CCCAGCTAGCATGAAAGG-3'	HptrpC-7 (–)	5'-TAAGCCCGCACACTTTATTTTCGCC-3'
HptrpC-6 (+)	5'-TAGAATGCAAAAAAGCATCGCCCTC-3'	HptrpC-6 (+)	5'-TAGAATGCAAAAAAGCATCGCCCTC-3'
		HptrpC-8 (–)	5'-GTCGTATTGGCG(CT)TTGAGCAAGCG-3'
Hp71S1 (+)	5'-CAATAAAGTCAGCTTGGCGCAACT-3'	Hp71S1 (+)	5'-CAATAAAGTCAGCTTGGCGCAACT-3'
Hp71S2 (+)	5'-GTTATTCGTAAGGTGCGTTTGTTG-3'	Hp71S2 (+)	5'-GTTATTCGTAAGGTGCGTTTGTTG-3'
Hp71S3 (+)	5'-GGCAATGCTAGGACTTGT-3'	Hp71S3 (+)	5'-GGCAATGCTAGGACTTGT-3'
Hp71AS1 (–)	5'-TCCCTTAGATTGCCAACTAAACGC-3'	Hp71AS1 (–)	5'-TCCCTTAGATTGCCAACTAAACGC-3'
yphC-F1	5'-CACTATTACCACGCCTATTTTTTTGAC-3'	yphC-F1	5'-CACTATTACCACGCCTATTTTTTTGAC-3'
yphC-F3	5'-CTTATGCGTTTTCTTCTTTTGG-3'	yphC-F3	5'-CTTATGCGTTTTCTTCTTTTGG-3'
yphC-R4	5'-AAGCAGCTGTTGTGATCACGGGGGC-3'	yphC-R4	5'-AAGCAGCTGTTGTGATCACGGGGGC-3'
yphC-R5	5'-TTTCTARGCTTTCTAAAATATC-3'	yphC-R5	5'-TTTCTARGCTTTCTAAAATATC-3'

B. *Nucleotide sequencing of PCR products*

The DNA fragments are purified either by column filtration through (e.g.) QIAquick (Qiagen, Crawley, UK) or ChromaSpin + TE100 (Clontech, Cambridge, UK) columns, or in an enzymatic reaction (7 μl of PCR product + 10 U exonuclease + 2 U shrimp alkaline phosphatase; Amersham Pharmacia Biotech, Little Chalfort, UK), as recommended by the respective manufacturers.

Sequencing reactions should be carried out on both strands, using the primers that were used for the initial PCR amplification (*S. pneumoniae*) or internal primers (*N. meningitidis, H. pylori*). Protocols recommended for the BigDye Ready Reaction Termination Mix (PE Applied Biosystems, Warrington, UK) are used, reduced to quarter-volumes. Samples are then electrophoresed on a Prism 377 Automated Sequencer (PE Applied Biosystems).

C. *Analysis of the sequence data*

After electrophoresis, the complementary strands must be aligned and the sequences edited so that they correspond exactly to the regions that are used to define the alleles. Various packages are suitable, including GCG, DNASTAR, AutoAssembler and Sequence Navigator Software (PE Applied Biosystems). For each strain, the consensus sequences for each gene fragment must be determined. The combination of alleles at the seven gene fragments identifies the allelic profile or ST of the strains, and the relatedness between isolates can be analysed in the same manner as for MEE data (i.e., genetic diversity and construction of a dendrogram from the matrix of pairwise differences between STs).

D. *Use of the database*

Meningococcal and pneumococcal MLST databases have been set up and are continuously expanding (*http://mlst.zoo.ox.ac.uk*). The meningococcal database currently contains 415 STs from isolates recovered from patients and healthy carriers, while the pneumococcal database contains 270 STs of isolates, obtained mostly from cases of serious invasive disease, together with various penicillin-resistant and multiply antibiotic-resistant isolates.

Consensus sequences of each gene fragment can be compared with those in the databases. The software checks that the sequences are the correct length and that they do not contain any undetermined characters. A check is also made to verify that the submitted sequence is at least 70% similar to another allele at that locus. Options are available to identify the allele at a single locus, to enter an allele profile, to find isolates in the database that match or nearly match an allele profile, or to browse the database. Consensus sequences not represented in the database can be submitted as a new allele. The database curator evaluates the traces of the sequence before assigning a number to the new allele and including it in the database.

E. Applications of MLST

MLST was first evaluated for *N. meningitidis* using sequences of *c*.470-bp fragments from 11 housekeeping genes in a reference set of 107 isolates from invasive disease and healthy carriers, selected on the basis of their multilocus genotypes as determined by MEE (Maiden *et al.*, 1998). The strain associations obtained by MLST were consistent with the clonal groupings determined by MEE. Most isolates from hyper-virulent lineages of meningococci belonging to serogroups A, B and C were identical for all loci or differed from the majority type at only a single locus. Six loci were selected that reliably identified the major lineages associated with epidemic disease. A seventh gene (*fumC*) was then added in an attempt to discriminate the new clone variant (ET-15) that was recently associated with serogroup C outbreaks in North America, Europe and Australia.

A similar study has been performed for *S. pneumoniae.* Among 274 isolates from recent cases of invasive pneumococcal disease in eight countries, 143 STs were resolved. Isolates of the same ST were recovered from cases of invasive disease in different countries, implying that strains with specific genotypes had an increased capacity to cause invasive disease. The relationship between STs and serotypes suggested that horizontal exchange of capsular genes was uncommon in the pneumococcal population associated with invasive disease (Enright & Spratt, 1998). A study of 74 penicillin-resistant isolates from Taiwan using MLST showed that 86% of the isolates belonged to one of three clusters, two of which were previously undescribed (Shi *et al.*, 1998). Coffey *et al.* (1999) provided new evidence by MLST that penicillin-resistant serotype 14 pneumococcal isolates were identical to the Spanish penicillin-resistant serotype 9V clone, and that they arose by recombinational replacement of the capsular locus and flanking regions.

A collection of 20 strains of *H. pylori* analysed by MLST for fragments of seven housekeeping genes revealed the existence of two weak clonal groups, in spite of extensive inter-strain recombination (Achtman *et al.*, 1999).

12.5 PERSPECTIVES

MLST is a powerful new approach for the characterisation of microorganisms, since it provides unambiguous molecular typing data that are electronically portable between laboratories, and which can be used in studies of global epidemiology. MLST schemes for *Staph. aureus*, *Strep. pyogenes*, *H. influenzae*, *Yersinia* spp. and other major pathogens are under development. MLST is a simple technique that requires only the ability to amplify specific DNA fragments by PCR and to sequence those fragments. While the method is still quite expensive, further technological developments in automated sequencing should soon render MLST accessible to all major public health laboratories, including those in the developing world. In line with the impact of MEE, MLST will further improve our understanding of the population and evolutionary biology of microbial pathogens, as well as other microorganisms. For the more stable microbial pathogens, sequence analysis of one or a few hypervariable loci, in addition to the MLST scheme based on housekeeping genes, will provide a versatile tool for studies of both global and local epidemiology.

ACKNOWLEDGEMENTS

This paper is dedicated to Prof. Robert K. Selander.

REFERENCES

Achtman, M. (1990). Molecular epidemiology of epidemic bacterial meningitis. *Reviews in Medical Microbiology* **1**, 29–38.

Achtman, M. (1994). Clonal spread of serogroup A meningococci. A paradigm for the analysis of microevolution in bacteria. *Molecular Microbiology* **11**, 15–22.

Achtman, M. (1995). Global epidemiology of meningococcal disease. In *Meningococcal disease,* Cartwright K., ed., pp. 159–175. Wiley, New York.

Achtman, M. (1997). Microevolution and epidemic spread of serogroup A *Neisseria meningitidis* – a review. *Gene* **192**, 135–140.

Achtman, M., Heuzenroeder, M., Kusecek, B., Ochman, H., Caugant, D.A., Selander, R.K., Väisänen-Rhen, V., Korhonen, T.K., Stuart, S., Ørskov, F. & Ørskov, I. (1986). Clonal analysis of *Escherichia coli* O2:K1 isolated from diseased humans and animals. *Infection and Immunity* **51**, 268–276.

Achtman, M., Azuma, T., Berg, D.E., Ito, Y., Morelli, G., Pan, Z.J., Suerbaum, S., Thompson, S.A., van Der Ende, A. & van Doorn, L.J. (1999). Recombination and clonal groupings within *Helicobacter pylori* from different geographical regions. *Molecular Microbiology* **32**, 459–470.

Aeschbacher, M. & Piffaretti, J.-C. (1989). Population genetics of human and animal enteric *Campylobacter* strains. *Infection and Immunity* **57**, 1432–1437.

Altwegg, M., Reeves, M.W., Altwegg-Bissig, R., & Brenner, D.J. (1991). Multilocus enzyme analysis of the genus *Aeromonas* and its use for species identification. *Zentralblatt für Bakteriologie* **275**, 28–45.

Amonsin, A., Wellehan, J.F., Li, L.L., Vandamme, P., Lindeman, C., Edman, M., Robinson, R.A. & Kapur, V. (1997). Molecular epidemiology of *Ornithobacterium rhinotracheale*. *Journal of Clinical Microbiology* **35**, 2894–2898.

Angen, O., Caugant, D.A., Olsen, J.E. & Bisgaard, M. (1997). Genotypic relationships among strains classified under the *(Pasteurella) haemolytica*-complex as indicated by ribotyping and multilocus enzyme electrophoresis. *Zentralblatt für Bakteriologie* **286**, 333–354.

Arbeit, R.D., Slutsky, A., Barber, T.W., Maslow, J.N., Niemczyk, S., Falkinham, J.O., O'Connor, G.T. & Von Reyn, C.F. (1993). Genetic diversity among strains of *Mycobacterium avium* causing monoclonal and polyclonal bacteremia in patients with AIDS. *Journal of Infectious Diseases* **167**, 1384–1390.

Ashton, F.E., Ryan, J.A., Borczyk, A., Caugant, D.A., Mancino, L. & Huang, D. (1991). Emergence of a virulent clone of *Neisseria meningitidis* serotype 2a that is associated with meningococcal group C disease in Canada. *Journal of Clinical Microbiology* **29**, 2489–2493.

Ausubel, F.M., Brent, R., Kingston, R.E., Moore, D.D., Seidman, J.G., Smith, J.A. & Struhl, K. (1994). *Current protocols in molecular biology*. Wiley, New York.

Balmelli, T. & Piffaretti, J.C. (1995). Association between different clinical manifestations of Lyme disease and different species of *Borrelia burgdorferi* sensu lato. *Research in Microbiology* **146**, 329–340.

Balmelli, T. & Piffaretti, J.C. (1996). Analysis of the genetic polymorphism of *Borrelia burgdorferi* sensu lato by multilocus enzyme electrophoresis. *International Journal of Systematic Bacteriol-*

ogy **46**, 167–172.

Baxter, F., Wright, F., Chalmers, R.M., Low, J.C. & Donachie, W. (1993). Characterization by multilocus enzyme electrophoresis of *Listeria monocytogenes* isolates involved in ovine listeriosis outbreaks in Scotland from 1989 to 1991. *Applied and Environmental Microbiology* **59**, 3126–3129.

Beltran, P., Musser, J.M., Helmuth, R., Farmer, J.J., Frerichs, W.M., Wachsmuth, I.K., Ferris, K., McWhorter, A.C., Wells, J.G., Cravioto, A. & Selander, R.K. (1988). Toward a population genetic analysis of *Salmonella*: genetic diversity and relationships among strains of serotypes *S. choleraesuis, S. derby, S. dublin, S. enteritidis, S. heidelberg, S. infantis, S. newport, and S. typhimurium. Proceedings of the National Academy of Sciences of the United States of America* **85**, 7753–7757.

Beltran, P., Plock, S.A., Smith, N.H., Whittam, T.S., Old, D.C. & Selander, R.K. (1991). Reference collection of strains of the *Salmonella typhimurium* complex from natural populations. *Journal of General Microbiology* **137**, 601–606.

Beltran, P., Delgado, G., Navarro, A., Trujillo, F., Selander, R.K. & Cravioto, A. (1999). Genetic diversity and population structure of *Vibrio cholerae. Journal of Clinical Microbiology* **37**, 581–590.

Bert, F., Picard, B., Lambert Zechovsky, N. & Goullet, P. (1995). Identification and typing of pyogenic streptococci by enzyme electrophoretic polymorphism. *Journal of Medical Microbiology* **42**, 442–451.

Beutin, L., Ørskov, I., Ørskov, F., Zimmermann, S., Prada, J., Gelderblom, H., Stephan, R. & Whittam, T.S. (1990). Clonal diversity and virulence factors in strains of *Escherichia coli* of the classic enteropathogenic serogroup O114. *Journal of Infectious Diseases* **162**, 1329–1334.

Bibb, W.F., Schwartz, B., Gellin, B.G., Plikaytis, B.D. & Weaver, R.E. (1989). Analysis of *Listeria monocytogenes* by multilocus enzyme electrophoresis and application of the method to epidemiologic investigations. *International Journal of Food Microbiol*ogy **8**, 233–239.

Bibb, W.F., Gellin, B.G., Weaver, R., Schwartz, B., Plikaytis, B.D., Reeves, M.W., Pinner, R.W. & Broome, C.V. (1990). Analysis of clinical and food-borne isolates of *Listeria monocytogenes* in the United States by multilocus enzyme electrophoresis and application of the method to epidemiologic investigations. *Applied and Environmental Microbiology* **56**, 2133–2141.

Blackall, P.J., Trott, D.J., Rapp-Gabrielson, V. & Hampson, D.J. (1997). Analysis of *Haemophilus parasuis* by multilocus enzyme electrophoresis. *Veterinary Microbiology* **56**, 125–134.

Blackall, P.J., Fegan, N., Chew, G.T. & Hampson, D.J. (1998). Population structure and diversity of avian isolates of *Pasteurella multocida* from Australia. *Microbiology* **144**, 279–289.

Boerlin, P. (1997). Applications of multilocus enzyme electrophoresis in medical microbiology. *Journal of Microbiological Methods* **28**, 221–231.

Boerlin, P. & Piffaretti, J.C. (1991). Typing of human, animal, food, and environmental isolates of *Listeria monocytogenes* by multilocus enzyme electrophoresis. *Applied and Environmental Microbiology* **57**, 1624–1629.

Boerlin, P. & Piffaretti, J.C. (1995). Multilocus enzyme electrophoresis. In *Methods in molecular biology,* Howard, J. & Whitcombe D.M., eds, vol. **46**, pp. 63–78. Humana Press, Totowa, NJ.

Boerlin, P., Rocourt, J. & Piffaretti, J.C. (1991). Taxonomy of the genus *Listeria* by using multilocus enzyme electrophoresis. *International Journal of Systematic Bacteriology* **41**, 59–64.

Boerlin, P., Peter, O., Bretz, A.-G., Postic, D., Baranton, G. & Piffaretti, J.-C. (1992). Population genetic analysis of *Borrelia burgdorferi* isolates by multilocus enzyme electrophoresis. *Infection and Immunity* **60**, 1677–1683.

Boerlin, P., Bannerman, E., Jemmi, T. & Bille, J. (1996). Subtyping *Listeria monocytogenes* isolates genetically related to the Swiss epidemic clone. *Journal of Clinical Microbiology* **34**, 2148–2153.

Boerlin, P., Boerlin Petzold, F., Bannerman, E., Bille, J. & Jemmi, T. (1997). Typing *Listeria monocytogenes* isolates from fish products and human listeriosis cases. *Applied and Environmental Microbiology* **63**, 1338–1343.

Boyd, E.F., Hiney, M.P., Peden, J.F., Smith, P.R. & Caugant, D.A. (1994). Assessment of genetic diversity among *Aeromonas salmonicida* isolates by multilocus enzyme electrophoresis. *Journal of Fish Diseases* **17**, 97–98.

Boyd, E.F., Wang, F.S., Whittam, T.S. & Selander, R.K. (1996). Molecular genetic relationships of the salmonellae. *Applied and Environmental Microbiology* **62**, 804–808.

Brenner, D.J., Steigerwalt, A.G., Gorman, G.W., Wilkinson, H.W., Bibb, W.F., Hackel, M., Tyndall, R.L., Campbell, J., Feeley, J.C., Thacker, W.L., Skaliy, P., Martin, W.T., Brake, B.J., Fields, B.S., McEachern, H.V. & Corcoran, L.K. (1985). Ten new species of *Legionella*. *International Journal of Systematic Bacteriology*, **35**, 50–59.

Carapetis, J., Robins-Browne, R., Martin, D., Shelby-James, T. & Hogg, G. (1995). Increasing severity of invasive group A streptococcal disease in Australia: clinical and molecular epidemiological features and identification of a new virulent M-nontypeable clone. *Clinical Infectious Diseases* **21**, 1220–1227.

Carlson, C.R., Caugant, D.A. & Kolsto, A.B. (1994). Genotypic diversity among *Bacillus cereus* and *Bacillus thuringiensis* strains. *Applied and Environmental Microbiology* **60**, 1719–1725.

Caugant, D.A. (1998). Population genetics and molecular epidemiology of *Neisseria meningitidis*. *Acta Pathologica Microbiologica Scandinavica* **106**, 505–525.

Caugant, D.A., Levin, B.R. & Selander, R.K. (1981). Genetic diversity and temporal variation in the *E. coli* population of a human host. *Genetics* **98**, 467–490.

Caugant, D.A., Levin, B.R., Ørskov, I., Ørskov, F., Svanborg, E.C. & Selander, R.K. (1985). Genetic diversity in relation to serotype in *Escherichia coli*. *Infection and Immunity* **49**, 407–413.

Caugant, D.A., Frøholm, L.O., Bøvre, K., Holten, E., Frasch, C.E., Mocca, L.F., Zollinger, W.D. & Selander, R.K. (1986). Intercontinental spread of a genetically distinctive complex of clones of *Neisseria meningitidis* causing epidemic disease. *Proceedings of the National Academy of Sciences of the United States of America* **83**, 4927–4931.

Caugant, D.A., Mocca, L.F., Frasch, C.E., Frøholm, L.O., Zollinger, W.D. & Selander, R.K. (1987). Genetic structure of *Neisseria meningitidis* populations in relation to serogroup, serotype, and outer membrane protein pattern. *Journal of Bacteriology* **169**, 2781–2792.

Caugant, D.A., Kristiansen, B.-E., Frøholm, L.O., Bøvre, K. & Selander, R.K. (1988). Clonal diversity of *Neisseria meningitidis* from a population of asymptomatic carriers. *Infection and Immunity* **56**, 2060–2068.

Caugant, D.A., Aleksic, S., Mollaret, H.H., Selander, R.K. & Kapperud, G. (1989). Clonal diversity and relationships among strains of *Yersinia enterocolitica*. *Journal of Clinical Microbiology* **27**, 2678–2683.

Caugant, D.A., Bol, P., Høiby, E.A., Zanen, H.C. & Frøholm, L.O. (1990a). Clones of serogroup B *Neisseria meningitidis* causing systemic disease in the Netherlands, 1958–1986. *Journal of Infectious Diseases* **162**, 867–874.

Caugant, D.A., Selander, R.K. & Olsen, I. (1990b). Differentiation between *Actinobacillus* (*Haemophilus*) *actinomycetemcomitans*, *Haemophilus aphrophilus* and *Haemophilus paraphrophilus* by multilocus enzyme electrophoresis. *Journal of General Microbiology* **136**, 2135–2141.

Caugant, D.A., Høiby, E.A., Magnus, P., Scheel, O., Hoel, T., Bjune, G., Wedege, E., Eng, J. & Frøholm, L.O. (1994). Asymptomatic carriage of *Neisseria meningitidis* in a randomly sampled population. *Journal of Clinical Microbiology* **32**, 323–330.

Caugant, D.A., Ashton, F.E., Bibb, W.F., Boerlin, P., Donachie, W., Low, C., Gilmour, A., Harvey, J. & Norrung, B. (1996). Multilocus enzyme electrophoresis for characterization of *Listeria monocytogenes* isolates: results of an international comparative study. *International Journal of Food Microbiology* **32**, 301–311.

Coffey, T.J., Daniels, M., McDougal, L.K., Dowson, C.G., Tenover, F.C. & Spratt, B.G. (1995). Genetic analysis of clinical isolates of *Streptococcus pneumoniae* with high-level resistance to expanded-spectrum cephalosporins. *Antimicrobial Agents and Chemotherapy* **39**, 1306–1313.

Coffey, T.J., Daniels, M., Enright, M.C. & Spratt, B.G. (1999). Serotype 14 variants of the Spanish penicillin-resistant serotype 9V clone of *Streptococcus pneumoniae* arose by large recombina-

of especially virulent clonal types of *Actinobacillus actinomycetemcomitans*. *Journal of Clinical Microbiology* **33**, 395–401.

Haubek, D., DiRienzo, J.M., Tinoco, E.M.B., Westergaard J., Lopez, N.J., Chung, C.-P., Poulsen, K. & Kilian, M. (1997). Geographic dissemination of a highly toxic clone of *Actinobacillus actinomycetemcomitans* associated with juvenile periodontitis. *Journal of Clinical Microbiology* **35**, 3037–3042.

Hauge, M., Jespersgaard, C., Poulsen, K. & Kilian, M. (1996). Population structure of *Streptococcus agalactiae* reveals an association between specific evolutionary lineages and putative virulence factors but not disease. *Infection and Immunity* **64**, 919–925.

Helgason, E., Caugant, D.A., Lecadet, M.M., Chen, Y., Mahillon, J., Lovgren, A., Hegna, I., Kvaloy, K. & Kolsto, A.B. (1998). Genetic diversity of *Bacillus cereus/B. thuringiensis* isolates from natural sources. *Current Microbiology* **37**, 80–87.

Helmig, R., Uldbjerg, N., Boris, J. & Kilian, M. (1993). Clonal analysis of *Streptococcus agalactiae* isolated from infants with neonatal sepsis or meningitis and their mothers and from healthy pregnant women. *Journal of Infectious Diseases* **168**, 904–909.

Holm, S.E., Norrby, A., Bergholm, A.M. & Norgren, M. (1992). Aspects of pathogenesis of serious group A streptococcal infections in Sweden, 1988–1989. *Journal of Infectious Diseases* **166**, 31–37.

Istock, C.A., Duncan, K.E., Ferguson, N. & Zhou, X. (1992). Sexuality in a natural population of bacteria – *Bacillus subtilis* challenges the clonal paradigm. *Molecular Ecology* **1**, 95–103.

Jackson, L.A., Schuchat, A., Reeves, M.W. & Wenger, J.D. (1995). Serogroup C meningococcal outbreaks in the United States. An emerging threat. *Journal of the American Medical Association* **273**, 383–389.

Jelfs, J., Jalaludin, B., Munro, R., Patel, M., Kerr, M., Daley, D., Neville, S. & Capon, A. (1998). A cluster of meningococcal disease in western Sydney, Australia initially associated with a nightclub. *Epidemiology and Infection* **120**, 263–270.

John, M.A. & Hussain, Z. (1994). Multilocus enzyme electrophoresis using ultrathin polyacrylamide gels. *Journal of Microbiological Methods* **19**, 307–313.

Kaczmarski, E.B. (1997). Meningococcal disease in England and Wales: 1995. *Communicable Diseases Report Review* **7**, R55-R59

Kapur, V., Kanjilal, S., Hamrick, M.R., Li, L.L., Whittam, T.S., Sawyer, S.A. & Musser, J.M. (1995a). Molecular population genetic analysis of the streptokinase gene of *Streptococcus pyogenes*: mosaic alleles generated by recombination. *Molecular Microbiology* **16**, 509–519.

Kapur, V., Sischo, W.M., Greer, R.S., Whittam, T.S. & Musser, J.M. (1995b). Molecular population genetic analysis of *Staphylococcus aureus* recovered from cows. *Journal of Clinical Microbiology* **33**, 376–380.

Klugman, K.P., Coffey, T.J., Smith, A., Wasas, A., Meyers, M. & Spratt, B.G. (1994). Cluster of an erythromycin-resistant variant of the Spanish multiply resistant 23F clone of *Streptococcus pneumoniae* in South Africa. *European Journal of Clinical Microbiology and Infectious Diseases* **13**, 171–174.

Kokka, R.P., Janda, J.M., Oshiro, L.S., Altwegg, M., Shimada, T., Sakazaki, R. & Brenner, D.J. (1991). Biochemical and genetic characterization of autoagglutinating phenotypes of *Aeromonas* species associated with invasive and noninvasive disease. *Journal of Infectious Diseases* **163**, 890–894.

Kolstad, J., Caugant, D.A. & Rørvik, L.M. (1992). Differentiation of *Listeria monocytogenes* isolates by using plasmid profiling and multilocus enzyme electrophoresis. *International Journal of Food Microbiology* **16**, 247–260.

Kristinsson, K.G., Hjalmarsdottir, M.A. & Steingrimsson, O. (1992). Increasing penicillin resistance in pneumococci in Iceland. *Lancet* **339**, 1606–1607.

Kriz, P., Giorgini, D., Musilek, M., Larribe, M. & Taha, M.K. (1999). Microevolution through DNA exchange among strains of *Neisseria meningitidis* isolated during an outbreak in the Czech Republic. *Research in Microbiology* **150**, 273–280.

Kumar, S., Tamura, K. & Nei, M. (1994). MEGA: molecular evolutionary genetics analysis software for microcomputers. *Computers in Applied Bioscience* **10**, 189–191.

Lagos, R., Avendano, A., Horwitz, I., Musser, J.M., Hoiseth, S.K., Maneval, D.R., Jones, M.J., Levine, M.M., Dattas, J.P., Prenzel, I., Enriquez, N., Topelberg, S., Olivari, F. & Morris, J.G. (1991). Molecular epidemiology of *Haemophilus influenzae* within families in Santiago, Chile. *Journal of Infectious Diseases* **164**, 1149–1153.

Lanser, J.A., Adams, M., Doyle, R., Sangster, N. & Steele, T.W. (1990). Genetic relatedness of *Legionella longbeachae* isolates from human and environmental sources in Australia. *Applied and Environmental Microbiology* **56**, 2784–2790.

Lawrence, L.M. & Gilmour, A. (1995). Characterization of *Listeria monocytogenes* isolated from poultry products and from the poultry-processing environment by random amplification of polymorphic DNA and multilocus enzyme electrophoresis. *Applied and Environmental Microbiology* **61**, 2139–2144.

Levin, M.H., Weinstein, R.A., Nathan, C., Selander, R.K., Ochman, H. & Kabins, S.A. (1984). Association of infection caused by *Pseudomonas aeruginosa* serotype O11 with intravenous abuse of pentazocine mixed with tripelennamine. *Journal of Clinical Microbiology* **20**, 758–762.

Li, J., Musser, J.M., Beltran, P., Kline, M.W. & Selander, R.K. (1990). Genotypic heterogeneity of strains of *Citrobacter diversus* expressing a 32-kilodalton outer membrane protein associated with neonatal meningitis. *Journal of Clinical Microbiology* **28**, 1760–1765.

Lomholt, H. (1995). Evidence of recombination and an antigenically diverse immunoglobulin A1 protease among strains of *Streptococcus pneumoniae*. *Infection and Immunity* **63**, 4238–4243.

Loos, B.G., Dyer, D.W., Whittam, T.S. & Selander, R.K. (1993). Genetic structure of populations of *Porphyromonas gingivalis* associated with periodontitis and other oral infections. *Infection and Immunity* **61**, 204–212.

Lymbery, A.J., Hampson, D.J., Hopkins, R.M., Combs, B. & Mhoma, J.R. (1990). Multilocus enzyme electrophoresis for identification and typing of *Treponema hyodysenteriae* and related spirochaetes. *Veterinary Microbiology* **22**, 89–99.

Maiden, M.C., Bygraves, J.A., Feil, E., Morelli, G., Russell, J.E., Urwin, R., Zhang, Q., Zhou, J., Zurth, K., Caugant, D.A., Feavers, I.M., Achtman, M. & Spratt, B.G. (1998). Multilocus sequence typing: a portable approach to the identification of clones within populations of pathogenic microorganisms. *Proceedings of the National Academy of Sciences of the United States of America* **95**, 3140–3145.

Marques, M.T., Bornstein, N. & Fleurette, J. (1995). Combined monoclonal antibody typing, multilocus enzyme electrophoresis, soluble protein profiles and plasmid analysis of clinical and environmental *Legionella pneumophila* serogroup 1 isolated in a Portuguese hospital. *Journal of Hospital Infection* **30**, 103–110.

Martin, P.R. & Hoiby, E.A. (1990). Streptococcal serogroup A epidemic in Norway 1987–1988. *Scandinavian Journal of Infectious Diseases* **22**, 421–429.

Martin, D.R. & Single, L.A. (1993). Molecular epidemiology of group A streptococcus M type 1 infections. *Journal of Infectious Diseases* **167**, 1112–1117.

Martin, C., Sibold, C. & Hakenbeck, R. (1992). Relatedness of penicillin-binding protein 1a genes from different clones of penicillin-resistant *Streptococcus pneumoniae* isolated in South Africa and Spain. *EMBO Journal* **11**, 3831–3836.

Martin, D.R., Walker, S.J., Baker, M.G. & Lennon, D.R. (1998a). New Zealand epidemic of meningococcal disease identified by a strain with phenotype B:4:P1.4. *Journal of Infectious Diseases* **177**, 497–500.

Martin, D.R., Walker, S.J., Glennie, A.C., Baker, M.G., Eyles, R.F., Lennon, D.R. & Roberts, A.P. (1998b). Continuation of meningococcal disease epidemic in New Zealand. In *Abstracts of the eleventh international pathogenic neisseria conference*, Nassif, X., Quentin-Millet, M.-J. & Taha, M.-K., eds, p. 8. Paris.

Martin, C., Boyd, E.F., Quentin, R., Massicot, P. & Selander, R.K. (1999). Enzyme polymorphism in *Pseudomonas aeruginosa* strains recovered from cystic fibrosis patients in France. *Microbiol-*

ogy **145**, 2587–2594.

Martinez, M.B., Whittan, T.S., McGraw, E.A., Rodrigues, J. & Trabulsi, L.R. (1999). Clonal relationship among invasive and non-invasive strains of enteroinvasive *Escherichia coli* serogroups. *FEMS Microbiology Letters* **172**, 145–151.

Maslow, J.N., Whittam, T.S., Gilks, C.F., Wilson, R.A., Mulligan, M.E., Adams, K.S. & Arbeit, R.D. (1995). Clonal relationships among bloodstream isolates of *Escherichia coli*. *Infection and Immunity* **63**, 2409–2417.

Maynard Smith, J., Smith, N.H., O'Rourke, M. & Spratt, B.G. (1993). How clonal are bacteria? *Proceedings of the National Academy of Sciences of the United States of America*, **90**, 4384–4388.

McDougal, L.K., Rasheed, J.K., Biddle, J.W. & Tenover, F.C. (1995). Identification of multiple clones of extended-spectrum cephalosporin-resistant *Streptococcus pneumoniae* isolates in the United States. *Antimicrobial Agents and Chemotherapy* **39**, 2282–2288.

McLaren, A.J., Trott, D.J., Swayne, D.E., Oxberry, S.L. & Hampson, D.J. (1997). Genetic and phenotypic characterization of intestinal spirochetes colonizing chickens and allocation of known pathogenic isolates to three distinct genetic groups. *Journal of Clinical Microbiology* **35**, 412–417.

Milkman, R. (1973). Electrophoretic variation in *Escherichia coli* from natural sources. *Science* **182**, 1024–1026.

Moore, P.S., Reeves, M.W., Schwartz, B., Gellin, B.G. & Broome, C.V. (1989). Intercontinental spread of an epidemic group A *Neisseria meningitidis* strain. *Lancet* **ii**, 260–263.

Munoz, R., Coffey, T.J., Daniels, M., Dowson, C.G., Laible, G., Casal, J., Hakenbeck, R., Jacobs, M., Musser, J.M., Spratt, B.G. & Tomasz, A. (1991). Intercontinental spread of a multiresistant clone of serotype 23F *Streptococcus pneumoniae*. *Journal of Infectious Diseases* **164**, 302–306.

Munoz, R., Musser, J.M., Crain, M., Briles, D.E., Marton, A., Parkinson, A.J., Sorensen, U. & Tomasz, A. (1992). Geographic distribution of penicillin-resistant clones of *Streptococcus pneumoniae*: characterization by penicillin-binding protein profile, surface protein A typing, and multilocus enzyme analysis. *Clinical Infectious Diseases* **15**, 112–118.

Musser, J.M. (1996). Molecular population genetic analysis of emerged bacterial pathogens: selected insights. *Emerging Infectious Diseases* **2**, 1–17.

Musser, J.M. & Kapur, V. (1992). Clonal analysis of methicillin-resistant *Staphylococcus aureus* strains from intercontinental sources: association of the *mec* gene with divergent phylogenetic lineages implies dissemination by horizontal transfer and recombination. *Journal of Clinical Microbiology* **30**, 2058–2063.

Musser, J.M. & Selander, R.K. (1990). Brazilian purpuric fever: evolutionary genetic relationships of the case clone of *Haemophilus influenzae* biogroup aegyptius to encapsulated strains of *Haemophilus influenzae*. *Journal of Infectious Diseases* **161**, 130–133.

Musser, J.M., Granoff, D.M., Pattison, P.E. & Selander, R.K. (1985). A population genetic framework for the study of invasive diseases caused by serotype b strains of *Haemophilus influenzae*. *Proceedings of the National Academy of Sciences of the United States of America* **82**, 5078–5082.

Musser, J.M., Barenkamp, S.J., Granoff, D.M. & Selander, R.K. (1986). Genetic relationships of serologically nontypable and serotype b strains of *Haemophilus influenzae*. *Infection and Immunity* **52**, 183–191.

Musser, J.M., Hewlett, E.L., Peppler, M.S. & Selander, R.K. (1986). Genetic diversity and relationships in populations of *Bordetella* spp. *Journal of Bacteriology* **166**, 230–237.

Musser, J.M., Bemis, D.A., Ishikawa, H. & Selander, R.K. (1987a). Clonal diversity and host distribution in *Bordetella bronchiseptica*. *Journal of Bacteriology* **169**, 2793–2803.

Musser, J.M., Rapp, V.J. & Selander, R.K. (1987b). Clonal diversity in *Haemophilus pleuropneumoniae*. *Infection and Immunity* **55**, 1207–1215.

Musser, J.M., Kroll, J.S., Moxon, E.R. & Selander, R.K. (1988a). Clonal population structure of encapsulated *Haemophilus influenzae*. *Infection and Immunity* **56**, 1837–1845.

Musser, J.M., Kroll, J.S., Moxon, E.R. & Selander, R.K. (1988b). Evolutionary genetics of the encapsulated strains of *Haemophilus influenzae*. *Proceedings of the National Academy of Sciences of the United States of America* **85**, 7758–7762.

Musser, J.M., Mattingly, S.J., Quentin, R., Goudeau, A. & Selander, R.K. (1989). Identification of a high-virulence clone of type III *Streptococcus agalactiae* (group B *Streptococcus*) causing invasive neonatal disease. *Proceedings of the National Academy of Sciences of the United States of America* **86**, 1731–1735.

Musser, J.M., Kroll, J.S., Granoff, D.M., Moxon, E.R., Brodeur, B.R., Campos, J., Dabernat, H., Frederiksen, W., Hamel, J. & Hammond, G. (1990a). Global genetic structure and molecular epidemiology of encapsulated *Haemophilus influenzae*. *Reviews of Infectious Diseases* **12**, 75–111.

Musser, J.M., Schlievert, P.M., Chow, A.W., Ewan, P., Kreiswirth, B.N., Rosdahl, V.T., Naidu, A.S., Witte, W. & Selander, R.K. (1990b). A single clone of *Staphylococcus aureus* causes the majority of cases of toxic shock syndrome. *Proceedings of the National Academy of Sciences of the United States of America* **87**, 225–229.

Musser, J.M., Hauser, A.R., Kim, M.H., Schlievert, P.M., Nelson, K. & Selander, R.K. (1991). *Streptococcus pyogenes* causing toxic-shock-like syndrome and other invasive diseases: clonal diversity and pyrogenic exotoxin expression. *Proceedings of the National Academy of Sciences of the United States of America* **88**, 2668–2672.

Musser, J.M., Gray, B.M., Schlievert, P.M. & Pichichero, M.E. (1992). *Streptococcus pyogenes* pharyngitis: characterization of strains by multilocus enzyme genotype, M and T protein serotype, and pyrogenic exotoxin gene probing. *Journal of Clinical Microbiology* **30**, 600–603.

Musser, J.M., Kapur, V., Kanjilal, S., Shah, U., Musher, D.M., Barg, N.L., Johnston, K.H., Schlievert, P.M., Henrichsen, J., Gerlach, D., Rakita, R.M., Tanna, A., Cookson, B.D. & Huang, J.C. (1993a). Geographic and temporal distribution and molecular characterization of two highly pathogenic clones of *Streptococcus pyogenes* expressing allelic variants of pyrogenic exotoxin A (scarlet fever toxin). *Journal of Infectious Diseases* **167**, 337–346.

Musser, J.M., Nelson, K., Selander, R.K., Gerlach, D., Huang, J.C., Kapur, V. & Kanjilal, S. (1993b). Temporal variation in bacterial disease frequency: molecular population genetic analysis of scarlet fever epidemics in Ottawa and in eastern Germany. *Journal of Infectious Diseases* **167**, 759–762.

Musser, J.M., Kapur, V., Szeto, J., Pan, X., Swanson, D.S. & Martin, D.R. (1995). Genetic diversity and relationships among *Streptococcus pyogenes* strains expressing serotype M1 protein: recent intercontinental spread of a subclone causing episodes of invasive disease. *Infection and Immunity* **63**, 994–1003.

Mwaniki, C.G., Robertson, I.D., Trott, D.J., Atyeo, R.F., Lee, B.J. & Hampson, D.J. (1994). Clonal analysis and virulence of Australian isolates of *Streptococcus suis* type 2. *Epidemiology and Infection* **113**, 321–334.

Møller, K., Nielsen, R., Andersen, L.V. & Kilian, M. (1992). Clonal analysis of the *Actinobacillus pleuropneumoniae* population in a geographically restricted area by multilocus enzyme electrophoresis. *Journal of Clinical Microbiology* **30**, 623–627.

Nei, M. (1977). F-statistics and analysis of gene diversity in subdivided populations. *Annals of Human Genetics* **41**, 225–233.

Nei, M. (1978). Estimation of average heterozygosity and genetic distance from a small sample of individuals. *Genetics* **89**, 583–590.

Nesbakken, T., Kapperud, G. & Caugant, D.A. (1996). Pathways of *Listeria monocytogenes* contamination in the meat processing industry. *International Journal of Food Microbiology* **31**, 161–171.

Ngeleka, M., Kwaga, J.K., White, D.G., Whittam, T.S., Riddell, C., Goodhope, R., Potter, A.A. & Allan, B. (1996). *Escherichia coli* cellulitis in broiler chickens: clonal relationships among strains and analysis of virulence-associated factors of isolates from diseased birds. *Infection and Immunity* **64**, 3118–3126.

Norton, R., Roberts, B., Freeman, M., Wilson, M., Ashhurst-Smith, C., Lock, W., Brookes, D. & La

Brooy, J. (1998). Characterisation and molecular typing of *Burkholderia pseudomallei*: are disease presentations of melioidosis clonally related? *FEMS Immunology and Medical Microbiology* **20**, 37–44.

Nouvellon, M., Pons, J.L., Sirot, D., Combe, M.L. & Lemeland, J.F. (1994). Clonal outbreaks of extended-spectrum beta-lactamase-producing strains of *Klebsiella pneumoniae* demonstrated by antibiotic susceptibility testing, beta-lactamase typing, and multilocus enzyme electrophoresis. *Journal of Clinical Microbiology* **32**, 2625–2627.

Nørrung, B. & Gerner Smidt, P. (1993). Comparison of multilocus enzyme electrophoresis (MEE), ribotyping, restriction enzyme analysis (REA) and phage typing for typing of *Listeria monocytogenes*. *Epidemiology and Infection* **111**, 71–79.

Nørrung, B. & Skovgaard, N. (1993). Application of multilocus enzyme electrophoresis in studies of the epidemiology of *Listeria monocytogenes* in Denmark. *Applied and Environmental Microbiology* **59**, 2817–2822.

Ochman, H. & Selander, R.K. (1984a). Evidence for clonal population structure in *Escherichia coli*. *Proceedings of the National Academy of Sciences of the United States of America* **81**, 198–201.

Ochman, H. & Selander, R.K. (1984b). Standard reference strains of *Escherichia coli* from natural populations. *Journal of Bacteriology* **157**, 690–693.

Ochman, H., Whittam, T.S., Caugant, D.A. & Selander, R.K. (1983). Enzyme polymorphism and genetic population structure in *Escherichia coli* and *Shigella*. *Journal of General Microbiology* **129**, 2715–2726.

Olyhoek, T., Crowe, B.A. & Achtman, M. (1987). Clonal population structure of *Neisseria meningitidis* serogroup A isolated from epidemics and pandemics between 1915 and 1983. *Reviews of Infectious Diseases* **9**, 665–692.

O'Rourke, M. & Spratt, B.G. (1994). Further evidence for the non-clonal population structure of *Neisseria gonorrhoeae*: extensive genetic diversity within isolates of the same electrophoretic type. *Microbiology* **140**, 1285–1290.

O'Rourke, M. & Stevens, E. (1993). Genetic structure of *Neisseria gonorrhoeae*: a non-clonal pathogen. *Jounal of General Microbiology* **139**, 2603–2611.

Ørskov, F. & Ørskov, I. (1983). Summary of a workshop on the clone concept in the epidemiology, taxonomy, and evolution of the enterobacteriaceae and other bacteria. *Journal of Infectious Diseases* **148**, 346–357.

Oxberry, S.L., Trott, D.J. & Hampson, D.J. (1998). *Serpulina pilosicoli*, waterbirds and water: potential sources of infection for humans and other animals. *Epidemiology and Infection* **121**, 219–225.

Perea Mejia, L.M., Stockbauer, K.E., Pan, X., Cravioto, A. & Musser, J.M. (1997). Characterization of group A *Streptococcus* strains recovered from Mexican children with pharyngitis by automated DNA sequencing of virulence-related genes: unexpectedly large variation in the gene (*sic*) encoding a complement-inhibiting protein. *Journal of Clinical Microbiology* **35**, 3220–3224.

Piffaretti, J.-C., Kressebuch, H., Aeschbacher, M., Bille, J., Bannerman, E., Musser, J.M., Selander, R.K. & Rocourt, J. (1989). Genetic characterization of clones of the bacterium *Listeria monocytogenes* causing epidemic disease. *Proceedings of the National Academy of Sciences of the United States of America* **86**, 3818–3822.

Pinero, D., Martinez, E. & Selander, R.K. (1988). Genetic diversity and relationships among isolates of *Rhizobium leguminosarum* biovar *phaseoli*. *Applied and Environmental Microbiology* **54**, 2825–2832.

Pinner, R.W., Schuchat, A., Swaminathan, B., Hayes, P.S., Deaver, K.A., Weaver, R.E., Plikaytis, B.D., Reeves, M., Broome, C.V. & Wenger, J.D. (1992). Role of foods in sporadic listeriosis. II. Microbiologic and epidemiologic investigation. *Journal of the American Medical Association* **267**, 2046–2050.

Pons, J.L., Mandement, M.N., Martin, E., Lemort, C., Nouvellon, M., Mallet, E. & Lemeland, J.F. (1996). Clonal and temporal patterns of nasopharyngeal penicillin-susceptible and penicillin-resistant *Streptococcus pneumoniae* strains in children attending a day care center. *Journal of*

Clinical Microbiology **34**, 3218–3222.

Popovic, T., Fields, P.I., Olsvik, O., Wells, J.G., Evins, G.M., Cameron, D.N., Farmer, J.J., Bopp, C.A., Wachsmuth, K. & Sack, R.B. (1995). Molecular subtyping of toxigenic *Vibrio cholerae* O139 causing epidemic cholera in India and Bangladesh, 1992–1993. *Journal of Infectious Diseases* **171**, 122–127.

Popovic, T., Kombarova, S.Y., Reeves, M.W., Nakao, H., Mazurova, I.K., Wharton, M., Wachsmuth, I.K. & Wenger, J.D. (1996). Molecular epidemiology of diphtheria in Russia, 1985–1994. *Journal of Infectious Diseases* **174**, 1064–1072.

Porras, O., Caugant, D.A., Gray, B., Lagergård, T., Levin, B.R. & Svanborg-Edén, C. (1986a). Difference in structure between type b and nontypable *Haemophilus influenzae* populations. *Infection and Immunity* **53**, 79–89.

Porras, O., Caugant, D.A., Lagergård, T. & Svanborg-Edén, C. (1986b). Application of multilocus enzyme gel electrophoresis to *Haemophilus influenzae*. *Infection and Immunity* **53**, 71–78.

Poulsen, K., Theilade, E., Lally, E.T., Demuth, D.R. & Kilian, M. (1994). Population structure of *Actinobacillus actinomycetemcomitans*: a framework for studies of disease-associated properties. *Microbiology* **140**, 2049–2060.

Pupo, G.M., Karaolis, D.K., Lan, R. & Reeves, P.R. (1997). Evolutionary relationships among pathogenic and nonpathogenic *Escherichia coli* strains inferred from multilocus enzyme electrophoresis and *mdh* sequence studies. *Infection and Immunity* **65**, 2685–2692.

Quentin, R., Goudeau, A., Wallace, R.J., Smith, A.L., Selander, R.K. & Musser, J.M. (1990). Urogenital, maternal and neonatal isolates of *Haemophilus influenzae*: identification of unusually virulent serologically non-typable clone families and evidence for a new *Haemophilus* species. *Journal of General Microbiology* **136**, 1203–1209.

Quentin, R., Martin, C., Musser, J.M., Pasquier Picard, N. & Goudeau, A. (1993). Genetic characterization of a cryptic genospecies of *Haemophilus* causing urogenital and neonatal infections. *Journal of Clinical Microbiology* **31**, 1111–1116.

Quentin, R., Huet, H., Wang, F.S., Geslin, P., Goudeau, A. & Selander, R.K. (1995). Characterization of *Streptococcus agalactiae* strains by multilocus enzyme genotype and serotype: Identification of multiple virulent clone families that cause invasive neonatal disease. *Journal of Clinical Microbiology* **33**, 2576–2581.

Rasmussen, O.F., Beck, T., Olsen, J.E., Dons, L. & Rossen, L. (1991). *Listeria monocytogenes* isolates can be classified into two major types according to the sequence of the listeriolysin gene. *Infection and Immunity* **59**, 3945–3951.

Rasmussen, O.F., Skouboe, P., Dons, L., Rossen, L. & Olsen, J.E. (1995). *Listeria monocytogenes* exists in at least three evolutionary lines: evidence from flagellin, invasive associated protein and listeriolysin O genes. *Microbiology* **141**, 2053–2061.

Reda, K.B., Kapur, V., Goela, D., Lamphear, J.G., Musser, J.M. & Rich, R.R. (1996). Phylogenetic distribution of streptococcal superantigen SSA allelic variants provides evidence for horizontal transfer of *ssa* within *Streptococcus pyogenes*. *Infection and Immunity* **64**, 1161–1165.

Reeves, M.W., Evins, G.M., Heiba, A.A., Plikaytis, B.D. & Farmer, J.J. (1989). Clonal nature of *Salmonella typhi* and its genetic relatedness to other salmonellae as shown by multilocus enzyme electrophoresis, and proposal of *Salmonella bongori* comb. nov. *Journal of Clinical Microbiology* **27**, 313–320.

Rodrigues, J., Scaletsky, I.C., Campos, L.C., Gomes, T.A., Whittam, T.S. & Trabulsi, L.R. (1996). Clonal structure and virulence factors in strains of *Escherichia coli* of the classic serogroup O55. *Infection and Immunity* **64**, 2680–2686.

Rørvik, L.M., Caugant, D.A. & Yndestad, M. (1995). Contamination pattern of *Listeria monocytogenes* and other *Listeria* spp. in a salmon slaughterhouse and smoked salmon processing plant. *International Journal of Food Microbiology* **25**, 19–27.

Sacchi, C.T., Pessoa, L.L., Ramos, S.R., Milagres, L.G., Camargo, M.C.C., Hidalgo, N.T.R., Melles, C.E.A., Caugant, D.A. & Frasch, C.E. (1992a). Ongoing group B *Neisseria meningitidis* epidemic in Sao Paulo, Brazil, due to increased prevalence of a single clone of the ET-5 complex.

Journal of Clinical Microbiology **30**, 1734–1738.

Sacchi, C.T., Zanella, R.C., Caugant, D.A., Frasch, C.E., Hidalgo, N.T., Milagres, L.G., Pessoa, L.L., Ramos, S.R., Camargo, M.C.C. & Melles, C.E.A. (1992b). Emergence of a new clone of serogroup C *Neisseria meningitidis* in Sao Paulo, Brazil. *Journal of Clinical Microbiology* **30**, 1282–1286.

Salih, M.A.M., Danielsson, D., Bäckman, A., Caugant, D.A., Achtman, M. & Olcén, P. (1990). Characterization of epidemic and non-epidemic *Neisseria meningitidis* serogroup A strains from Sudan and Sweden. *Journal of Clinical Microbiology* **28**, 1711–1719.

Schill, W.B., Phelps, S.R. & Pyle, S.W. (1984). Multilocus electrophoretic assessment of the genetic structure and diversity of *Yersinia ruckeri*. *Applied and Environmental Microbiology* **48**, 975–979.

Scholten, R.J.P.M., Poolman, J.T., Valkenburg, H.A., Bijlmer, H.A., Dankert, J. & Caugant, D.A. (1994). Phenotypic and genotypic changes in a new clone complex of *Neisseria meningitidis* causing disease in the Netherlands, 1958–1990. *Journal of Infectious Diseases* **169**, 673–676.

Schwartz, B., Hexter, D., Broome, C.V., Hightower, A.W., Hirschhorn R.B., Porter, J.D., Hayes, P.S., Bibb, W.F., Lorber, B. & Faris D.G. (1989). Investigation of an outbreak of listeriosis: new hypotheses for the etiology of epidemic *Listeria monocytogenes* infections. *Journal of Infectious Diseases* **159**, 680–685.

Schwartz, B., Facklam, R.R. & Breiman, R.F. (1990). Changing epidemiology of group A streptococcal infection in the USA. *Lancet* **336**, 1167–1171.

Segovia, L., Pinero, D., Palacios, R. & Martinez Romero, E. (1991). Genetic structure of a soil population of nonsymbiotic *Rhizobium leguminosarum*. *Applied and Environmental Microbiology* **57**, 426–433.

Selander, R.K. & Levin, B.R. (1980). Genetic diversity and structure in *Escherichia coli* populations. *Science* **210**, 545–547.

Selander, R.K., McKinney, R.M., Whittam, T.S., Bibb, W.F., Brenner, D.J., Nolte, F.S. & Pattison, P.E. (1985). Genetic structure of populations of *Legionella pneumophila*. *Journal of Bacteriology* **163**, 1021–1037.

Selander, R.K., Caugant, D.A., Ochman, H., Musser, J.M., Gilmour, M.N. & Whittam, T.S. (1986a). Methods of multilocus enzyme electrophoresis for bacterial population genetics and systematics. *Applied and Environmental Microbiology* **51**, 873–884.

Selander, R.K., Korhonen, T.K., Väisänen-Rhen, V., Williams, P.H., Pattison, P.E. & Caugant, D.A. (1986b). Genetic relationships and clonal structure of strains of *Escherichia coli* causing neonatal septicemia and meningitis. *Infection and Immunity* **52**, 213–222.

Selander, R.K., Caugant, D.A. & Whittam, T.S. (1987a). Genetic structure and variation in natural populations of *Escherichia coli*. In *Escherichia coli and Salmonella typhimurium cellular and molecular biology,* Neidhardt, F.C., Ingraham, J.L., Low, F.C., Magasanik, B., Schaechter, M. & Umbarger, H.E., eds, vol.II, pp. 1625–1648. American Society for Microbiology, Washington, D.C.

Selander, R.K., Musser, J.M., Caugant, D.A., Gilmour, M.N. & Whittam, T.S. (1987b). Population genetics of pathogenic bacteria. *Microbial Pathogenesis* **3**, 1–7.

Selander, R.K., Beltran, P., Smith, N.H., Barker, R.M., Crichton, P.B., Old, D.C., Musser, J.M. & Whittam, T.S. (1990a). Genetic population structure, clonal phylogeny, and pathogenicity of *Salmonella paratyphi* B. *Infection and Immunity* **58**, 1891–1901.

Selander, R.K., Beltran, P., Smith, N.H., Helmuth, R., Rubin, F.A., Kopecko, D.J., Ferris, K., Tall, B.D., Cravioto, A. & Musser, J.M. (1990b). Evolutionary genetic relationships of clones of *Salmonella* serovars that cause human typhoid and other enteric fevers. *Infection and Immunity* **58**, 2262–2275.

Selander, R.K., Smith, N.H., Li, J., Beltran, P., Ferris, K.E., Kopecko, D.J. & Rubin, F.A. (1992). Molecular evolutionary genetics of the cattle-adapted serovar *Salmonella dublin*. *Journal of Bacteriology* **174**, 3587–3592.

Shah, H.N. & Gharbia, S.E. (1992). Biochemical and chemical studies on strains designated *Pre-*

votella intermedia and proposal of a new pigmented species, *Prevotella nigrescens* sp. nov. *International Journal of Systematic Bacteriology* **42**, 542–546.

Shi, Z.Y., Enright, M.C., Wilkinson, P., Griffiths, D. & Spratt, B.G. (1998). Identification of three major clones of multiply antibiotic-resistant *Streptococcus pneumoniae* in Taiwanese hospitals by multilocus sequence typing. *Journal of Clinical Microbiology* **36**, 3514–3519.

Sibold, C., Wang, J., Henrichsen, J. & Hakenbeck, R. (1992). Genetic relationships of penicillin-susceptible and -resistant *Streptococcus pneumoniae* strains isolated on different continents. *Infection and Immunity* **60**, 4119–4126.

Sneath, P.H.A. & Sokal, R.R. (1973). *Numerical taxonomy*. Freeman, San Francisco.

Soares, S., Kristinsson, K.G., Musser, J.M. & Tomasz, A. (1993). Evidence for the introduction of a multiresistant clone of serotype 6B *Streptococcus pneumoniae* from Spain to Iceland in the late 1980s. *Journal of Infectious Diseases* **168**, 158–163.

Souza, V., Rocha, M., Valera, A. & Eguiarte, L.E. (1999). Genetic structure of natural populations of *Escherichia coli* in wild hosts on different continents. *Applied and Environmental Microbiology* **65**, 3373–3385.

Spratt, B.G. (1999). Multilocus sequence typing: molecular typing of bacterial pathogens in an era of rapid DNA sequencing and the internet. *Current Opinion in Microbiology* **2**, 312–316.

Stanton, T.B., Trott, D.J., Lee, J.I., McLaren, A.J., Hampson, D.J., Paster, B.J. & Jensen, N.S. (1996). Differentiation of intestinal spirochaetes by multilocus enzyme electrophoresis analysis and 16S rRNA sequence comparisons. *FEMS Microbiology Letters* **136**, 181–186.

Stockbauer, K.E., Grigsby, D., Pan, X., Fu, Y.X., Mejia, L.M., Cravioto, A. & Musser, J.M. (1998). Hypervariability generated by natural selection in an extracellular complement-inhibiting protein of serotype M1 strains of group A *Streptococcus*. *Proceedings of the National Academy of Sciences of the United States of America* **95**, 3128–3133.

Struelens, M.J. & the Members of the European Study Group on Epidemiological Markers. (1996). Consensus guidelines for appropriate use and evaluation of microbial epidemiologic typing systems. *Clinical Microbiology and Infection* **2**, 2–11.

Takala, A.K., Vuopio-Varkila, J., Tarkka, E., Leinonen, M. & Musser, J.M. (1996). Subtyping of common pediatric pneumococcal serotypes from invasive disease and pharyngeal carriage in Finland. *Journal of Infectious Diseases* **173**, 128–135.

Tekle Haimanot, R., Caugant, D.A., Fekadu, D., Bjune, G., Belete, B., Frøholm, L.O., Høiby, E.A., Rosenqvist, E., Selander, R.K. & Bjorvatn, B. (1990). Characteristics of serogroup A *Neisseria meningitidis* responsible for an epidemic in Ethiopia, 1988–89. *Scandinavian Journal of Infectious Diseases* **22**, 171–174.

Thurm, V. & Ritter, E. (1993). Genetic diversity and clonal relationships of *Acinetobacter baumannii* strains isolated in a neonatal ward: epidemiological investigations by allozyme, whole-cell protein and antibiotic resistance analysis. *Epidemiology and Infection* **111**, 491–498.

Tomayko, J.F. & Murray, B.E. (1995). Analysis of *Enterococcus faecalis* isolates from intercontinental sources by multilocus enzyme electrophoresis and pulsed-field gel electrophoresis. *Journal of Clinical Microbiology* **33**, 2903–2907.

Trott, D.J., Robertson, I.D. & Hampson, D.J. (1993). Genetic characterisation of isolates of *Listeria monocytogenes* from man, animals and food. *Journal of Medical Microbiology* **38**, 122–128.

Trott, D.J., Atyeo, R.F., Lee, J.I., Swayne, D.A., Stoutenburg, J.W. & Hampson, D.J. (1996). Genetic relatedness amongst intestinal spirochaetes isolated from rats and birds. *Letters in Applied Microbiology* **23**, 431–436.

Trott, D.J., Jensen, N.S., Saint, G.I., Oxberry, S.L., Stanton, T.B., Lindquist, D. & Hampson, D.J. (1997a). Identification and characterization of *Serpulina pilosicoli* isolates recovered from the blood of critically ill patients. *Journal of Clinical Microbiology* **35**, 482–485.

Trott, D.J., Oxberry, S.L. & Hampson, D.J. (1997b). Evidence for *Serpulina hyodysenteriae* being recombinant, with an epidemic population structure. *Microbiology* **143**, 3357–3365.

Trott, D.J., Mikosza, A.S., Combs, B.G., Oxberry, S.L. & Hampson, D.J. (1998). Population genetic analysis of *Serpulina pilosicoli* and its molecular epidemiology in villages in the eastern High-

Author index

Keyword index